T0296565

CAMBRIDGE MONOGRAPHS ON PHYSICS

X-RAY MICROSCOPY

X-RAY MICROSCOPY

BY

V. E. COSSLETT, M.A., Ph.D.
University Lecturer in Physics, Cavendish Laboratory, Cambridge

AND

W. C. NIXON, M.A., Ph.D.
Assistant Director of Research, Engineering Laboratory, Cambridge
formerly A.E.I. Research Fellow, Cavendish Laboratory, Cambridge

CAMBRIDGE
AT THE UNIVERSITY PRESS
1960

CAMBRIDGE
UNIVERSITY PRESS

University Printing House, Cambridge CB2 8BS, United Kingdom

Published in the United States of America by Cambridge University Press, New York

Cambridge University Press is part of the University of Cambridge.

It furthers the University's mission by disseminating knowledge in the pursuit of education, learning and research at the highest international levels of excellence.

www.cambridge.org
Information on this title: www.cambridge.org/9781107654655

© Cambridge University Press 1960

First published 1960
First paperback edition 2014

A catalogue record for this publication is available from the British Library

ISBN 978-1-107-65465-5 Paperback

CONTENTS

CHAPTER 4

Reflexion X-Ray Microscopy: Mirror Systems

CHAPTER 5

Reflexion X-Ray Microscopy: Curved Crystals

CHAPTER 6

X-Ray Absorption Microanalysis

CHAPTER 7

X-Ray Emission Microanalysis

CHAPTER 8

Production of X-Rays

CHAPTER 9

Specimen Preparation Techniques

CHAPTER 10

Techniques of Contact Microradiography

CHAPTER 11

Techniques of Projection Microscopy

CHAPTER 12

Applications of X-Ray Microscopy in Biology and Medicine

CHAPTER 13

Inorganic Applications of X-Ray Microscopy

CHAPTER 14

Microdiffraction

CHAPTER 15

Some New Experimental Methods

APPENDIX

Absorption and Emission Data

PREFACE

In writing this monograph our intention has been to provide a connected account of the use of X-rays for microscopical investigation, commencing with the physical principles and including some microanalytical applications. Although its possibilities as a supplement to the established optical methods have long been clear, X-ray microscopy has become a practicable research tool only in recent years. A number of technological advances have combined to remove the major difficulties, which indeed largely stem from the properties of short wavelength and high penetration which make the method so attractive. A dioptric system, analogous to the compound light microscope, remains unrealized, but reflexion focusing at near grazing incidence has been thoroughly explored. Its aberrations are now well understood, if not yet more than partially correctable. On the other hand, the difficulties of focusing X-rays have been avoided by means of the two-wavelength procedure, in which a primary image is recorded with X-rays and the final image with visible light, all or part of the magnification being obtained by optical enlargement. The contact method, in which a one-to-one image is first obtained on a fine-grained emulsion, relies upon an optical microscope for the whole of the magnification, whereas in the projection method an image is formed directly at moderate magnification by a point source of X-rays.

These three methods of X-ray microscopy are now finding many practical applications, especially since they allow microanalysis of specific chemical elements to be carried out in a non-destructive manner, on the basis either of differential absorption or of the emission of characteristic line spectra. By the use of television techniques the latter method can be made to give a visible image of the distribution of a particular element in a specimen. These possibilities of microanalysis make the X-ray microscope a valuable complement to optical methods, even though its resolving power is so far no better.

The basic principles and limitations of the different methods have been fully investigated during the past decade, and their fields of application are now being actively explored. A review of the subject should therefore be timely. Short surveys have appeared in year-books and elsewhere, but no comprehensive treatment has hitherto been published. The Symposium held at Cambridge in 1956, under the auspices of the International Union of Pure and Applied Physics, brought together for the first time the various workers in the subject, and gave the opportunity for an assessment of its state of development and future prospects. The present text owes a great deal to the inspiration of that meeting, as also to the advice and encouragement of many who contributed papers to it; nevertheless, the division of the subject-matter adopted here, and the views expressed on unsettled issues, are those of the authors themselves.

The book falls roughly into three parts: the principles of the main methods of X-ray microscopy (chapters 1–5), the practical details of their use for qualitative microradiography and quantitative microanalysis (chapters 6 and 7, 9–11), and examples of applications in biology, medicine, metallurgy and technology (chapters 12 and 13). In addition, it seemed useful to include some account of the production of X-rays (chapter 8), of microdiffraction procedures (chapter 14) and of certain recent developments of promise for X-ray microscopy (chapter 15). An appendix contains data about the absorption coefficients and the characteristic emission lines of the elements.

In a text covering so wide a field, in much of which the authors have no personal experience, it is only too possible that errors will exist. We shall be grateful to have them brought to our notice, and thanks are tendered in advance for this and for other comments which may help to improve the text of any future edition.

V.E.C.
W.C.N.

CAMBRIDGE
January 1959

Postcript: As publication was delayed by a printing dispute, we have been able to add some reference to recent work and especially to the papers given at the Symposium on X-ray Microscopy and X-ray Microanalysis held in Stockholm in June 1959.

ACKNOWLEDGEMENTS

The authors wish to thank their colleagues Dr P. Duncumb, Dr N. A. Dyson and Dr J. V. P. Long, who have read various parts of the manuscript and offered valuable comment and suggestions. They are also greatly indebted to Miss M. Clayden for drawing most of the diagrams, to Mr B. K. Harvey for photographic assistance, and to Miss J. A. L. Beckett and Miss J. D. P. D. Gardiner for typing the manuscript.

Our thanks are also due to the following authors, publishers and learned bodies for permission to reproduce Text-figures and Plates from books and journals. Academic Press: Figs. 1.2, 2.1, 2.5, 3.6, 3.8, 6.10, 6.11, 6.12, 6.13, 6.14, 6.17, 7.6, 7.11, 7.13, 8.7, 8.8, 8.12, 9.1, 9.2, 10.3, 10.5, 11.8, 14.4, 14.13, 15.3, 15.4, and Pls. IB, II, III, VA, VIA, XIIB, XIIIA, XIX, XXB, XXIA, XXIB, XXIIIB, XXVA, XXVIIB, XXVIIC. *Acta Orthopaed. Scand.* and Dr K. Holmstrand: Fig. 6.7b. *Acta Paediatr. Stockh.* and Dr G. Wallgren: Fig. 8.10. *Acta Radiol., Stockh.*: Figs. 6.3, 6.4, 6.5, 6.6, 6.8. Dr K. W. Andrews: Pls. XIX, XXB. *Appl. Sci. Res.*: Fig. 15.5 and Plate XXXIIA. Prof. A. V. Baez: Fig. 7.13. Dr S. Bellman: Fig. 6.8. *Brit. J. Radiol.*: Pls. VIIIB, XXA, XXIV. *Cancer* and Dr B. Engfeldt: Pl. XVB. *Curr. Sci.*: Fig. 5.7. Dr P. Duncumb: Figs. 7.2, 7.3, 7.4, 7.8, 7.9, 7.10, 8.5, 15.1, 15.2. Dr J. Dyson: Figs. 4.11, 4.15, 4.16. Dr N. A. Dyson: Figs. 6.10, 6.11, 6.12 and Table A. 1. Dr W. Ehrenberg: Fig. 14.1. Prof. A. Engström: Figs. 2.1, 2.5, 6.4, 6.5, 6.13, 9.2, 10.4, 10.5 and Pls. XIIB, XVIA, XVII. *Exp. Cell Res.*: Fig. 10.4. *Fortschr. Phys.* and Dr G. Hildenbrand: Figs. 4.2, 4.3, 4.4, 4.13, 4.14, 4.18, 4.19. Dr A. Franks: Fig. 14.3. General Electric Co., Milwaukee: Pl. IXB. Prof. L. von Hámos: Figs. 5.2, 5.3, 5.4 and Pl. IV. Dr B. L. Henke: Figs. 8.12, 14.2 and Table A. 2. Mr K. Hooper: Pl. XXVIIA. Dr W. Hoppe and Dr H. J. Trurnit: Fig. 5.8. Prof. H. Hydén: Pl. VA. Mr C. K. Jackson: Pls. XVIB, XXIA. *J. Appl. Phys.*: Figs. 10.2, 11.3, 14.2 and Table A. 2. *J. Inst. Metals* and Prof. R. W. K. Honeycombe: Pl. XXVIII. *J. Sci. Instrum.*: Figs. 5.2, 5.3, 5.4, 6.15, 6.16. *J. Ultrastructure Res.*: Pl. XVIA. Prof. P. Kirkpatrick and Dr H. H. Pattee: Figs. 1.2, 4.12, 4.17, 14.4 and Pls. II, III. Dr G.

xiv ACKNOWLEDGEMENTS

Langner: Fig. 3.8. Prof. J. B. Le Poole: Fig. 15.5 and Pls. XXVB, XXXIIA. Dr B. Lindström: Figs. 6.3, 6.6. Dr J. V. P. Long: Figs. 6.15, 6.16, 6.17, 6.18, 7.12 and Pls. XXVI, XXXA. Dr L. Marton: Fig. 11.8. Dr J. F. McGee: Pl. IB. *Metallurgia* and Dr D. A. Melford: Pl. VII. Prof. G. Möllenstedt: Figs. 15.3, 15.4. Mr T. Mulvey: Fig. 7.7. *Nature, Lond.*: Pl. IA. Martin Nijhoff, Delft, and Dr R. Castaing and Prof. A. Guinier: Fig. 7.5. Dr S. B. Newman: Pl. XXXA. *Norelco Reporter* and Dr I. Bessen: Fig. 8.11 and Pl. X. *Norelco Reporter* and Dr P. W. Zingaro: Tables A. 3, A. 4. Dr H. H. Pattee: Pl. XXXIIB. N. V. Philips, Eindhoven: Pl. VIIIA. *Philips Tech. Rev.* and Dr J. E. de Graaf and Dr W. J. Oosterkamp: Fig. 8.9. *Phil. Trans.*: Pl. XVIII. *Proc. Phys. Soc.*: Figs. 4.11, 4.15, 4.16, 14.1, 14.3. *Proc. Roy. Soc.*: Fig. 11.5, and Pl. XIIIB. *Progress in Biophysics* and Miss S. M. Clark and Dr J. Iball: Pl. VB. Dr G. N. Rama-chandran: Fig. 5.7. Dr A. Rich and Dr H. Davies: Pl. XXXB. *J. R. Micr. Soc.*: Pls. IXA, XXII, XXIIIA. Prof. R. L. de C. H. Saunders: Pls. XI, XIIA, XIVA, XIVB, XVA. Dr J. Sawkill and Tube Investments Research Laboratories: Pls. VIB, XXIX. Dr H. E. See-mann and Dr H. R. Splettstosser: Fig. 10.2. Dr R. S. Sharpe: Fig. 9.1 and Pls. XXIB, XXIIIB, XXVA. Dr D. S. Smith: Pl. XIIIA. Springer Verlag (Berlin): Figs. 4.12, 4.17, 7.7, 7.10, 15.6. Springer Verlag (Berlin) and Prof. R. Glocker: Fig. 6.9. Springer Verlag (Vienna) and Dr K. Michel: Fig. 6.7a. *Z. Naturforsch.*: Figs. 5.8, 5.14. *Z. Physik* and Dr L. Y. Huang: Pl. XXXI. *Brit. J. Appl. Phys.*: Figs. 7.8, 7.10. Dr S. L. van den Broek: Fig. 10.3.

CHAPTER I

INTRODUCTION

1.1. The use of X-rays for microscopy

The chief distinguishing characteristics of X-rays have special value for microscopy. Their short wavelength compared with visible radiation offers a correspondingly higher resolution than that of the optical microscope. Their relatively great penetration into matter gives the possibility of investigating internal structure, in both biological and inorganic specimens. The simple line-structure of X-ray emission, and the rapid variation of absorption coefficient with atomic number, provide two independent methods of analysis of the elements present. Unfortunately, however, there are serious difficulties in the way of focusing X-rays and it is only comparatively recently that some of these prospective advantages have been realised, with the development of methods of X-ray microscopy having a resolving-power approaching that of the best optical microscope.

The discovery of X-rays was made by Röntgen on 9 November 1895, and attempts were made at once to explore their use in microscopy. Röntgen (1895) himself recognized, in his original communication, the difficulties involved in attempting to focus X-rays with any sort of lens; he was unable to detect any refractive effect and concluded that it must, at best, be very small. The possibility of constructing an X-ray microscope was, therefore, dismissed and only some thirty years later was it realized that a mirror-system might be practicable, following upon detailed investigations of the reflexion of X-rays from polished surfaces. On the other hand, within a few months of Röntgen's discovery, X-rays were being used for the study of fine internal structure, by enlarging contact radiographs: in botanical specimens by Burch (1896) and by Ranwez (1896), and in alloys by Heycock & Neville (1898). From these beginnings the subject of microradiography grew, slowly until 1930, but with increasing refinement and width of application immediately before the Second World War and in the decade since its end. The point projection method, in which the specimen is as close as possible to a fine focus, was not suggested until 1939 and has been realised in practice only since 1950. The development of both these non-focusing systems has suffered from technical limitations: the contact method requires a very fine-grained emulsion, the projection method a powerful fine-focus X-ray tube. In the meantime the possibilities of focusing X-rays, with curved crystals as well as with polished mirrors, have been thoroughly explored.

1.2. Contact microradiography

The simplest method of recording fine detail with X-rays is that of contact microradiography. The specimen is placed as close as possible to' a photographic emulsion of very fine grain, at some distance from an X-ray source (Fig. 1.1), so that a one-to-one

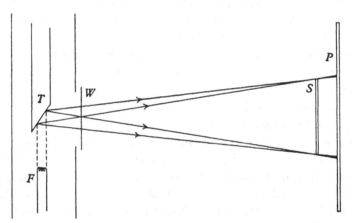

Fig. 1.1. Contact microradiography. *F*, filament; *T*, target; *W*, window; *S*, specimen; *P*, photographic emulsion.

image is obtained. The resulting negative is examined under a high-power optical microscope and selected regions are subsequently enlarged photographically. Being the easiest it was also the first method to be used for X-ray microscopy. After the early experiments, the first detailed investigation of its value as a research technique was made by Goby (1913a), who introduced the term 'microradiography'. For more than two decades its development

INTRODUCTION 3

was almost entirely confined to France and mainly to biological applications. In the late 1930's the commercial availability of ultra-fine-grain emulsions led to a rapid extension of its use, particularly to metallography. Since 1947 it has been developed by Engström and his school as a powerful method of microanalysis in biology and especially in cytology.

For many purposes useful results can be obtained, at a resolution down to a few microns, with a normal type of X-ray tube and an emulsion of moderately fine grain. The resolution is determined by the penumbra width, as well as by grain size; to minimize the penumbra the specimen must be as close to the emulsion as possible and the angular size of the source must be small. For high-resolution investigations, therefore, a source of small size ($\sim$0·1 mm.) and high intensity is desirable, and suitable X-ray tubes have recently been developed. The specimen, usually a section a few microns thick, is laid directly on the photographic emulsion; a thin layer of nitrocellulose may be placed under it, or sometimes a strip of cellulose tape upon it, to facilitate its removal before the emulsion is processed. Great care is needed at all stages to avoid scratches or the deposit of dust on the negative; fine-grained emulsions are fortunately slow enough for all the procedures, including specimen mounting and removal, to be carried out under a safelight.

The advantages and limitations of the contact method are discussed in detail in chapter 2. Its apparent simplicity is complicated by the demands of specimen mounting and photographic processing in obtaining the original negative, and of high-resolution photomicrography in producing the final enlarged positive picture. The exposure-time is somewhat longer than in the projection method at the same resolution, but the field of view is much larger. The X-ray equipment is not so elaborate and costly as that needed for projection imaging, but the very best optical microscope must be available. The ultimate limitation to the contact method is in fact set by the resolving-power of the microscope used for the final enlargement, since the grain size is smaller than this in the best emulsions. With visible light the resolution limit is about 0·2μ, and it has already been closely approached in some of Engström's microradiographs (cf. Plate XVII). The use of the electron microscope for enlarging the image would require drastic treatment of the

1-2

4 X-RAY MICROSCOPY

emulsion, unless a radically different method of recording the image can be developed (§ 15.6). The greatest utility of the contact method promises to be for applications requiring only moderate resolution, of the order of a micron; that is to say, essentially as an extension of routine procedures in metallurgical and biological macroradiography.

1.3. Reflexion X-ray microscopy

The reflexion coefficient is very small for X-rays incident normally on a polished surface, since the refractive index for radiation of such short wavelength is very close to unity. However, the index is *less* than one, and so total reflexion takes place in the less dense medium, not in the denser as with light. This phenomenon of total external reflexion makes itself evident in a high reflexion coefficient for X-rays incident at low angle on a polished surface. Its use for image-formation was first suggested, and experimentally investigated, by Jentzsch in 1929, but the practical realization of a reflexion X-ray microscope was delayed until Kirkpatrick & Baez (1948) found an effective means of overcoming the strong astigmatism inherent in the process. By mounting two mirrors at right-angles to each other (Fig. 1.2) the astigmatism of the first is cancelled by that of the second. As focusing is very much stronger in the plane of incidence than in that normal to it, cylindrical instead of spherical surfaces can be used without loss of power and with gain in simplicity. The critical angle θ is between 89° and 90°, so that the glancing-angle for total reflexion ϕ is of order 30′; in Fig. 1.2 ϕ has been greatly exaggerated for clarity. The smallness of this angle makes the setting up of the mirrors highly critical.

The aberrations peculiar to such a non-centred system have been evaluated by Dyson (1952), Pattee (1953a) and Montel (1953, 1954). It is found that obliquity of the field can be corrected by suitably placed stops and spherical aberration by figuring the two mirrors of a pair, but coma only by resorting to a four-mirror system, again suitably figured. With a single pair of mirrors McGee (1957) has obtained a resolution of order 0.5μ, at the low primary magnification of ×6 (in the interests of short exposure) and with aluminium K radiation (8·3 Å), in the interests of larger glancing-angle and higher image contrast.

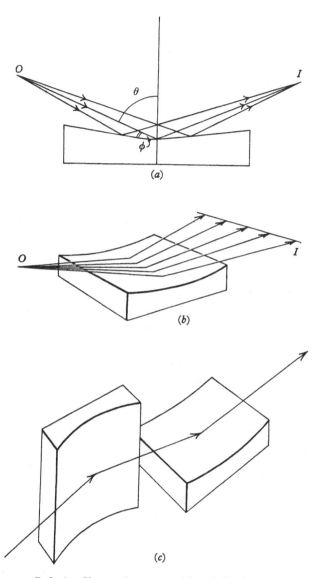

Fig. 1.2. Reflexion X-ray microscopy. (*a*) and (*b*) X-rays diverging from a source *O* are focused by a cylindrical surface to form an astigmatic image *I*; (*c*) arrangement of two cylindrical mirrors for eliminating astigmatism. (Kirkpatrick & Pattee, 1953.)

The development of the reflexion method is hampered more by practical difficulties than by any theoretical limitation. The ultimate resolution set by diffraction depends on the wavelength of the X-rays used and on the angular aperture, which can at maximum equal the glancing-angle for total reflexion. At a wavelength of 1 Å and critical glancing-angle of 0·4°, the diffraction limit is 85 Å. However, this would be attainable only over a very small field with the mirrors available owing to the effect of other aberrations. The first requirement is to find an acceptable compromise between width of field and resolution. Computation of the properties of a four-mirror system using elliptical and paraboloidal surfaces (Pattee, 1957a) predicts a resolution of 500 Å over a field of some 10μ at a wavelength of 4 Å. The technical difficulty then remains of figuring the surface to the required accuracy, which is of the order of 50 Å according to Pattee; in addition the surface must be free of local irregularities greater than 10 Å, to reduce X-ray scattering. These are much more stringent requirements than those encountered in optical technology. It can be expected that the necessary techniques of working and testing surfaces will gradually be developed, but it is likely to be some time before the reflexion X-ray microscope approaches its theoretical limit of performance. The problems involved are discussed in detail in chapter 4.

In principle an X-ray microscope can also be based on Bragg diffraction, the reflecting elements being curved lattices formed by bending thin crystals (Cauchois, 1946). Reflexion at normal or near-normal incidence is then possible. Although reasonably good images have been obtained at low magnification with a single mirror (Ramachandran & Thathachari, 1951), high-resolution microscopy is hindered by the lack of perfection even of single crystals over a sufficiently large volume and by the finite angular tolerance for Bragg diffraction. The various systems proposed, and their relation to the curved crystal systems used in focusing spectrometers, are described in chapter 5. An instrument devised by von Hámos (1934, 1953) combines some of the features of both spectrometer and microscope, and allows a partial qualitative analysis of the specimen to be carried out by fluorescent excitation (cf. § 5.2.2).

1.4. Point projection with X-rays

The simplest way of forming enlarged images with X-rays is to place the object close to a point source (Fig. 1.3); from the geometry of point projection, the magnification is the ratio of source–image to source–object distance. The resolution depends primarily on the diameter of the source, the exposure time on its intensity. One method of obtaining a point source is to place a pinhole in front of a macroscopic focal spot, as in the camera obscura. Early attempts were disappointing (Czermak, 1897; Uspenski, 1914) on account of the low intensity of the X-ray tubes then available, but the method was applied with greater success by

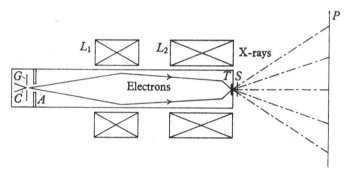

Fig. 1.3. Projection X-ray microscopy. The electron lenses L_1L_2 form a reduced image at T of the cathode C; the X-rays emitted from T project an image of a specimen S on to the screen (or plate) P.

Sievert (1936) and more recently by Rovinsky & Lutsau (1957) using the modern type of tube developed for crystallographic analysis. A similar arrangement, without an object, has been used for many years for inspecting the shape of the focal spot and the distribution of X-ray intensity across it.

The use of a fine focal spot, instead of a limiting-pinhole was first proposed by Malsch in 1933. In 1939 Ardenne and Marton independently suggested that the newly developed electron optical techniques would make it possible to concentrate an electron beam into a minute focus, 1μ or less in diameter, of great X-ray intensity. Experiments by Cosslett (1940) confirmed that a projection microscope of this type, with magnetic electron lenses, was a practicable instrument but that the available intensity was severely limited by the lens aberrations. A first, low-power, model was constructed by Cosslett & Taylor (1948), and has since been developed for high magnification and a resolution approaching that of the ultra-violet microscope by Cosslett & Nixon (1951, 1952a, b, 1953).

The ultra-fine focus T is formed by two-stage demagnification of a cathode C by the lenses L_1, L_2 (Fig. 1.3). By setting the target in the wall of the tube and making it from thin metal foil the emitted X-rays can be utilized most efficiently. The specimen can be brought close to the focus, allowing a high magnification ($\times$ 50–500) and short exposure (0·5–5 min.) in a camera-length of a few centimetres. The size and intensity of the focus can be varied by changing the excitation of the electron lenses. In principle, it can be made as small as a few Ångström units in diameter, if the final lens aperture is stopped down to limit the effect of spherical aberration. At the same time the accelerating voltage must be lowered, to limit penetration of the electron beam into the target, which would enlarge the spot by elastic and inelastic scattering processes. Both measures diminish the X-ray intensity, so that the ultimate resolution attainable by point projection is closely related to the tolerable exposure time, at least until lenses can be corrected for spherical aberration.

The experimental resolution is also affected by Fresnel diffraction in the specimen, though this can in principle be reduced below any assigned limit by bringing it close enough to the source. In practice, this separation cannot be brought much below 1 μ, corresponding to a resolving-power of about 100 Å. A limit of the same order is set by the weakness of the absorption of X-rays in thin specimens: resolution is only of value so long as the resolved details produce visible contrast. These considerations set a limit to the useful resolving-power of the projection method, at least for the foreseeable future, of 100–250 Å. The best value so far obtained lies between 1000 and 2000 Å (Nixon, 1955a, b).

The projection method thus offers the prospect of better ultimate resolution than either the contact or reflexion methods. It also gives shorter exposure-time at given resolution (§ 3.8.3.) It shares with the reflexion method the advantage of readily allowing treatment of the specimen during microscopy and conversion to microdiffraction operation. Stereophotographs can be obtained, and microanalysis by absorption or emission carried out, more easily than in the contact method. The main disadvantage of the projection method is that it requires a special fine-focus tube, with highly stabilized electrical equipment to counteract the chromatic aberration of the electron lenses. On the other hand, photography of the image is simple and little or no subsequent enlargement is needed. The method is fully described in chapter 3.

1.5. Scanning methods

In all the methods described above the whole of the field of view is illuminated and its image recorded simultaneously. It is equally possible to explore it point by point, by scanning it with a fine electron beam ('probe'). The X-rays emitted (or transmitted, if the beam first falls on a target) are detected with a counter, the amplified signal from which is conveyed to a cathode-ray tube operated synchronously with the scan, so that an image of the area scanned is displayed on its screen (Fig. 7.8). Such a flying-spot system has the advantage of better heat dissipation in the target and of allowing electronic control of both magnification and contrast in the image. It was proposed by Pattee (1953b) for high resolution imaging by X-ray absorption, with the specimen in contact with a thin window-target, and by Cosslett (1952b) for microanalysis by emission. As discussed later (§ 15.3), the experimental realization of the former system is closely bound up with the availability of a high-intensity source of electrons (Pattee, 1957b).

In the alternative method the electron probe falls directly on the specimen and the X-rays thus excited are recorded. As originally proposed by Castaing (1951) the specimen was moved mechanically under a stationary probe. Greater speed and flexibility is gained if the specimen is fixed and is scanned by the probe (Cosslett & Duncumb, 1956, 1957). Since each element has a simple and characteristic X-ray line spectrum, direct microanalysis of the composition of a surface or thin foil can be carried out, point by point (§ 7.1).

1.6. Comparison with optical and electron microscopy

X-ray microscopy provides information supplementary to that obtained with the established optical and electron methods. The

physical respects in which it differs from them define its special
sphere of usefulness: short wavelength, small but specific absorption, and a simple emission spectrum. In addition it shares with
electron microscopy, in contrast to the light microscope, the possibility of microdiffraction on crystalline specimens and of stereophotography at high resolution on those of openwork structure.

1.6.1. *Penetration and absorption in matter.* The penetration of
X-rays is high compared with visible light because of their much
shorter wavelength (cf. Fig. 1.4) and it is high compared with that
of an electron beam because of the absence of charge. Only those
materials can be observed in the optical microscope which have

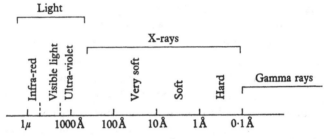

Fig. 1.4. Wavelength spectrum (1 Å $= 10^{-8}$ cm.; $1\mu = 10^{-4}$ cm.).

transmission bands between the wavelengths 0.1 and 0.7μ, that is,
which are transparent in the usual sense of the word. In electron
microscopy there are no specifically transparent substances, and
at the highest voltage normally used (100 kV) the limiting thickness
is about 0.2μ for compact material, but nearer 1μ for biological
specimens. For X-rays of wavelength 1 Å, elements of medium
atomic number are effectively transparent (35–40% transmission)
up to a thickness of 20–50μ. Images can be obtained through much
thicker specimens, as in medical radiography, but with poor
definition on account of X-ray scattering. In the thin specimens
used in X-ray microscopy, scattering is negligible in comparison
with absorption, except in certain metallurgical applications
(cf. § 2.2).

As with visible light, the absorption of X-rays in matter is
exponential and is expressed in terms of a linear absorption
coefficient μ (cf. § 2.1), the value of which varies with the wave-

length λ and with the atomic number Z of the absorber. In the wavelength region used for X-ray microscopy $(1-10 \text{ Å})\mu$ is approximately proportional to λ^3 and Z^3. At given wavelength, therefore, contrast in the recorded image will strongly depend on local differences in elementary composition of the specimen. Adequate contrast can be obtained even with specimens of low atomic number, such as biological tissues, by using soft enough X-rays.

Contrast in the X-ray image is thus due to physical processes different from those effective in electron microscopy (elastic scattering of electrons) and in optical microscopy (primarily molecular absorption). By differential absorption measurements on micrographs obtained at known X-ray wavelengths, it is possible to determine the amount of a given element present in different regions of the specimen, or, in biological tissues, the variation in dry mass. The accuracy is of order 1% if a series of determinations is made at wavelengths on either side of the absorption edge of the element to be estimated. An area a few microns in diameter in a specimen a few microns thick can be analysed, with a detection sensitivity of about 10^{-12} g. These microanalytical methods are described in chapter 6; they are analogous to optical absorption spectrometry, but since the X-ray line system is much simpler than optical spectra they are easier to apply on the microscale. No equivalent technique exists in electron microscopy as yet.

1.6.2. *X-ray emission analysis.* In analogy with the detection of elements from the presence of their lines in visible radiation, elementary microanalysis can be carried out by detecting the specific X-ray wavelengths excited when electrons strike a specimen. The scanning system devised for this purpose, mentioned above, is described in § 7.3. Alternatively emission may be excited by an X-ray beam (fluorescence analysis). In either case the analysis proceeds without appreciable temperature rise of the specimen, in contrast to the related technique of electron emission microscopy (Rathenau & Baas, 1951, 1956) in which the specimen must be raised to 800–1000° C. The information given by the two methods is again complementary: electron emission shows differences in phase or crystallographic orientation within a single metal, whereas X-ray emission shows the location and amount of the different elements in a mixture or alloy. In its present state of development, the X-ray method can detect 1 part in 10,000, in a volume of approximately $1\mu^3$, corresponding to a detection limit of 10^{-14} to 10^{-15} g.

1.6.3. *Microdiffraction*. X-rays share with electrons the property of being diffracted from regular arrays of atoms and, by use of a small aperture, individual small crystals in an alloy or mixture can be investigated. From the pattern obtained in a beam of known wavelength the crystal structure can be determined. Selective area diffraction has become a valuable auxiliary technique in electron microscopy and it can readily be practised with the reflexion and projection methods of X-ray microscopy. As a throw of some distance is required to spread the diffraction pattern to observable size, the contact method cannot be used. In the projection method, where a point source of great intensity is available and can be closely approached by the crystal, diffraction patterns can be recorded in one-tenth to one-hundredth the time normally needed in crystallographic analysis. The technique is discussed in detail, with examples of its applications, in chapter 14.

1.6.4. *Stereomicroscopy*. Stereophotography is practicable only if the depth of object space in sharp focus is sufficiently great, as otherwise no three-dimensional impression will be gained from the reproduction. In the optical microscope the depth of focus decreases in the same proportion as resolving-power increases, when the objective aperture is increased, until it is of the same order as the resolving-power, both being a fraction of a micron. It is thus possible to obtain stereo-images only at low magnification with visible light. In the reflexion X-ray microscope, as in the electron microscope, the usable aperture is so small that the depth of focus is very great indeed, being of the order of one hundred to one thousand times the resolving-power. In the contact and the projection methods, where no focusing elements are involved, the whole of the object is in focus at once; the fact that the magnification varies from the front to the back surface aids rather than hinders interpretation. Thus an X-ray microscope image contains more useful information than does an optical micrograph taken at the same magnification. By tilting, or translating, the specimen between two exposures by an amount dependent on the magnification (cf. §§ 10.3 and 11.11), stereo-pairs can be obtained which give excellent three-dimensional reconstruction. An insect, or part of a larger organism that has to be sectioned and examined piecemeal in the optical microscope, will be seen in sharp detail in a single stereo-pair of microradiographs. The method is particularly easy to practise in the projection X-ray microscope; in the contact method the specimen must be transferred from one emulsion to the other between exposures.

1.6.5. *Resolving-power.* In resolving-power each of the three main methods of X-ray microscopy is at present approaching the limit of the optical microscope (1000 Å), set by the wave nature of light. Various practical factors hinder further advance towards the ultimate limit set by the wave-nature of X-rays. In the reflexion method this limit, according to the Abbe theory of imaging, is set by the ratio of the wavelength to the aperture of the objective. The smallness of the aperture that can be used in grazing reflexion restricts resolution, in the best case, to about one hundred times the wavelength, that is, to about 100 Å. In the contact and projection methods there exists an analogous limit, set by Fresnel diffraction, with the angular size of the source taking the place of the numerical aperture of the objective (cf. § 3.3). The ultimate limit, set by practical considerations of intensity, is again of order 100 Å. Such a resolution is probably attainable with the projection method; the contact method would require an effectively grainless recording emulsion and its re-enlargement in an electron microscope instead of with light.

The useful range of X-ray microscopy, so far as resolution is concerned, thus runs across the boundary between optical and electron microscopy. It can already equal the former; it will never be able to compete with the latter. In practice its performance is likely to be limited more by lack of contrast or length of exposure than by resolving-power. Details as small as 100 Å will give very little contrast, even in the case of heavy elements; and the use of longer wavelengths, in pursuit of higher absorption, is hindered by the low efficiency of sources of soft X-rays.

1.7. Scope and applications of X-ray microscopy

The types of investigation in which X-ray microscopy may be usefully applied follow from the above discussion of the special properties of X-rays. For the most part they can be considered as logical extensions of macroscopic radiology on the one hand and on the other of optical microscopy, with the added potentialities of elementary microanalysis. Only an outline is given here, since practical applications are dealt with at greater length in chapters 12 and 13; details of specimen preparation are described in chapter 9.

1.7.1. *Materials opaque to light.* There is a considerable number of materials which cannot be examined optically because of their opacity: metals, ores, rocks, bone, teeth, wood. With these may be included biological specimens, such as insects, which can be rendered transparent only by a laborious process of sectioning and clearing, often supplemented by staining. X-ray microscopy both reduces the possibility of artefact formation and facilitates study of the morphology of the specimen as a whole. Bulk tissues still have to be sectioned (§§ 9.1.2. and 9.2.1.), since the resolution is in some degree dependent on the thickness of the object, but the sections can be much thicker than would be needed for optical examination at the same resolution, especially if stereophotographs are taken.

1.7.2. *Structure of thick specimens.* The spatial structure of materials that are transparent or semi-transparent to light is more readily explored with X-ray microscopy, owing to the limited focal depth of the optical microscope. This is particularly true of the fine details of the circulation of the blood, and microangiography has become an important branch of anatomy, following the pioneering work of Barclay (1951). In order to make the network easily visible in the microradiograph, a contrast medium such as thorium dioxide may be injected into the system (§ 9.1.3). Stereophotography is again a valuable aid in tracing the three-dimensional reticulation.

1.7.3. *Living organisms.* It is impossible to observe living material in the electron microscope and only possible under severe limitations in the optical microscope. X-ray microscopy allows larger organisms to be studied, but only at limited magnification and resolution owing to the lethal nature of the radiation on the one hand and to its being scattered in the tissue on the other hand. Techniques developed for macroscopic radiology, and especially in radiotherapeutics, have been utilized for microscopy. Studies have been made of the circulation in the ear of a living rabbit, for instance, and of the changes induced in it by mechanical shock or drug action (§ 12.1). The possibilities of following the progress of radiation damage, after either whole body exposure or highly localized irradiation with a point focus tube, have not yet been properly explored.

1.7.4. *Quantitative microanalysis in biology.* The determination of the amount and localization of mineral constituents in biological tissues is a straightforward matter of differential absorption measurement at wavelengths appropriate to the element to be estimated. Studies of the calcium, phosphorus and sulphur content of bone and other tissues have been made in this way (§ 6.3). The lower the atomic number of the element the greater the experimental difficulties, as the characteristic absorption edges occur at longer wavelengths. Silicon and magnesium can be determined by present methods, but it will be some time before the problem of greatest interest in cytology is fully solved: the determination of carbon, nitrogen and oxygen. As an essential step in this direction, methods of high accuracy have been developed by Engström for the determination of the dry weight of cell components by X-ray absorption measurements in the continuous spectrum (§ 6.2).

1.7.5. *Quantitative microanalysis of inorganic specimens.* Similar methods may be applied to the micro-estimation of elements present in non-biological specimens, such as alloys, rock sections or dust samples. Differential absorption (§ 6.1) and primary excitation (§ 7.1) allow $0 \cdot 1 - 1 \%$ of a foreign element to be detected with a localization of 1μ or better. Secondary excitation or fluorescent analysis (§ 7.4), gives still higher detection sensitivity at the cost of poorer localization, unless very long exposure times can be tolerated. Many problems arise in metallurgy (both ferrous and non-ferrous), petrology, mining technology and health protection where minute inclusions, precipitates or dust deposits have to be identified and analysed. Microdiffraction plays a part also, in determining the crystal structure and hence the mode of combination of the elements present, subsequent to their individual detection.

1.8. Other X-ray methods

A number of other methods of investigating micro-structure with X-rays have been proposed. They have often been called X-ray microscopes, although in fact only one of them, the X-ray lens, produces a real image of the specimen in the usual sense of the term. As the X-ray lens remains a paper project, it is discussed only briefly in § 4.2. The other three procedures, outlined below, have all been investigated experimentally with varying degrees of success.

1.8.1. *'Darkfield' imaging by the Berg–Barrett method.* When an X-ray beam falls on a surface at arbitrary incidence, it will be coherently reflected only by crystalline regions which are in suitable

Fig. 1.5. The Berg–Barrett method of X-ray imaging.

orientation for Bragg diffraction. It was pointed out by Berg (1930, 1934) that information about local variations in structure could be obtained at grazing incidence with the arrangement shown in Fig. 1.5. The photographic plate *P*, placed almost parallel to the surface *S* and protected from the direct beam, records the locally diffracted beams. Contrast in the image will depend on lattice orientation in the surface and thus can reveal the grain and sub-grain structure. The technique, as refined by Barrett (1945), Honeycombe (1951) and Weissman (1956), has found increasing use in metallography. Coyle, Marshall, Auld & McKinnon (1957) find it advantageous to use much larger glancing angles (20–45°) and a longer specimen–film distance; they obtained striking detail in micrographs of deformed aluminium single crystals.

The Berg–Barrett technique, in optical terms, is a darkfield method since the image is formed by the scattered beam, the direct beam being deliberately avoided. There is very little primary magnification and in this sense it is closely related to the contact method of X-ray microscopy. In practice, however, emulsions of only moderately fine grain are used, in the interest of short exposure, with subsequent enlargement of 10–100 times. Applications of the method are described in § 14.2.3.

1.8.2. Optical reconstruction of lattice structure. Normal methods of crystallographic analysis record as many as possible of the diffracted orders, including the direct X-ray beam (zero order), in contrast to the technique discussed above. It was pointed out by Bragg (1939, 1942) that the diffracting lattice could be reconstructed by forming the diffraction-pattern of the diffraction-pattern, that is, by the experimental equivalent of a Fourier transformation. The recorded negative is used to prepare a mask in the form of a corresponding pattern of holes in an opaque plate, each of which represents one of the Fourier components of the crystal structure. The plate is then illuminated by a parallel beam of monochromatic light and the resulting Fraunhofer diffraction pattern is photographed. The method has recently been developed into a routine research tool in crystallographic analysis (Buerger, 1950; Lipson & Taylor, 1951; Hanson, Lipson & Taylor, 1953). The reconstruction of the lattice so obtained is not an image in the true optical sense. It will show the *regular* features of the arrangement of atoms in the original crystal, but not small local variations such as a vacant site or a foreign atom; it is a statistically smoothed picture of the lattice, not a one-to-one image. However, since it utilizes the basic physical principles of microscopic image formation, its inclusion in a discussion of X-ray microscopy is justifiable.

1.8.3. Diffraction microscopy. A more general application of the reconstruction principle, in which information recorded with one type of radiation is extracted by use of another type, has been put forward by Gabor (1949). He proposed to record the Fresnel diffraction fringes around the images of objects illuminated by a point source of light. A photographic reversal of this image is

C & N

introduced into the analogue of the original optical system, at the position of the image, but with the direction of illumination reversed. The reversed image being of the nature of a generalized zone plate, its diffraction-pattern will form a 'reconstruction' of the object at its original place, which can be observed through a telescope. Gabor was concerned with reconstructing in visible light an image recorded in the electron microscope, but it is also feasible to do the same with X-ray fringes. Baez (1952a, b) suggested that this would give a better ultimate resolving power than that attainable with the X-ray microscope by direct imaging, but Haine & Mulvey (1952) have pointed out that the improvement will be limited to a factor of two, at best. In practice, serious difficulties have been encountered in recording multiple Fresnel fringes in the conditions of X-ray microscopy (Baez & El-Sum, 1957; cf. § 3.3), although they had been obtained earlier at low magnification by Kellström (1932); a single fringe is not difficult to obtain (Nixon, 1955a, b), but the Gabor method only gives a satisfactory reconstruction from multiple fringes. Its value for X-ray microscopy remains doubtful; if successful, however, it would give a true image of the original object, in contrast to the Bragg method, since it works from a fringe pattern which is closely related to the detailed form of the object.

CHAPTER 2

CONTACT MICRORADIOGRAPHY

2.1. Introduction

Attempts to investigate fine structure were made in the very early days of X-rays, by enlarging radiographs taken with the object almost in contact with the photographic plate. Burch (1896) and Ranwez (1896) examined botanical specimens, and Heycock & Neville (1898) alloys, by this means. The first thorough investigation of the technique was made by Goby (1913 a), who was primarily concerned with the structure of plants, insects and foraminifera (Goby, 1913 b, 1914). He published images enlarged up to × 25, but gave no details of the operating voltage of his X-ray tube nor of his exposure-time (see also Barnard, 1915). Later he developed a method of obtaining stereo-photographs from comparatively thick specimens (Goby, 1925).

The pioneering work of Goby was continued by Dauvillier (1927) and Lamarque (1936) in biology, and later by Fournier (1938), Glocker & Schaaber (1939) and Trillat (1940) in metallurgy. Indeed, until the 1940's, microradiography was largely developed in French laboratories. The introduction of emulsions of the Lippmann type, in which the grain size is below 1μ, made possible useful magnifications nearer to those of everyday optical microscopy. Dauvillier (1927, 1930), who was primarily interested in pathology and histology, obtained enlargements up to × 600 from some of his microradiographs. He observed that the contrast of biological specimens improved as the X-ray wavelength was increased, and constructed a special tube to produce the 8·3 Å characteristic radiation of aluminium. The applied voltage was 3–7 kV and an aluminium window was used to filter and minimize absorption of the emergent beam; absorption was further reduced by filling the space between window and specimen with hydrogen. These trends were carried further by Lamarque (1936), who invented the term 'historadiography' for the application of the technique in histology. To obtain adequate contrast from thin sections (4μ), a long wave-

length X-ray tube was constructed which operated at voltages from
500 to 5000 V, the very soft X-rays emerging through a lithium
window (Lamarque & Turchini, 1936). Borrowing the technique
developed in medical radiology, salts of heavy elements such as
gold were injected to increase contrast and results of very high
resolution were obtained (Lamarque, Turchini & Castel, 1937;
Lamarque, 1938). After 1940, biological and medical applications
increased rapidly (Bohatirchuk, 1942, 1944), especially in arteriography (Barclay, 1947, 1951; Bellman, 1953). Mitchell (1954) and
Engström (1955a) have recently reviewed progress in these fields.
  The obvious application to the study of alloys, and to metal
specimens generally which contain more than one component, was
already explored before 1940 by Fournier, Trillat and, in the
United States, by G. L. Clark (1939). Since it is more difficult to
prepare thin specimens of metals than of tissues, more penetrating
X-rays must be used and voltages up to 30 kV were early employed.
With the increasing importance of special alloys and the growth
of the study of the solid state in general, the period since 1945 has
seen a gradual extension of the use of microradiography in
metallurgy. Some of these applications will be noticed below, but
it has to be said that developments, especially in quantitative
estimation, have been slower than in biology.
  Whilst the use of contrast media enables great progress to be
made in microhistology, they hinder rather than aid the main aim
of many cytological studies: the determination of the mass (or
density), and if possible the chemical identification, of individual
components of the cell or tissue. By refinement of differential
absorption techniques Engström and his collaborators have made
great strides in the last few years towards this goal (Engström,
1946, 1956). A special X-ray tube has now been commercially
developed for this work (Combée, 1955; van den Broek, 1957),
which can be operated in the range 1–5 kV. Results obtained with
it show a resolving-power of about $0\cdot5\mu$ on maximum-resolution
plates (Combée & Recourt, 1957). Engström has succeeded in
determining the dry mass of minute biological structures of the
order of 10^{-12} g., to an error of 10%. The estimation of individual
elements has been carried out for sulphur, phosphorus and
calcium. It is believed that determination of carbon, nitrogen and

oxygen should be possible, to an accuracy of 5-10 % (Lindström, 1957). If this aim is achieved, microradiography will become a most powerful tool in cytochemistry.

The contact method essentially requires an X-ray tube to produce a suitable beam of radiation, a fine-grain photographic emulsion to record the image, and an optical system of high resolving-power to enlarge the image details to visible size. The resolution in the X-ray image depends primarily on the grain-size (§ 2.5) and on the sharpness of the shadows formed (§ 2.4). The sharpness is mainly determined by the width of the penumbra ('geometrical blurring'), which in turn depends on the size of the X-ray source and the relative distances of specimen and source from the emulsion. Fresnel diffraction (§ 3.3) and mechanical instability in the apparatus ('kinetic blurring') may also play a part in some circumstances. The possibility of distinguishing specific features in the microradiograph depends on relative contrast, and thus on the differential absorption of X-rays in materials of differing density and atomic constitution. Absorption will also depend on the wavelength of the radiation used.

These determining factors of the contact method will be discussed in the present chapter, with emphasis on the underlying physical principles. Experimental details of qualitative microradiography are given in chapter 10, but those relating to quantitative microanalysis are separately treated in chapter 6. Particular applications in biology, medicine, metallurgy and other fields of research are described in chapters 12 and 13.

2.2. Absorption of X-rays in matter

The degree of absorption of X-rays of given wavelength depends on a high power of the atomic number Z of an element and on the amount of it in the path of the beam. For a given element, absorption increases with a high power of the wavelength, with interruptions at the characteristic absorption edges corresponding to its emission lines. Image contrast will thus depend on the experimental conditions and can be markedly altered by change of illuminating wavelength. Although the same fundamental considerations are common to all methods of X-ray microscopy, it is convenient to discuss them here in connexion with the contact method.

2.2.1. *Absorption coefficients.* When a collimated beam of X-rays passes through a thickness t of a particular element it is found that its intensity is reduced from the initial value I_0 to a value I according to an exponential relation (Lambert's Law):

$$I = I_0 e^{-\mu t}, \qquad (2.1)$$

where μ is a constant, the value of which depends on the atomic number of the element and on the wavelength of the X-rays. It has the dimensions of a reciprocal length (cm.$^{-1}$) and is known as the linear absorption coefficient. The form of the relationship implies that, in passing through successive elementary layers of thickness dt, the beam loses a constant fraction of its intensity in each layer:

$$dI/I = -\mu \, dt. \qquad (2.2)$$

The value of the linear absorption coefficient depends on the number of atoms in the path of the beam and hence on the state of aggregation of the element. Division by its density (ρ) gives a more practically useful quantity (μ/ρ), which is independent of the physical state. It is known as the mass absorption coefficient and has dimensions cm.^{2} g.$^{-1}$ Absorption is thus referred to the amount of matter in the path of the beam, measured as the mass per unit area or 'mass-thickness' m, and since $m = \rho t$, (2.1) becomes

$$I = I_0 e^{-(\mu/\rho) m}. \qquad (2.3)$$

Similarly, division of μ by the number of atoms per c.c. ($\rho N/A$) defines the absorption coefficient per atom

$$\mu_a = \mu/\rho \, . \, A/N,$$

where A is the atomic weight of the absorbing element and N is Avogadro's number. Since absorption is primarily concerned with the inner electron shells of the atom, μ_a is almost entirely independent of the physical state and mode of chemical combination of the absorbing element.

Discussion of the variation of the absorption coefficient with atomic number and wavelength is complicated by the fact that loss of beam intensity occurs both by absorption (which transfers energy to the absorber) and by scattering (which deflects it out of the recorded beam). The absorbed energy is re-emitted either as secondary (fluorescent) radiation or as photo-electrons. The scattering may be either coherent, including Bragg diffraction, or incoherent, with change of wavelength (Compton scattering). The experimentally determined mass absorption coefficient will thus be composed of two terms,

$$\mu/\rho = \tau/\rho + \sigma/\rho,$$

where τ/ρ represents absorption by photoelectric emission and fluorescence and σ/ρ is the mass scattering coefficient. In the wavelength range usually used for microradiography scattering is almost entirely coherent and in practice its value is nearly always negligible compared with that for absorption.

The value of τ/ρ varies greatly with wavelength λ and with the atomic number Z of the absorbing element. Away from characteristic absorption edges, which are considered below, the approximate relation

$$\tau/\rho = k\lambda^m Z^n \qquad (2.4)$$

is found to apply. For wavelengths in the range 1–10 Å, the value of both m and n is about 3; k is a constant only within a narrow range of λ and Z. Tables of the 'mass absorption coefficient', such as those in the Appendix, in fact give values of the photo-electric coefficient, based on (2.4). It will be seen that it is of order 10^2–10^3 for most elements at the wavelengths used in microradiography, and even at 1 Å falls below 10 only for elements of Z less than 12.

On the other hand, the mass scattering coefficient σ/ρ depends not on Z alone but on the ratio Z/A, which is almost constant through the periodic table. The value of σ/ρ increases slowly with wavelength, being almost 50% higher at 10 Å than at 0·1 Å. For all elements, except hydrogen, its value is below 0·25 at all X-ray wavelengths. Hence the scattering coefficient can be neglected as compared with the photo-electric coefficient in almost all microradiographic circumstances. For very hard X-rays in light elements, however, they become comparable and scattering must then be taken into account. A quite different effect enters when the specimen contains particles of regular form and size, if the incident wavelength is of the same order as the particle size, as in the scattering of ultra-soft X-rays by latex particles (Henke & DuMond, 1955).

The absorption coefficient, defined as above, gives a measure of the attenuation of a beam of given wavelength in its passage through matter. For many purposes it is convenient to have an index of its penetrating power, and this is provided by the 'half-value layer' (H.V.L.): the thickness of an absorber which reduces the intensity to one-half its initial value. From (2.1), with $I = I_0/2$, we have for

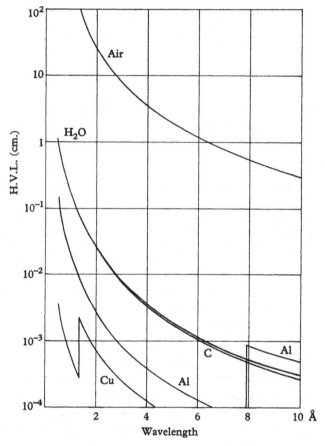

Fig. 2.1. Half-value layer (H.V.L.) of various absorbers, in function of wavelength. (Engström, 1956.)

the thickness of the half-value layer: $t_{\frac{1}{2}} = 0.69/\mu$. The variation of $t_{\frac{1}{2}}$ with λ for a number of elements and compounds is shown in Fig. 2.1. For wavelengths longer than 1 Å it is less than 1 mm., and beyond 6 Å less than 10μ, for all elements higher in the periodic table than carbon.

2.2.2. *Fluorescence radiation.* The energy absorbed from the X-ray beam will be carried off by ejection of electrons from the atoms involved (photo-electrons) and by raising of other electrons to higher levels within the atom (excitation). The photo-electrons may ionize other atoms in the absorber or may escape from it. In the latter event they may enter the photographic emulsion, especially in the contact method, and cause blurring of the image; this limitation on resolution is discussed below (§ 2.6.2). The excited atoms in due course return to the normal state, emitting in the process X-radiation of wavelength characteristic of the element. This secondary, or 'fluorescent', radiation is emitted in all directions from the absorber and can thus also have a blurring effect on the image. In the case of microanalysis by absorption, it can complicate the measurements by adding to the intensity of the direct beam, particularly in the contact method where the photographic plate is necessarily very close to the specimen. In the projection method its intensity at the recording system is very much less. Fortunately the fluorescent intensity is in any case small for a thin specimen, since it depends on the amount of absorption taking place. Its strength increases with atomic number, for a given spectral line, and for a given element decreases as the wavelength of successive lines increases. For elements higher in the periodic table than zinc ($Z > 30$), one-half or more of the energy absorbed in a thick specimen appears as fluorescent radiation. Most of the remainder is carried off by secondary ('Auger') photo-electrons, produced by absorption of fluorescent quanta in the outer shells of the atom.

The emission of fluorescent radiation may be turned to positive use, for elementary microanalysis, since it may be recorded separately from the primary radiation by a detector placed to one side of and shielded from the direct X-ray beam. Owing to the sharpness of the fluorescence lines, uncomplicated by the continuous radiation emitted by a normal X-ray source, such an arrangement allows excellent discrimination between elements which are close together in the periodic table (see § 7.4). For the same reason, a fluorescent source may be preferred for absorption microanalysis when exposure time can be sacrificed for detection sensitivity.

2.2.3. *Absorption edges.* The regular increase in absorption coefficient with wavelength is interrupted by more or less sharp absorption edges. These occur close to the wavelengths of the corresponding characteristic emission lines K, L, M, etc., because the mechanism of absorption and emission are closely related. The general nature and relative magnitudes of the absorption steps are shown in Fig. 2.2 for two elements of widely differing atomic number, platinum ($Z = 78$) and copper ($Z = 29$). With increasing Z the characteristic edges move towards shorter wavelengths, as do the emission lines, and decrease in height.

The energy of a quantum of radiation E_q is proportional to its frequency ν:

$$E_q = h\nu, \qquad (2.5)$$

where h is Planck's constant. The more energetic the quantum the better chance it has of penetrating matter without being absorbed. Absorption results in the transfer of energy to an electron in the atom, so that a quantum of given energy can cause an electron energy transition E_e of any value equal to or less than its own value: $E_e \leqslant E_q = h\nu$. As the frequency is increased (wavelength reduced) a point will be reached where the quantum of radiation has just enough energy to eject electrons in the next lowest level, or inner shell, of the atom. Extra energy is now transferred to the absorber, with a corresponding jump in the value of the mass absorption coefficient. Since the energy needed for ejection is greater the deeper the electron shell, the K-edge shows a bigger drop (measured as a percentage) than do the L- and M-edges; but as the value of the mass absorption coefficient increases rapidly with wavelength, the *absolute* value of the drop is less at the K than at the L-edge (cf. Fig. 6.4). Also, since the L, M and higher shells have multiple energy levels, whereas K is single valued, a number of substeps occur on all but the K-edge. A still finer substructure can be demonstrated by refined analysis, determined by the state of chemical combination of the element (Hanawalt, 1932; Kurylenko, 1955). Although this does not enter into account in microradiography, it can be investigated by the projection method of absorption analysis described in § 6.5 (Long, 1959, 1960b).

The values of the critical absorption wavelengths have been accurately measured for most elements, at least for the higher energy transitions (K-, L-, M-edges) and are tabulated by Compton & Allison (1935) and by Cauchois & Hulubei (1947). The magnitudes of the jumps are less well known, owing to lack of accurate measurements of the absorption coefficients (cf. Lindström, 1955). The critical absorption limit always occurs at a wavelength slightly less than that of the hardest of the corresponding emission lines. This follows from the fact that the

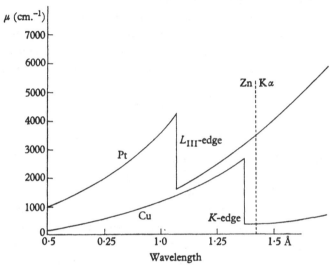

Fig. 2.2. Variation in absorption coefficient with wavelength for copper and platinum, in the neighbourhood of the Cu-K and Pt-L_{III} absorption edges.

absorption edge indicates the energy needed to eject an electron of a given shell entirely from the atom, whereas the emission lines arise from transitions from outer to inner shells only (cf. chapter 8). Hence photo-absorption involves rather more energy than that possessed by a quantum from the corresponding emission line, i.e. the wavelength of the K absorption edge (say) must be shorter than that of any of the $K\alpha$ or $K\beta$ lines of a given element. An important consequence is that an element is always a poor absorber of its own characteristic radiation, as this will lie on the long wavelength side of the edge, where the absorption coefficient is relatively low.

2.3. Optimum conditions for microradiography

2.3.1. *Wavelength and specimen thickness.* The optimum wavelength for microradiography will be that which gives maximum contrast between the region containing the element (or compound) of interest and the matrix in which it is situated. When two wavelengths offer much the same contrast, that which is available in greatest intensity will usually be preferred. The characteristics of the X-ray tube and the photographic material available will enter into the final choice of wavelength, as well as the desire to maximize the contrast differentiation in the microradiograph. The chosen wavelength λ (in Å) is related to the minimum voltage needed to excite it (V_m) by

$$V_m = \frac{12{,}398}{\lambda} \quad \text{(volts)}. \qquad (2.6)$$

Normally, an operating-voltage two or three times greater than V_m will be employed, in the interests of greater intensity (cf. § 8.5).

In the usual microradiographic arrangement (Fig. 1.1) the visible effect in the photographic emulsion is determined by the locally incident X-ray intensity and by the response of the emulsion to radiation of the wavelength used. The X-ray intensity depends initially on the output of the target T and, secondly, on the local absorption in the specimen S. The efficiency of X-ray production, in dependence on wavelength, is discussed in § 8.5, and that of photographic recording in § 6.4. We assume here that the X-ray source and photographic emulsion are given, and that at the given wavelength the emulsion characteristic is linear, so that the blackening produced (and hence the visible contrast) is directly proportional to the incident X-ray intensity. Attention is thus confined to the ratio of the intensity I transmitted by any part of the specimen to the incident intensity I_0, as given by (2.3) in terms of the local values of the mass absorption coefficient (μ/ρ) and the mass per unit area (m). When the emulsion characteristic is not linear, an optimum may exist for the density of blackening and hence for the exposure (Henke, Lundberg & Engström, 1957), as discussed in § 6.4.

When the specimen contains two components of mass absorption coefficients $(\mu/\rho)_1$, $(\mu/\rho)_2$, in amount m_1 and m_2 (g. cm.$^{-2}$), the ratio of the intensities of the transmitted beams will be, from (2.3):

$$I_2/I_1 = e^{-\{(\mu/\rho)_2 m_2 - (\mu/\rho)_1 m_1\}}. \tag{2.7}$$

The visible effect in the microradiograph depends on the differential contrast $(I_1 - I_2)/I_1 = \Delta I/I_1$. The values of m_1 and m_2 being fixed, contrast will be determined by the respective mass absorption coefficients. As is evident from Fig. 2.2, these increase rapidly with wavelength apart from sharp drops at the absorption edges. The right-hand side of (2.7) can be made to give almost any desired degree of contrast by taking the microradiograph with radiation of suitable wavelength. The degree of contrast sought will depend on the accuracy of discrimination required, and possibly also on the output of the source if exposure time threatens to be unduly long. In quantitative work, as discussed in chapter 6, the highest possible contrast is desirable. In qualitative microradiography, however, a contrast difference of 0·2 is ample and 0·1 is adequate for many purposes, since the eye can detect differences as small as 5 % in photographic blackening. A wavelength should thus be used at which the difference in absorption coefficients for the two components will just be enough to give easily detectable contrast. Use of a longer wavelength is in general not advisable owing to the attendant decrease in X-ray intensity.

The problem of wavelength selection is simplified when the specimen is of uniform thickness. The result will then depend on whether we want to distinguish a feature from the matrix in which it is embedded, for example, a precipitate in a metal, or two features one from the other with or without a matrix, as in a biological section. In the latter case an optimum specimen thickness at given wavelength and an optimum wavelength for given thickness are found to exist (Engström, 1947b, 1955b; Engström & Greulich, 1956). In the former case, contrast continues to increase for longer wavelengths than that giving the minimum detectable degree of contrast.

(i) We assume first that the feature of interest is a homogeneous inclusion of thickness t in a matrix of thickness D (Fig. 2.3a). From (2.7), or direct from (2.1), the ratio between the X-ray beam intensities transmitted by this feature (I_2) and by *the same thickness* of the matrix (I_1) is given by

$$I_2/I_1 = e^{-(\mu_2 - \mu_1)t}. \tag{2.8}$$

As the difference in absorption is assumed small, we can write in first approximation

$$(\mu_2 - \mu_1)t = 1 - I_2/I_1 = \Delta I/I_1. \qquad (2.9)$$

This expression is accurate to 10 % or better for values of the left-hand side of 0·20 and less, corresponding to an intensity ratio of 0·82 and more, that is, a differential contrast of 0·18 and less. Since this is roughly the range of $\Delta I/I$ suitable for observation, (2.9) is useful for giving a rapid estimate of the minimum wavelength required. Note that the calculation proceeds here in terms of the linear absorption coefficient μ, not of (μ/ρ).

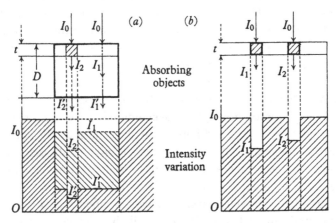

Fig. 2.3. Relative absorption of X-rays (a) by an inclusion of thickness t and a matrix of thickness D; and (b) by two particles side by side in free space.

The differential contrast, given by (2.9), increases in proportion to the thickness t of the inclusion. The optimum thickness of a section will thus be that which includes the whole of such an inclusion. For given t, the overall thickness D of the matrix is immaterial, since the ratio I_2/I_1 is unaltered by subsequent absorption in the remaining thickness of the matrix $(D-t)$. The two beams are equally affected, so that on emergence $I_2'/I_1' = I_2/I_1$, and the differential contrast $\Delta I/I_1$ remains unaltered. The effect of the additional material is simply to weaken the beam as a whole, making a longer exposure necessary. Further, it is evident that the same contrast will result whether the second component is locally concentrated or is dispersed in depth, so long as the comparison beam passes through the same total thickness of pure matrix. In such a case it is more useful to express the result of the estimation as a mass per unit area than as a thickness of the second component.

(ii) The contrast conditions are different if no matrix is present. When there are two particles of different composition side by side in free space (Fig. 2.3b), we wish to maximize the difference in the respective transmitted intensities I_1 and I_2. We assume the two features to be of equal thickness t, as in a section, and completely isolated, so that the beam suffers no further absorption after passing through them. The reference beam is now the uninterrupted incident radiation of intensity I_0, so that the difference in transmitted intensity, from (2.1), is

$$I_2 - I_1 = I_0(e^{-\mu_2 t} - e^{-\mu_1 t}).$$

This difference, which defines the contrast, at first increases with thickness but then falls again when the majority of the radiation is absorbed. It is a maximum for the thickness

$$t_x = \frac{\log_e(\mu_1/\mu_2)}{\mu_1 - \mu_2}. \tag{2.10}$$

In biological objects the elementary chemical composition is essentially constant, differences in absorption depending mainly on small variations in density. Hence $\mu_1 = \mu_2$ and in this limit t_x becomes

$$t_x \doteq 1/\mu. \tag{2.11}$$

Consequently optimum contrast is obtained when

$$I_1/I_0 \doteq I_2/I_0 \doteq 1/e.$$

The corresponding optimum thickness can be calculated at once from the linear absorption coefficient.

When the two inclusions are embedded in a matrix of absorption coefficient μ_m, instead of being in free space, the conditions for optimum contrast between their images (not between image and matrix), will be

$$t_x' = \frac{\log_e\{(\mu_1 - \mu_m)/(\mu_2 - \mu_m)\}}{\mu_1 - \mu_2} \doteq 1/(\mu_1 - \mu_m). \tag{2.12}$$

The optimum thickness now involves the matrix absorption

coefficient, and is not independent of it as stated by Engström, Lagergren and Lundberg (1957).

(iii) If the thickness of the specimen is given, then optimum contrast will be obtained at a particular wavelength (and hence X-ray tube voltage V) which can be found from the relations between μ and λ, and λ and V. Setting $\tau = \mu = 1/t$ in (2.4) and substituting for λ in (2.6), we find for the minimum voltage needed to excite the optimum wavelength:

$$V_m = 12\cdot4 \times 10^{-5}(k\rho t Z^n)^{1/m}.$$

For soft X-rays and elements of low atomic number, $n \doteq m \doteq 3$, and $k \doteq 5 \times 10^{21}$, so that

$$V_m = 2\cdot2 \times 10^3(t\rho)^{1/3}Z. \qquad (2.13)$$

In dehydrated biological tissue, the average value of Z is 6·6 and of ρ about 0·3, so that the minimum required voltage for a 4μ section would be about 750 V. Since it is usual to run an X-ray tube at about 1·5 times the minimum voltage, the optimum for biological work will be about 1000 V (Engström & Greulich, 1956; Engström, 1957). With more highly absorbing material the optimum voltage will be greater; for the mineral content of bone, for instance, the optimum voltage for a 30μ section would be about 6 kV.

The above discussion has assumed that the mass absorption coefficients of the specimen components are known, as will be so (within certain limits of accuracy) so long as they consist of local concentrations of elements. When mixtures or chemical compounds are concerned, an integrated value for the mass absorption coefficient must be used:

$$(\mu/\rho)_c = \sum_{k=1}^{n} a_k(\mu/\rho)_k/100, \qquad (2.14)$$

where n is the number of elements present, and $(\mu/\rho)_k$ the mass absorption coefficient and a_k the percentage by weight of a given element. The corresponding expression in place of (2.3), when m_k is the mass per unit area of a given element, is then

$$I_1 = I_0 \exp\left(-\sum_{k=1}^{n}(\mu/\rho)_k \cdot m_k\right).$$

2.3.2. *Choice of characteristic, continuous or fluorescent radiation.* It has been implicitly assumed above that radiation of a single well-defined wavelength should be used. Although an absolutely pure monochromatic beam cannot be obtained, it is desirable that it be as monochromatic as possible in order to allow direct comparison of the observed contrast with calculation or, more importantly, quantitative estimation of an element. Once the minimum wavelength needed to give a desired degree of contrast has been approximately evaluated, as described, an X-ray source will be chosen with a target that has a characteristic line at or on the longer wave side of this minimum. For most purposes the line radiation will be monochromatic enough, especially if use is made of a suitable filter (cf. § 8.7). But for the most exacting quantitative work use may be made of the fact that fluorescent X-radiation consists of extremely sharp lines, and some of the apparatus devised for this purpose will be described below (§§ 7.4 and 10.1). On the other hand the low intensity of fluorescent radiation entails a long exposure-time, so that its range of worthwhile application is limited.

In many applications it is possible to take advantage of the jump in absorption coefficient at an absorption edge, since the edge for one or other of the elements present may well fall in the neighbourhood of the minimum useful wavelength. The working wavelength will then be chosen on the short wavelength side of the edge of the element of lower Z, or on the long wave side of that of the element of higher Z. As an illustration, Fig. 2.2 shows the difference between the absorption coefficients of copper and platinum for wavelengths in the neighbourhood of the Cu K-edge. Immediately below the L_{III}-edge of platinum (1·07 Å) and again beyond the K-edge of copper (1·38 Å) the difference in μ is particularly high. With Zn $K\alpha$ radiation (1·43 Å), for instance, $\Delta\mu = 3300$ and (2.9) shows that 10 % contrast difference would be given by a thickness of 0·3μ of copper in platinum.

At the same time there are many applications for which the continuous X-ray spectrum may be satisfactorily used. As is evident from Fig. 2.2, the mass absorption coefficients of two elements usually differ radically over the whole range of wavelengths, except near absorption edges. The unfiltered continuous

C & N

radiation from almost any X-ray tube is thus adequate for most qualitative work in microradiography; up to a certain point it also has a place in quantitative work (cf. § 6.2). However, where a deep absorption step for one of the main component elements occurs in the wavelength range accessible with the voltage available, sharper differentiation will be obtained if line radiation of suitable wavelength is employed.

2.4. Geometrical blurring of the micro-image

In the contact method the specimen must be as close as possible to the recording emulsion. Analysis of the geometry of image formation shows also that the ratio of the source size to the source distance (its angular subtension) must be below a certain limit if a given resolution is to be obtained. The physical limitation lies in the formation of a penumbra and is usually referred to in radiography as 'geometrical blurring' to distinguish it from other image-marring effects. In other methods of X-ray microscopy it is also a factor in determining image sharpness, but in contact microradiography it is the most important limitation, next to grain size of the emulsion. 'Kinetic blurring' can also occur, due to relative motion of object or focal spot and emulsion; this can be largely eliminated by firmly fixing the camera to the X-ray tube and the specimen to the emulsion, but difficulty occurs with living subjects.

We suppose a source S of width s to form an image of an object O of width w on an emulsion E at distance a from S (Fig. 2.4a). O is distant c from E; for clarity it is shown well separated from it. An umbra (total shadow) is formed of width u, bordered by a penumbra of width p in which the illumination rises linearly to the background intensity. The widths of umbra and penumbra are evidently:

$$u = \frac{aw - cs}{a - c},$$
(2.15)

$$p = \frac{cs}{a - c}.$$
(2.16)

The umbra will become zero if $s/a \geqslant w/c$, that is, if the angular subtension of the source is greater than that of the object at E.

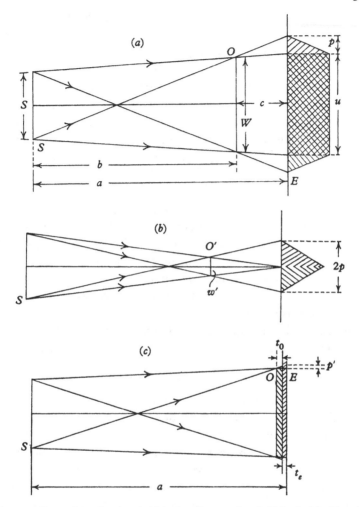

Fig. 2.4. Formation of umbra (width u) and penumbra (width p); (a) object O larger than source S; (b) object of critical width w', giving zero umbra; (c) object of thickness t_0 in contact with emulsion of thickness t_e.

This will be so if the source is too large or the object detail (w) too small; or, alternatively, if the source distance (a) is too small or the object distance (c) too large. But even with u zero an object detail will still be detected (Fig. 2.4b), on account of the penumbra.

The penumbra width p cannot be made zero, that is, the object outline cannot be perfectly sharp, since neither c nor s can be made zero in (2.16). However, it can be made smaller by moving O closer to E or by moving S farther away. The latter move has the disadvantage that illuminating intensity is rapidly reduced in accordance with the inverse square law; the former reaches a limit when the object is in contact with the emulsion (Fig. 2.4c). The object and emulsion will in fact be of finite thickness t_0 and t_e respectively, so that a feature on the surface of the object nearest to the source will be imaged at the back of the emulsion with a penumbra of width

$$p' = \frac{(t_0+t_e)s}{a-(t_0+t_e)} \doteqdot \frac{(t_0+t_e)s}{a}. \tag{2.17}$$

In radiography it is usual to measure image resolution in terms of edge sharpness and therefore primarily by penumbra width. We discuss later how far this coincides with the more usual optical definition of resolving-power in terms of point-point separation (§ 3.2). If, for contact microradiography, we now set p' equal to the minimum size of object detail it is sought to resolve w_m, (2.17) shows that it is determined by the ratio s/a, since (t_0+t_e) is already minimal. s/a is the angular size of the source, which thus will have a maximum permissible value for any required resolving-power, given by

$$s/a = w_m/(t_0+t_e). \tag{2.18}$$

For instance, if a specimen 10μ thick is in direct contact with an emulsion 15μ thick and a resolution of 0·5μ is sought, then s/a must be less than 1/50 (approx. 1°). A source of width 2 mm. would have to be more than 10 cm. distant; if the specimen were 100μ thick, it would have to be about 50 cm. from the source. It is of interest to note that equating penumbra to object size is equivalent to requiring vanishingly small umbra, in the contact method, as may be seen by substituting p from (2.16) for w in (2.15) and treating c as negligibly small; it is also obvious from Fig. 2.4b.

When the source of X-rays is given, it is immaterial (apart from considerations of experimental convenience), whether the value of s/a required by (2.18) is obtained by using the full focal spot at a great distance from the object or by stopping it down with a lead aperture and lessening the distance correspondingly. The amount of radiation depends on the area of source used (s^2) and on the inverse square of the distance ($1/a^2$), and so will be constant for a constant ratio s/a. However, the conditions for target cooling are more favourable for small than for large focal spots so that a greater emission intensity can be obtained from a fine-focus tube (cf. § 8.2). Thus when high resolution and short exposures are required, there is advantage in using the ultra-fine-focus tubes devised for projection microradiography, which have a specific loading of 1 megawatt/cm.^{2} and more.

2.5. Photographic and optical requirements

The successful practice of contact microradiography depends to a great extent on proper choice of photographic material and on its correct treatment. The problem is twofold, since the micronegative obtained in the X-ray beam must subsequently be enlarged and a print made at a magnification great enough to reveal the resolved detail. An emulsion of minimum grain size, and as free as possible from pinholes or foreign particles, is thus needed for initial recording of the image, and another of suitable contrast properties for making the final enlargement, in an optical system of high resolving-power. Alternatively, for quantitative analysis the optical density of the primary negative may be determined point by point in a microphotometer without enlargement (§ 6.3). The present state of knowledge of fine-grain emulsions is discussed in this section; practical details of processing them and the optical requirements for final enlargement are described in § 10.2.

2.5.1. *High resolution emulsions.* The fineness of image detail that can be unambiguously recorded by an emulsion depends on the distribution of grains and on their behaviour during processing, as well as on the size of the silver bromide grains in isolation. For this reason the resolution has to be treated as an empirical quantity; only values obtained by actual photography of objects of known dimensions under controlled conditions have any reliability. The standard photographic test-object is a ruled grating and the resolution of an emulsion is quoted in lines per millimetre. The wavelength and nature of the illumination should also be given, since the resolution may well be different for X-rays or electrons (for instance) from what it is for light. It must be noted that the criterion of resolving-power used in microscopy, the separation of neighbouring points, will be more conservative by a factor of 2–3, as it is much easier to detect a line than a point.

Hitherto the finest grained emulsions have been prepared for spectroscopic use and the line resolution quoted for them will be that determined in visible or ultra-violet light. Very few detailed measurements of their resolution for X-rays have been made. The most reliable are those of Engström & Lindström (1951), who compared three fine-grain emulsions as to speed and resolution with two of coarse grain, of the type used in medical radiography. Their results are summarized in Table 2.1. The Lippmann and the Eastman Kodak 548 and 649 Spectroscopic plates are some 5000–10,000 times slower than ordinary X-ray film to radiation of wavelength 3–4 Å, whilst their resolving power is 20–100 times greater. A slow response is to be expected from fine-grain material, since less target area is presented to the beam and each grain must absorb a quantum before becoming developable (for radiation of these wavelengths). On a simple picture of the process, as occurring in a layer containing a given thickness of grains, the sensitivity should decrease in proportion to the surface area of grain and so with the square of its linear extension. Assuming the resolving-power to reflect the grain size, we have multiplied the relative sensitivity by the square of the resolving-power and the third column of Table 2.1 shows the product to be roughly constant. Using this as a figure of merit, the 548 plate has a low rating and the 649 plate a high rating, a result which agrees with general experience.

TABLE 2.1

Type of emulsion	Relative sensitivity S	Resolving-power R (lines/mm.)	$S \times R^2$
Kodak No-Screen	1	10	100
Agfa Printon	⅓	50	150
Kodak 548	1/10,000	500	25
Lippmann	1/10,000	1000	100
Kodak 649	1/5,000	1000	200

In fact a number of other factors play a part in the relative performance of different emulsions, such as thickness and the concentration of silver halide. An ideal emulsion, that would possess both maximum sensitivity and resolution, should contain a high concentration of very fine grains uniformly distributed in a layer just thick enough to absorb effectively all the X-rays of the wavelength employed. The Lippmann-type emulsions seem to have the smallest grain size ($\sim$0·05μ) but there are appreciable inhomogeneities in thickness and concentration of the emulsion; also there is considerable variation from one batch to another, so that some are suitable for microradiography and others are not (Engström & Lindström, 1951). After careful comparative measurements of granularity in the developed plate, Lindström (1955) concluded that both the Kodak High Resolution and the Kodak Maximum Resolution plates were to be preferred to Lippmann film, being highly homogeneous over large areas and having a minimum of artefacts. The Eastman Kodak no. 649 plate has similar characteristics; it has a silver halide concentration of 40–50% and an average developed grain size of 0·1μ. This is well beyond the resolving-power of the optical microscope used to enlarge the negative, but even so a resolving-power better than 0·5μ has rarely been claimed in contact microradiography. Ehrenberg & White (1957) have suggested that the thickness of the emulsion may be the limiting factor: the depth of focus of a high-power objective is so small that it can focus only on a very thin layer of grains. The thickness of the Maximum Resolution (M.R.) emulsions is 10–15μ, whereas the focal depth of an objective of aperture 0·90 is only 0·4μ. The experiments of Ehrenberg and White indicated that the resolution was better with a specially prepared emulsion 5μ thick than with a normal M.R. emulsion. Working with still thinner coatings, White (1958) has shown that a resolution of 0·1–0·2μ can be obtained with an emulsion 0·9μ thick (0·5μ after processing). Combée & Recourt (1957), on the other hand, found no significant increase in resolution with an emulsion ($\sim$3μ) containing a higher concentration of silver halide; the sensitivity, however, was appreciably improved (cf. § 6.4).

It seems probable, nevertheless, that photographic material better suited to the special needs of contact microradiography can be developed. Since at the wavelengths now most used for biological specimens (8–20 Å) the X-rays penetrate less than 1μ in silver bromide and little more than 2μ into the gelatine (Engström, Greulich, Henke & Lundberg, 1957), there is likely to be little loss in sensitivity with a thinner emulsion than those now available.

For metallurgical work with hard X-rays, however, the need is for a higher concentration of silver bromide rather than for a thinner emulsion.

2.5.2. *Optical enlarging system.* Even with the emulsions already available, the contact method is faced with the problem of the finite resolving-power of the optical microscope which has to be used for examining the negative (cf. § 10.2.2). An objective of numerical aperture 1·32 will show a point resolution of about 0·25 μ with visible light. To go beyond this an illumination of shorter wavelength must be used. Ultra-violet microscopy is available, but has its own technical difficulties. Recourt (1957 a) has tried to make use of the much higher resolving-power of the electron microscope. He dissolved the celluloid base of his film in acetone and then digested away the gelatine with an enzyme preparation, so that only the silver skeleton of the microradiograph remained (cf. Comer & Skipper, 1954). This was mounted and photographed in the electron microscope at a magnification of × 4000. The resulting negative showed greater detail than did the enlargement of the original microradiograph. The dangers of artefact formation in such a procedure might be reduced by simply stripping the emulsion from its backing and viewing it in the electron microscope without further treatment, other than desiccation. Liquier-Milward (1956) has successfully used this method for autoradiography. The electron microscope should give a resolution of the order of 500–1000 Å from such a dried-down emulsion, owing to the great difference in contrast between the silver and the gelatine. With thinner emulsions a still better resolution could be obtained. An alternative proposal for utilizing the electron microscope, involving registration of the X-ray image on a plastic film instead of by photography, is described in § 15.6.

2.6. Ultimate resolving-power of the contact method

As has been explained, the resolution in a contact microradiograph depends, first, on the unsharpness inherent in the imaging X-ray beam; secondly, on the spread in the emulsion of the X-rays and the photo-electrons produced by them; thirdly on the size and degree of aggregation ('clumping') of the grains in the emulsion;

and fourthly, on the resolving-power of the optical system used to enlarge the primary negative. These limitations are all peculiar to the contact method. In common with other methods of X-ray microscopy, it is also limited ultimately by lack of contrast when the object details fail to absorb enough of the incident beam to produce a visible differentiation in the photographic image.

2.6.1. *Unsharpness in image formation.* Geometrical unsharpness can be brought below any desired resolution limit, for given specimen thickness t, by keeping the angular size of the source below the value given by (2.17); this requires either a small width of source s or a large distance a, which can be achieved only at the cost of long exposure. Secondly, the connexion between tube and camera must be rigid enough to prevent the emulsion from moving, with respect to the focal spot, by more than the distance p during exposure. A third cause of unsharpness is the diffraction of the X-ray beam at the edge of the specimen, giving rise to Fresnel fringes. As shown in § 3.3, the fringe width f is given with sufficient accuracy by the expression

$$f = (c\lambda)^{\frac{1}{2}}, \qquad (2.19)$$

where λ is the wavelength of the incident X-rays and c is the separation of the object edge from the imaging plane. In the usual conditions of contact microradiography the effect is negligible in comparison with other causes of unsharpness. Indeed, f is only $0 \cdot 1\mu$ for a specimen $0 \cdot 1$ mm. thick and a wavelength of 1 Å, as in metallurgy, or equally for a section 10μ thick at a wavelength of 10 Å, as in biology. It is, however, of importance for a thick specimen, or for any specimen not in close contact with the emulsion (cf. § 3.3).

2.6.2. *Image spread.* The image in the emulsion of a point in the specimen may also be laterally blurred if the X-ray beam is scattered in the emulsion, or if the ejected photo-electrons have a range greater than the grain separation. Scattering is negligible except when very hard or very soft X-rays are used. The ejected photo-electrons, however, constitute a serious limitation to resolving-power since they have an appreciable range even at low X-ray energy, as shown in Fig. 2.5 (Bellman & Engström, 1952). For Cu $K\alpha$ radiation (~ 8 kV) the range in the emulsion is 2μ, and for Al $K\alpha$ ($\sim$ 1·5 kV) it still is greater than 0·1 μ. It is thus of the same order as the grain separation in the concentrated emulsions necessary for high-resolution work. Ehrenberg & White (1957) have taken comparative microradiographs which show clearly the limiting effect of photo-electrons on resolution when copper radiation is used; with aluminium radiation no unsharpness due to this cause could be substantiated, but in any case it would

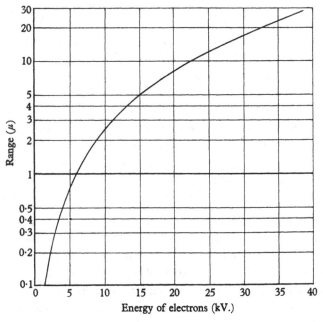

Fig. 2.5. Variation with energy of the range of photo-electrons in a photographic emulsion. (Engström, 1956.)

have been beyond the optical resolving-power of the enlarger. The attainment of a resolution better than 1000 Å by the contact method will only be possible if X-ray wavelengths of order 20 Å and longer are employed. (See also Hoh & Lindström 1959).

2.6.3. *Size and aggregation of photographic grains.* It follows from what has been said in § 2.5, that there is no technical difficulty in preparing emulsions with adequately small grains in very high concentration. Probably processing methods can be found which would avoid the aggregation of grains which results from the use

of rapid developers. Very little appears to be known about aggregation on this very small scale, but one may expect still further
progress in the production of ultra-fine-grain developers once the
need for them is realized. It seems likely that the properties of the
photographic material will not become the limiting factor until
resolutions below 1000 Å are sought. Even then an alternative
means of recording may well be found in the solid state effects of
X-rays investigated by Ladd & Ladd (1957) (cf. § 15.6) or in the
image converter devised by Möllenstedt & Huang (1957) (cf. § 15.3).

2.6.4. *Resolving-power of the enlarging system.* As explained in
§ 2.5, the most important limitation on resolution lies in the
optical system used to enlarge the primary negative: first, in its
small depth of focus, and secondly, in its finite resolving-power
owing to diffraction effects. To obtain an optical resolution of
0.25μ requires a first-class microscope and long experience in
setting it up. Beyond this limit it may be expected that the
electron microscope will be called upon, but it has to be shown
that appreciable artefacts are not introduced in the process of
separating the emulsion from its base and preparing it for insertion
in the vacuum of the electron microscope.

2.6.5. *Contrast requirements.* The diminution in the initial
intensity due to absorption in the specimen is given by (2.1):

$$I/I_0 = e^{-\mu t}.$$

The visibility of an object detail thus depends on its thickness t
and its linear absorption coefficient μ. If we accept 5 % as the
practical minimum of the change in I_0 needed for its detection, the
object must have a value for the product $\mu t \geqslant 0.05$. From the
values of μ given in the Appendix, it follows that at 10 Å wavelength the limiting thickness of uranium is 80 Å, of osmium 100 Å,
of copper 110 Å, and of carbon 2000 Å (see Table 2.2). It is seen
that biological tissues, which are essentially composed of carbon
and elements near to it in the periodic table, would have to be
heavily stained with elements high in the table, such as lead or
osmium, if details smaller than $\sim$ 1000 Å are to be made visible.

2.6.6. *Ultimate resolution.* Summing up, it can be said that the
immediate limit to the resolving-power of the contact method is

set by the optical system used for enlarging the micronegative, if proper precautions have been taken in selecting and processing the recording emulsion. By use of the electron microscope the resolution may be extended beyond the optical limit of about $0.25\,\mu$, but it remains to be shown that artefacts are not produced in the process. A resolution below 1000 Å will require the use of very soft X-rays, to limit the range of photo-electrons, and a still further improvement in photographic materials and processing control. Until such a stage is reached, contrast in the image and Fresnel diffraction in the object do not limit the attainable resolution. Taking into account all the difficulties, it remains unlikely that the contact method of X-ray microscopy will exceed the resolving-power of the best optical microscope for some time to come. Its range of application within this limit, however, is wide and ever increasing, especially for quantitative microanalysis in biology (chapters 6 and 12).

TABLE 2.2

Atomic Number	Element	Density	$\lambda = 2.5$ Å		$\lambda = 10$ Å		$\lambda = 20$ Å	
			μ/ρ	t	μ/ρ	t	μ/ρ	t
12	C	2·25	19·0	11·5μ	1120	2000 Å	7300	310 Å
29	Cu	8·9	200	0·28μ	5100	110 Å	4900	115 Å
76	Os	22·5	550	0·040μ	2200	100 Å	—	—
92	U	18·7	850	0·031μ	3400	80 Å	—	—

2.7. Associated techniques of microradiography

The contact method has been combined with other optical or radiographic techniques for particular applications.

2.7.1. Stereo-microradiography. An advantage of the contact method (shared by the projection method) over the optical microscope is that the depth of focus is very great. As the image is formed by geometrical projection, with no focusing, all planes in the specimen are sharply imaged, apart from penumbra and Fresnel edge-effects. It is thus possible to obtain three-dimensional pictures: the X-ray source is translated or the specimen is tilted by the appropriate amount between two exposures, and the finally enlarged pair of negatives (or prints) are inspected in a stereo-viewer. The complete internal structure is then visible and, if need be, quantitative measurement of spatial relationships can be made. In an optical microscope at the same final magnification the depth of focus is so small that stereophotography is valueless. The stereo method finds many applications where specimens possess a relatively openwork structure. Since comparatively thick sections can be used, it obviates the labour of reconstructing the three-dimensional arrangement from a series of pictures of thin sections. The contact method can be used for X-ray stereomicroscopy (cf. § 10.3) but it has two limitations: geometrical blurring increases rapidly for planes further from the emulsion, and, more importantly, the specimen must be transferred from one emulsion to another to obtain the two images. In the projection method these two difficulties do not arise and stereophotography can be practised readily even at high resolution (§ 11.10).

2.7.2. *Microfluoroscopy.* As the mounting and photography of specimens in the contact method is somewhat time consuming, direct observation of the X-ray image on a fluorescent screen is often preferable. For given resolution, contact conditions allow greater visual intensity at the screen than does direct magnification (cf. § 3.9). The specimen is placed in close contact with the side of the screen coated with fluorescent material and the image is viewed from the farther side with an optical microscope. The resolution is then determined mainly by the grain size of the screen. Hitherto this has been so coarse as to limit seriously the application of the method, but the ultra-fine-grained screens now available promise to make it a most valuable technique for microradiography (cf. § 10.4).

2.7.3. *Electron microradiography.* Methods of electron radiography have been devised which are analogous to the usual X-ray methods and complementary in so far as the absorption coefficient for electrons depends in a different way on the atomic constants of matter. The measurements may be made with a primary electron beam, as in electron microscopy (cf. Hall, 1955; Hall & Inoue, 1957; Zeitler & Bahr, 1957; Engström, Bergendahl, Björnerstedt & Lundberg, 1957), or with β-rays from radioactive material incorporated in the specimen, when the term autoradiography is used (Taylor, 1956; Pelc, 1957). Alternatively, secondary electrons produced by X-radiation may be used, as in the method of electron microradiography devised by Trillat (1940). Images may be obtained either by 'reflexion' or in transmission, contrast depending on the secondary emission intensity or the electron absorption coefficient respectively.

The 'reflexion' method of Trillat is illustrated in Fig. 2.6a. A beam of X-rays from a high-voltage tube T passes through a sandwich comprised of black paper B, the photographic film F and the specimen S. Electrons photo-electrically emitted from the surface of S produce an image on F. In order to ensure that this

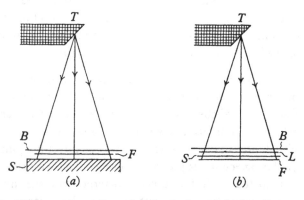

Fig. 2.6. Electron microradiography: image formed (a) by electrons ejected from specimen S by X-rays from target T; and (b) by electrons ejected from lead foil L (F, photographic film; B, black paper).

image is due primarily to electrons, and not to the X-rays from the initial beam or from fluorescent emission, the tube must be operated at very high voltage (150–200 kV) and the emulsion must be thin. A filter is used to remove the soft component from the beam. The film must be in close contact with the specimen and must be of fine grain, to allow subsequent enlargement. Lippmann films have been used, but also standard fine-grain emulsions as used in amateur photography or even copying papers (Velox). Exposure times of a few minutes were obtained at a tube–film distance of 45 cm. A number of applications to metals, paper, and some biological specimens are recorded. The photographic density produced was found to increase only slowly with atomic number, so that the method is best used where the constituents differ widely in Z.

The transmission method (Fig. 2.6b) employs a secondary emitter in the form of a lead foil, 0·2 mm. thick (L), the electrons from which pass through a specimen S of similar thinness to the film F. Contrast now depends on differential absorption of electrons by the elements in S, which again varies only slowly over the periodic table. In addition to the applications described by Trillat and his collaborators (see § 13.1.4), results were obtained by similar methods (and independently) by Seemann (1937, 1949) and by Tasker & Towers (1945).

Electron microradiography is an interesting and useful complement to X-ray methods, in spite of its limited applicability. For most purposes, however, it would be preferable to use a primary rather than a secondary beam of electrons, owing to the low efficiency of photo-electric emission, and to obtain a direct enlargement by projection or with electron lenses rather than use the contact method, owing to the slow speed of fine-grain film. The main advantages of excitation by X-radiation lie in the ready availability of high-voltage equipment, as used in medical or industrial radiography, and in the possibility of examining the surfaces of solid specimens, which is difficult with the electron microscope.

48

CHAPTER 3

MICROSCOPY BY POINT PROJECTION

3.1. Introduction

The simplest way of magnifying an object is to project an image of it on to a distant plane by means of a point source close to the object, the magnification being equal to the ratio of source–image to source–object distance. The definition in the image is determined primarily by the source size. In the early days of X-rays attempts at this method of microradiography were made by placing a pinhole before the anti-cathode, but the 'brightness' of the focal spot was too low to provide an acceptably short exposure time. More recently, special point-focus tubes have been used in which the specific loading of the target is very much higher than in a spot of normal size, giving a correspondingly shorter exposure and allowing direct magnifications as high as × 1000.

In his second communication on the 'new type of radiation', Röntgen (1896) mentions experiments on forming pinhole images, but shows no results, as also does Dorn (1896). The first published pinhole pictures were due to Czermak (1897), being simple images of the cathode, anti-cathode, anode and interposed rings. The method was revived, apparently in ignorance of the earlier work, by Uspenski in 1914 and again by Sievert in 1936. With a slit 2–3 mm. wide in front of a target carrying 3 mA, the former obtained images at an unstated distance with an exposure of 15 min. He also succeeded in getting a recognizable image of a composite rod of tin, wood and lead from the secondary X-rays emitted when it was placed near the target, but with a 'very much greater' exposure. Sievert first suggested projection as an alternative to the contact method for high definition radiography. He used a pinhole only 5μ in diameter and obtained fairly sharp images at a direct magnification of × 5, reproduced at a total magnification of × 80. The exposure time was 15 min. for a target–pinhole distance of 35 mm. and a pinhole–film distance of 10 mm., with a tube operating at 15 kV and 20 mA. He suggested that, with a pinhole of 1μ diameter, a direct magnification of $\times 10$ or more, and an overall useful magnification of $\times 250$–500, would be obtainable so long as very fine-grained emulsions were used which would allow a high final optical enlargement. The increase in exposure time with such a small pinhole could be offset at least in part by scaling down the camera.

With the advent of electron lenses, it became possible to form an X-ray source of great intensity by focusing an electron beam to a point on the target. The potentialities of such a tube for projection radiography appear to have been first realized by Malsch in 1933. He later designed a tube (Malsch, 1939) which was commercially produced for the detection of small flaws in castings. It operated at 150 kV and $1 \cdot 5$ mA, with a focus of diameter $0 \cdot 1$ mm. produced by a magnetic electron lens. The projected images were 'still sharp' at a direct magnification of $\times 10$. Without reference to this industrial work, Ardenne (1939) suggested using an ultra-fine focus to obtain radiographs with a resolution of the same order as that of the best optical microscope. He pointed out the usefulness of a transmission target in allowing a very much shorter focus–object distance than is possible with the normal type of tube, in which the window is at some distance from the target. He published a design for a microfocus tube using electrostatic lenses (Ardenne, 1940), which were apparently replaced by magnetic lenses sometime later (Ardenne, 1956), but no results from either have so far been published. Marton (1939) independently explored the same type of system, but abandoned it in the belief that the exposure-time at any usefully high magnification would be prohibitively long (Marton, 1954). Similar investigations by Cosslett (1940) and later by Cosslett & Jones (1946) proved to be more promising and led to a practicable point projection X-ray microscope (Cosslett & Nixon, 1951, 1952a, b, 1953, 1954). This instrument employed magnetic lenses to form a focus of diameter less than 1μ, giving exposure-times of the order of a few minutes at direct magnifications up to $\times 250$ in a camera length of a few centimetres. It has since been improved by Nixon (1955b) to give a spot of size $0 \cdot 1\mu$ and a useful magnification of $\times 2000$–3000, approaching the resolving limit of the ultra-violet microscope. The practical limit to its further development is set primarily by lack

4

of X-ray intensity as the focal spot is made smaller (Cosslett, 1952*a*, 1956).

The main principles of the projection method will be discussed in this chapter. The geometry of image formation and the complicating effect of Fresnel diffraction lead to a discussion of resolving-power. The intensity problem involves the brightness of the electron source, the aberrations of the electron lenses and the thermal properties of the target. Both resolution and intensity enter into the discussion of the best photographic emulsion to use, as also into a comparison of the merits of the contact and the projection methods. The intermediate method which operates at a magnification of × 2, proposed by Le Poole & Ong (1956), is also critically considered. The more practical aspects of projection microscopy are discussed in chapter 11, in particular details of the construction and operation of fine-focus tubes.

3.2. The geometry of projection

The intensity at the image and the limit to the detail resolvable in it, are in the first place determined by the geometrical relations of X-ray source, object and image (Fig. 3.1). These are in principle the same as in the contact method, apart from the position of the object (cf. Fig. 2.4*a*). A focal spot S of lateral extension s illuminates an opaque object O of width w at distance b, such that a projected image is received on a plane I, normal to the axis at a distance a from S. The magnification M is given by a/b. The image will be characterized by a penumbra of width p and an umbra of width u, given by

$$\left.\begin{array}{l} p = s(a-b)/b, \\ u = \{aw - s(a-b)\}/b. \end{array}\right\} \tag{3.1}$$

This relation is identical with (2.15–16) apart from the replacement of c by $(a-b)$.

In macroscopic radiography the definition in the image is discussed in terms of the penumbra width p (cf. § 2.4), whilst in optics it is usual to speak of the resolution at the object. The equivalent (or 'reduced') penumbra width at the object, p_0, will be

$$p_0 = p/M = s(a-b)/a. \tag{3.2}$$

The detectability of an object will depend also on the local change in intensity (contrast) to which it gives rise in the image. The limit of detectability is assumed to be reached when zero umbra exists at the image plane, so that the penumbra extends to the axis from both sides (Fig. 3.1, right). From (3.1), $u = 0$ when

$$w' = s(a-b)/a = p_0. \tag{3.3}$$

So that at the contrast limit, the penumbra width in object space is equal to the width of the object itself (cf. Fig. 2.4b). For given source size and object position, therefore, the minimum resolvable size of object is fixed and is equal to the reduced penumbra width,

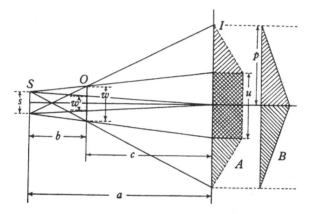

Fig. 3.1. Projection of image I of object O by finite source S; and intensity distribution (shaded) across image, (A) for finite umbra, of width u, and (B) when umbra falls to zero and penumbra has width p.

so that p_0 may be taken as a suitable measure of the edge resolution δ_E. Properly speaking it is a limit of detection, not of resolution, and we discuss below its relation to the classical definition of the latter in terms of the separability of neighbouring objects.

In projection microscopy, where b is negligibly small, (3.3) gives the edge resolution as

$$\delta_E \doteqdot s \tag{3.4}$$

confirming that the resolution is effectively given by the size of the source. The error in assuming $p_0 = s$ is only 10 % at a magnification of × 10 and it becomes negligible at the high direct magnifications usual in practice. When the object is midway between S and I, we have $a = 2b$ and so $\delta_E = s/2$ and $M = 2$; this is the condition in the 'times-two' method (§ 3.9). In the contact method, on the other hand, $b \doteq a$ and $M \doteq 1$, so that from (3.2)

$$\delta_E = p/M \doteq p, \qquad (3.5)$$

which confirms that the resolution then equals the actual penumbra width, as discussed in § 2.4.

As the object is moved from O towards I the reduced penumbra width p_0 decreases (cf. 3.2) as well as the actual penumbra width, so that all object planes to the right of O are imaged with increasing sharpness. Thus, although there is no true depth of focus in the optical sense, a three-dimensional object will be imaged with high definition, by contrast to the light microscope in which the depth of focus is of the same order as the resolved distance. For this reason it is readily possible to obtain stereographs at high magnification by projection microscopy; the plane nearest the source can be identified as it will be most magnified and least sharp.

The concept of reduced penumbra width, related by (3.3) to the detection limit of an individual object, allows a treatment of resolving-power along the same lines as in optics, in terms of the closest distance between two opaque objects for which the optical system can still form separate images of them. Such a definition also involves a detection limit, however, since some criterion of visibility of separation has to be adopted. It is usual to regard two (bright) images as resolved if the level of intensity between them falls by not more than 25 % of the peak intensity (Fig. 3.2a). Light optics is concerned with diffraction images in which intensity varies in a complex manner, whereas in shadow projection we have overlapping penumbras in which the variation is linear. To adapt the definition of resolving-power to these circumstances, we need only require the centres of two objects to be far enough apart for the rise in intensity between their (dark) image points to be 25 % of the maximum. From simple geometry (Fig. 3.2b) this will be so when they are separated by 1·25w, if w is the width of each object, and this may be taken as the limit of resolution δ.

On the other hand, the detection limit of a single object will be equal to w when the source size is such that $w = p_0$, by (3.3). So the criterion of detection as equal to reduced penumbra width is thus simply related to the more exact definition of resolution in terms of separability of neighbouring points. Since the difference is only that between w and $1 \cdot 25w$, and since in practice the eye can detect an intensity change of much less than the 25 % assumed above, it is justifiable to take the simpler result as the practical measure of resolution: $\delta = w = \delta_E$. In most applications of projection microscopy the direct magnification will be so high that

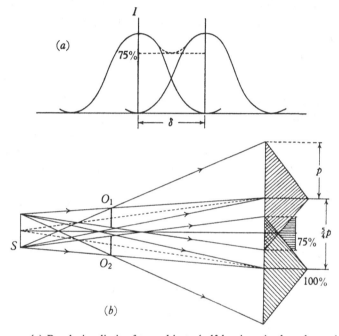

Fig. 3.2. (a) Resolution limit of two objects (self-luminous) when determined by diffraction: intensity falls to 75 % of peak value. (b) Resolution limit of two objects, when determined by penumbra formation: the composite intensity is 25 % above zero level when the gap between objects is one-quarter of the diameter of either.

the further step of convenience may be taken of equating the point-point resolution with source width, as already done in (3.4) for edge resolution δ_E. Since in many experiments the source size (s) is not known with any exactitude, it is possible to use this relationship in reverse, that is, to estimate s from the resolution actually obtained. By this means it is also possible to draw conclusions about the astigmatism of the electron optical system, or of the source itself, from the observed asymmetry of the image (Dyson, 1957).

3.3. Fresnel diffraction

The image may also be blurred by Fresnel diffraction, which results in edge-effects which may obscure detail and limit the resolving-power of the X-ray microscope. It occurs also in the reflexion and contact methods, but it is discussed here because it can be a more immediate limiting factor in projection conditions. The usual wave optical treatment can be applied to the diffraction of X-rays at an opaque straight-edge (Fig. 3.3). The positions of the fringes are given by the relation

$$x = \nu(ac\lambda/2b)^{\frac{1}{2}}, \tag{3.6}$$

where x is the distance of a point on the screen from the geometrical shadow of the edge, λ is the wavelength of the illumination and a, b and c specify the relative separations of source, obstacle and viewing screen. The dimensionless number ν specifies the variation with x of the amplitude of the resultant disturbance, its value being given by the Fresnel integrals or graphically by the Cornu spiral. At the first maximum it is approximately $\sqrt{2}$, so that (3.6) becomes

$$x_1 = (ac\lambda/b)^{\frac{1}{2}} \tag{3.7}$$

and x_1 is termed the width of the first fringe. Referred to the object plane, the 'reduced' fringe width will be

$$x_0 = x_1/M = (bc\lambda/a)^{\frac{1}{2}}. \tag{3.8}$$

In contact microradiography, where the object is close to its image, we have $b \doteqdot a$ and so

$$x_{0c} \doteqdot (c\lambda)^{\frac{1}{2}}. \tag{3.9}$$

If the fringe width is to be less than the resolution, as determined by the penumbra width (2.17), the requisite object–image distance c may be found from (3.9) for given wavelength of the illuminating X-ray beam. For $\lambda = 1$ Å, for instance, the fringe width will be 1μ when $c = 1$ cm. If a resolution of 0.2μ at a wavelength of 10 Å is sought, then c must be 40μ or less, a condition which is automatically satisfied by the usually adopted technique of placing the specimen directly on the recording emulsion. However, it sets an upper limit to the thickness of specimen that can be examined at this resolution, even if a source of small angular size is used in order to keep the penumbra width below the limit set by (2.18). From this viewpoint, Fresnel diffraction has to be taken into account in high-resolution work with the contact method as well as in projection microscopy (cf. also § 3.9).

When the object is close to the source, as it is for projection, we have $a \doteqdot c$ and (3.8) becomes

$$x_{0p} \doteqdot (b\lambda)^{\frac{1}{2}}, \tag{3.10}$$

so that similar limits are set to specimen thickness and position as in contact microradiography. But, in projection microscopy,

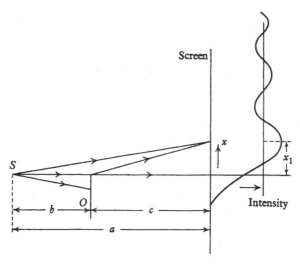

Fig. 3.3. Formation of Fresnel fringes by diffraction of illumination from source S at straight-edge O. Right: the variation of intensity across any subsequent plane of observation.

resolution is determined by the source-diameter s and not simply by the penumbra width (cf. 3.4); therefore the Fresnel effect will limit resolving-power unless the condition $x_{0p} \leqslant s$ is satisfied. For hard X-rays ($\lambda \sim 1$ Å) and moderate resolution ($s \sim 1\mu$), this is not difficult to ensure, since (3.10) requires only that $b \leqslant 1$ cm. But for a source as small as $0 \cdot 1\mu$, b would have to be less than 100μ, which already appreciably limits the specimen thickness. For soft X-rays (~ 10 Å), which must be used to get adequate contrast in biological material, the restriction to $b \leqslant 10\mu$ begins to be serious, so that Fresnel diffraction may set the ultimate limit to resolving-power in the projection method.

We have considered so far the diffraction effects at one edge of an opaque object only. When the dimensions of the object are so small as to be comparable with the fringe width, complications arise because interference can take place between the waves diffracted from different sides of it. The intensity variation in the fringes then depends somewhat on the shape of the object, whereas the geometrical penumbra is independent of either shape or dimensions of the object. The limiting resolving-power, involving as it does the level of intensity between the images of two objects, will thus vary slightly with their shape—if they are bars or disks, for instance.

In the case of a single bar object, the fringe pattern outside the geometrical shadow is almost identical with that at a semi-infinite straight-edge, but within the shadow equally spaced fringes are now formed. The central fringe is always bright and the spacing is the same as that of Young's fringes formed by two slits. Reduced to the object plane, and for b small, as in projection, the internal spacing is

$$x_{0c} \doteqdot b\lambda/w. \qquad (3.11)$$

As the bar width w is reduced, the separation of the internal fringes increases and their number decreases rapidly. The limit of detectability may conveniently be taken as that width of bar for which one internal fringe remains, of width equal to the bar width. With $x_{0c} = w$, we have from (3.11):

$$w_x \doteqdot (b\lambda)^{\frac{1}{2}} \qquad (3.12)$$

so that the internal fringe width then equals that of the first external maximum (cf. (3.10)). It may indeed be possible to infer the existence of a line object from the interference pattern, even when its width is less than that of the first fringe, but for practical purposes we may take the reduced width of the first fringe (3.10) as setting the limit of detection of a bar object, in conditions where the Fresnel effect is the limiting factor, i.e. when $x_{0p} > s$. This is directly analogous to the situation when penumbra formation predominates and the limit of detection is equal to the reduced penumbra width (cf. 3.3).

The case of two parallel bar objects can be treated in the same way as in penumbra formation above. We assume them to be just resolved when the rise in intensity between their images reaches 25 % of the maximum. The rise in the first fringe is steep and, over the central region, linear, so that as before the combined intensity will reach the limit of 25 % when the centre to centre separation of the bars is approximately $1 \cdot 25 x_0$, if x_0 is the reduced width of the first fringe as given by (3.10). Thus the resolving-power δ, in the strict sense of separability, is given by

$$\delta = 1 \cdot 25 (b\lambda)^{\frac{1}{2}}. \tag{3.13}$$

At the same time, the individual bars will be discernible only if the width of each is at least equal to that of the first fringe, by (3.12). If α_0 is the semi-angle subtended at the source by such a bar, then

$$\alpha_0 = w/b = (\lambda/b)^{\frac{1}{2}}$$

from (3.12). Substituting in (3.13), we have:

$$\delta = 0 \cdot 625 \lambda / \alpha_0, \tag{3.14}$$

which is the analogue in projection conditions of the well-known expression for resolving-power in optics. There is now no image-forming system, and the angular width of the beam intercepted by the object replaces that of the beam accepted by the objective.

Disk objects may be discussed in the same way as bar objects. The numerical results are slightly different, but for all practical purposes we may take (3.12) as the limit of detectability of a single disk object and (3.13) as giving the resolving-power for two neighbouring disks. A subjective influence comes into play, however, in that it is easier for the eye to pick out line fringes than rings, against a more or less confused background. For this reason the line resolution is always better than the point resolution in X-ray images, as in optical microscopy. The factor of advantage may be roughly estimated at between 2 and 5 times, depending on the opacity of the object and on the nature of the background against which it is seen.

To summarize the above discussion: in conditions in which Fresnel diffraction is the limiting effect, both the limit of detection of an isolated object and the true resolving-power for two neighbouring objects may be taken as equal to the width of the first fringe. In the conditions of projection, this width is equal to the square root of the product of the X-ray wavelength and the source–object distance, by (3.10). It should be noted that the first fringe only need be considered. In special conditions, Kellström (1932) was able to obtain several orders of fringes around the X-ray shadow image of a wire, but a thorough investigation of the factors influencing fringe formation (Baez & El-Sum, 1957) has shown that it is difficult to obtain more than a single fringe in the conditions of microscopy. The higher orders are blurred out, owing to the partial transparency of most objects and the finite size and lack of monochromaticity of the normal X-ray source. Even the first fringe is less well defined than theory predicts.

In sum, therefore, Fresnel diffraction is a serious limitation in projection microscopy only when very high resolution is sought (cf. Pl. IA). In these conditions Nixon (1955b) has proposed that positive use be made of it, as a direct measure of the resolution obtained.

3.4. Experimental systems

Three methods of producing point sources have been developed in recent years. In practice only that which uses electron lenses has found wide application.

3.4.1. *Camera obscura.* The use of a pinhole in front of an X-ray source of normal dimensions is the simplest way of projection microscopy (Fig. 3.4), and the only means not requiring a special tube. The effective intensity of the pinhole is given by the product of its area and the radiation per unit area from the target. Since the radiation per unit area in a microfocus tube depends inversely on the size of the focal spot, owing to the conditions of thermal dissipation (§ 8.2), it is better to form a spot equal to the pinhole in size. A physical pinhole can then be dispensed with altogether, except when microdiffraction or microabsorption measurements are to be made. Apart from intensity limitations, it is not easy to make holes smaller than a few microns in diameter in a plate. The only work reported with a camera obscura system in recent times is that by Rovinsky & Lutsau (1957). They employed a pinhole 1μ in diameter with a high intensity X-ray tube (10 kV and 5 mA) and obtained images at a direct magnification of × 100 with an exposure of $\frac{1}{2}$–1 hr.

3.4.2. Point anode tube. A point source is directly obtained if the target is made in the form of a needle-point, faced by a ring cathode (Fig. 3.5). Such a point anode tube has been designed by Drenck & Pepinsky (1951) for microdiffraction, and operated intermittently with heavy current pulses. A more elaborate type was

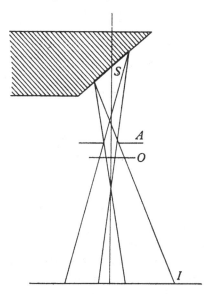

Fig. 3.4. Image projection with finite source S, through pinhole aperture A in front of object O (camera obscura).

made especially for projection microscopy by Rovinsky, Lutsau & Avdeyenko (1957, 1960), incorporating a focusing electrode. It has a cathode loop 0·25 mm. in diameter and a tungsten anode of 0·1 μ radius at the tip, sharpened electrolytically, giving a focal spot about 0·5 μ in diameter. The X-rays issue along the axis of the cathode, through a window into the camera. At 8 kV and 3–5 μA the exposure time was 3–15 min. at a direct magnification of × 30–100, depending on the specimen, in a camera length of 28 cm.

The point anode tube provides a compact microfocus X-ray source, utilizing a much simpler focusing system than is required to form a point focus on an extended target. Its disadvantage lies in the poor heat dissipation properties of a needle as compared with a block or even a foil target (§ 8.2). The target loading, and therefore the X-ray intensity, are correspondingly limited. In the operating conditions quoted, the foil target can dissipate roughly three times as much energy.

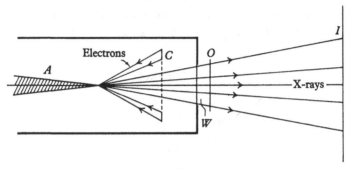

Fig. 3.5. Image projection with point source, formed by electrons from a ring cathode *C* exciting X-rays at the tip of a needle anode *A*; the radiation issues through a window *W* and illuminates the object *O* (after Rovinsky, Lutsau & Avdeyenko, 1957).

3.4.3. *Window target tube.* A point focus of almost any desired size can be formed on a target by demagnification of a point cathode. One or more electron lenses form a reduced image of the latter (Fig. 3.6), in accordance with the usual optical laws. If the cathode is a hairpin filament thick enough to have a reasonably long life, a magnification of order 1/100 is needed, and the filament–lens distance must be large and the tube correspondingly lengthy (20–40 cm.). This type of tube, employing magnetic lenses, has been most used for projection microscopy. It has the great advantage that the object can be brought very close to the X-ray source whilst remaining in atmospheric conditions, giving high direct magnification in short camera length with correspondingly short exposure. In tubes of normal construction, with massive target, the window is usually at least 1–2 cm. from the source, so that the higher target loading allowed on a block is more than offset by the effect of the inverse square law at the camera. However, the semi-microfocus tube with electrostatic lens, designed by Ehrenberg & Spear (1951a) for microdiffraction, is useful for projection microscopy at moderate resolution (~ 40μ), although it is better used for contact microradiography at high resolution (Ehrenberg & White, 1957).

The resolution obtained with the point focus tube is primarily limited by the size of the X-ray source and, secondarily, by Fresnel diffraction in the specimen. The incident electron spot can be made as small as a few Ångström units, aside from intensity limitations, but it is spread by scattering in the target to a lateral extent of the order of the electron range. When very small spots are needed, the target must thus be made thinner than the range at the given operating voltage. The factors involved, and the practically attainable resolution, are discussed in § 3.6.3.

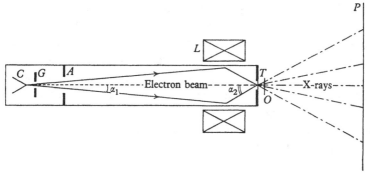

Fig. 3.6. Image projection with point source formed on window target T by electrons issuing from cathode C, through grid G and anode A, and focused by electron lens L; the X-rays throw an image of the object O on to the screen P. (Cosslett, 1957.)

The main limitation to the performance of the projection microscope is the available X-ray intensity. It depends on a number of factors: the specific emission from the cathode and the aberrations of the electron lenses; the efficiency of X-ray production, the thermal dissipation, electron scattering and X-ray absorption in the target; and the sensitivity of the photographic emulsion or other recording-system. These influences, and the compromise between them that is reached in practice, are discussed below. At the present time the lens aberrations and the cathode brightness are the chief limiting factors. Until these can be improved, heat-dissipation in the target is of secondary importance.

3.5. Electron optical limitations

3.5.1. *Cathode emission.* The electron intensity at the target depends in the first place on the intensity emitted from the cathode and then on the gathering power of the focusing system, which is severely limited by spherical aberration. The discussion starts from the concept of brightness β, defined as the electron flux per unit area per unit solid angle or the current density (ρ) per unit solid angle:

$$\beta = I/\pi^2 r^2 \sin^2\alpha = \rho/\pi\sin^2\alpha, \qquad (3.15)$$

where I is the current through an area of radius r within the angular aperture α. In analogy with the Helmholtz–Lagrange law of optics, β is invariant through any electron optical system (Langmuir, 1937). As recently pointed out by Haine (1957) it holds even in an aberrated image so long as α is properly measured as the angle of the illuminating cone at a given image point, not as the beam angle as a whole. Account must be taken, however, of changes in refractive index, which depends in electron optics on the local electrostatic potential and so on the voltage. Thus the brightness (β_1), in the beam that has been accelerated to the full anode voltage V_1 and is now contained in a cone of semi-angle α_1, will be

$$\beta_1 = \frac{I}{\pi^2 r_1^2 \sin^2\alpha_1} = \beta_c \frac{V_1}{V_c} = \frac{I}{\pi^2 r_c^2 \sin^2\alpha_c} \frac{V_1}{V_c}, \qquad (3.16)$$

where β_c is the brightness at the cathode, α_c is the angle and r_c the radius over which emission takes place; r_1 is the radius of the virtual cathode (that is, of the image formed by producing backwards the rays issuing through the anode) and V_c is the effective velocity with which electrons leave the cathode. $\sin\alpha_c$ can be taken as unity, since electrons are emitted at all angles up to $90°$ to the cathode surface, and $\sin\alpha_1 \doteq \alpha_1$ since the final beam is very narrow. The current is constant along the beam, and so the relationship between the observable quantities is

$$(r_1/r_c)^2 \, \alpha_1^2 = V_c/V_1. \qquad (3.17)$$

In most electron guns r_1 is of the same order as r_c and V_c is a small fraction of a volt, so that α_1 is of order 10^{-2} to 10^{-3} and the beam emerging from the anode is contained in a very small cone. The main exception to this working rule is the point cathode operating by cold emission (§ 11.1).

In a perfect focusing-system in which there is no further change in potential, and irrespective of the number of lenses, the brightness at the image β_2 will be equal to that of the virtual cathode β_1, so that

$$r_2^2\sin^2\alpha_2 = r_1^2\sin^2\alpha_1 \qquad (3.18)$$

$$\rho_2 = \rho_1(\sin\alpha_2/\sin\alpha_1)^2$$

$$= \rho_1/M^2,$$

where the magnification M is defined in the usual way as r_2/r_1 and r_2, α_2 and ρ_2 specify the conditions at the image. It is sometimes useful to refer the final to the initial conditions: if ρ_c is the cathode current density and M_c the overall magnification, r_2/r_c,

$$\rho_2 = \rho_c/M_c^2$$

$$= \rho_c(\sin\alpha_2/\sin\alpha_c)^2 V_1/V_c. \qquad (3.19)$$

But $\alpha_c = \frac{1}{2}\pi$ and the maximum value of α_2 is also $\frac{1}{2}\pi$, so that the maximum possible current density at the image will be

$$\rho_x = \rho_c(V_1/V_c) \qquad (3.20)$$

at a minimum magnification (maximum demagnification) given by

$$M_x = (V_c/V_1)^{\frac{1}{2}}. \qquad (3.21)$$

For an anode voltage of 10 kV a perfect system would give a focused current density of order 10^4 A cm.$^{-2}$ from a cathode of specific emission 1 A cm.$^{-2}$ The required magnification of 10^{-2} could be readily obtained with two stages of demagnification of 1 to 10.

3.5.2. *Lens aberrations.* Electron lenses, whether electrostatic or magnetic, suffer from a large number of aberrations, three of which are severe: astigmatism, chromatic and spherical aberration. Astigmatism is due mainly to mechanical shortcomings in making the lens, particularly to lack of circularity of the bore. It can be considerably reduced by care in design and construction, and any residual astigmatism may be compensated by auxiliary fields possessing astigmatism of opposite sign (§ 11.3). Chromatic aberration may be nullified by using a highly monochromatic beam, that is, by stabilizing the anode voltage and lens supplies to the order of 1 part in 10^3 or 10^4, depending on the size of focal spot required. These limits of stability are less severe than those for electron microscopy and the necessary techniques are well known. Spherical aberration is much more difficult to minimize. It is due to the outer zones of a lens having higher refractive power than the paraxial zone and remains severe even for a point on the axis, for which all other field aberrations vanish. It is thus of great importance in the production of a point focus.

In the presence of spherical aberration a vanishingly small point source will be imaged as a disk of confusion of radius r_s proportional, as in optics, to the third power of the angular aperture of the lens α_2 and to a factor C_s which depends on the form of the lens and the strength at which it is operated:

$$r_s = 1/4(C_s\alpha_2^3).\tag{3.22}$$

The factor of $1/4$ is absent in the geometrical optical treatment, but arises in the wave theory and corresponds to the Rayleigh limit of resolution in light optics. The effect of the aberration is to limit the permissible demagnification: α_2 cannot be larger than allowed by (3.22). This smaller value takes the place of $\tfrac{1}{2}\pi$ in (3.19), so that the current density is correspondingly limited, i.e. the whole of the current available from the cathode can no longer be focused into the spot. The limit to α_2 usually adopted is that which makes r_s, found from (3.22), equal to r_2 as found from the electron optical magnification by (3.18); a physical aperture of appropriate size is inserted in the lens to ensure this equality. The maximum current density in the image, from (3.19) with $\sin\alpha_c = 1$, is then

$$\rho'_x = \rho_c\alpha_2^2(V_1/V_c)$$
$$= \rho_c(4r_2/C_s)^{\frac{2}{3}}(V_1/V_c)\tag{3.23}$$

and the maximum target current I'_2 is

$$I'_2 = \{4\pi\rho_c r_2^{\frac{8}{3}}\}/\{(2C_s)^{\frac{2}{3}}(V_1/V_c)\}.\tag{3.24}$$

The target current in the point focus tube thus decreases rapidly (cf. Fig. 3.7) as the spot size is reduced below the size at which C_s becomes the limiting factor (Cosslett, 1952a).

In principle, means of correcting spherical aberration in electron lenses are already known (Scherzer, 1947; Burfoot, 1953; Archard, 1955), but none of them has yet been developed to a stage of practical usefulness. Until this is achieved, the aperture of the strongest electron lens in the imaging system has to be limited to the value given by the required spot size and the aberration constant of the lens, according to (3.22). In the usual type of point

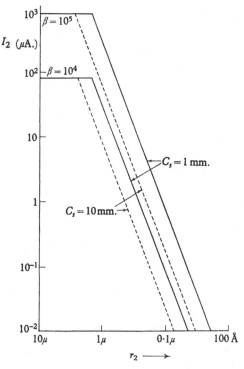

Fig. 3.7. Variation of focused current I_2 with electron spot radius r_2, for two values of source brightness β (in A cm.$^{-2}$ sterad.$^{-1}$) and two values of lens spherical aberration coefficient C_s.

focus tube this will be the final lens. The value of C_s depends on a high power of the focal length so that it is desirable to operate the final lens at maximum strength, i.e. with very high demagnification. In these circumstances the depth of focus of the electron spot is very small (a few microns) so that the position of the target with respect to the lens is critical and the control of focal length even more so (cf. § 11.9).

3.6. Target limitations

When the focused electrons strike the target, they are slowed down and scattered by interaction with the atoms in it and, except in very thin targets, finally come to rest after losing all the energy imparted to them in the electron gun. In the process X-rays are emitted, but by far the greater part of the energy is dissipated as heat. The output of X-rays may, therefore, be limited by the amount of energy which can be continuously removed by cooling, as determined by the geometry and the thermal properties of the target. Apart from this, the intensity of X-rays available for image formation will depend on the efficiency of their production in a given target material at a given voltage, and on the amount lost by absorption within the target. The volume from which X-rays are emitted, and so the size of the source for projection purposes, will be influenced by the extent of scattering, which again depends on voltage and on target material.

3.6.1. *Production and absorption of X-rays.* The factors which govern the production of X-rays both in the characteristic lines and over the continuous spectrum, and also their absorption in the target, are considered in detail in chapter 8. For the continuous spectrum the energy efficiency is of order 10^{-3} and depends directly on voltage and on the atomic number Z of the element bombarded. The intensity in the characteristic spectrum depends in a more complicated way on the voltage and is again proportional to Z. In general it is much less than the total energy in the continuous spectrum, although the ratio improves in the soft X-ray region.

In X-ray microscopy, and even more so in microanalysis, the working voltage has to be chosen primarily with regard to the absorption properties of the specimen to be examined. In general, absorption increases rapidly with wavelength, so that soft X-rays will be preferred in the interest of image contrast or element detectability, whereas efficiency of production demands high voltage and harder X-rays. In practice a compromise is reached by using the highest voltage that will still give the required contrast (§ 2.3). The choice of target material also is restricted when absorption measurements are to be made at particular wavelengths on either side of the edge for the element under analysis. In other circumstances X-ray output will be served by using a target material of high Z, such as gold or tungsten, so long as its absorption coefficient for the peak wavelength in the continuous, at the operating voltage, is not so high as to produce serious attenuation of the X-rays passing out through the target. With a window target the thickness must be kept so small, to minimize electron diffusion (see below), that self-absorption is usually negligible except for soft X-rays. When a characteristic line is to be employed, the self-absorption has a beneficial monochromatizing effect (§ 8.7), since an element is highly transparent to its own radiation. In most work, however, the choice of target material has to be made as much for its thermal properties as for efficiency of X-ray production and absorption.

3.6.2. *Thermal dissipation.* The conditions of thermal dissipation in X-ray targets have been thoroughly analysed by Oosterkamp (1948) for focal spots of differing size and shape. For spots of macroscopic size the rate of dissipation, even under the most favourable conditions of forced cooling, is less than the maximum possible rate of energy supply from the incident electron beam (see § 8.2). As the spot size is reduced, the supply of energy (at constant current density) decreases with the square of its radius (r), whereas loss by conduction falls only in direct proportion to r since, in a thin target, dissipation is almost entirely in the radial direction. For very small focal spots, therefore, the maximum potential rate of loss may be greater than the maximum practicable rate of supply, so that the limitation on X-ray output comes to be set by the properties of the electron gun and lens system, as discussed in § 3.5, and not by the thermal properties of the target (Cosslett, (1952 a). The critical spot size, below which more energy could be dissipated than can be supplied, will depend on the specific emission from the cathode, the anode voltage and the spherical aberration coefficient of the lens on the one hand, and on the thickness, thermal conductivity and melting-point of the target material on the other hand. The maximum permissible temperature in the spot is usually taken as three-quarters of the melting-temperature.

The outer radius of the target, at 'sink' temperature, is of secondary importance because it enters only logarithmically into (8.9).

On the basis of their thermal properties, the order of preference for target materials is then found to be tungsten, copper, silver, gold, aluminium, in descending order. In typical experimental conditions the critical spot radius on a tungsten foil is about 1μ, with $V = 10$ kV and $I = 25\mu$A, the corresponding specific loading W_e being 100 kW mm.$^{-2}$ The point of energy balance can only be moved towards smaller spot sizes if either cathodes of higher specific emission or lenses at least partially corrected for spherical aberration can be made available. The prospects for using point cathodes, operating under field emission conditions, to achieve the first goal have been discussed by Cosslett & Haine (1956) and by Drechsler, Cosslett & Nixon (1960).

3.6.3. *Electron scattering.* The resolution attainable in projection depends primarily on the size of the X-ray source. Its diameter s is determined in the first place by the size of the electron focal spot, but also by the extent to which the electrons spread laterally in the target in the process of coming to rest. This scattering effect in practice makes it only possible to obtain an X-ray source smaller in diameter than the range of electrons in the target at the cost of serious loss in intensity. The experimental limit on resolving-power is thus set more by the inconvenience of long exposure times than by any electron optical difficulty in forming very minute focal spots.

The diameter of the incident electron spot can be calculated from the size of the virtual cathode and the demagnification produced by the electron lenses (§ 3.5). Provided that the aperture of the last lens is not larger than that specified by (3.22), the enlargement of the spot by spherical aberration will be negligible compared with that due to scattering. The effect of scattering will depend on the thickness of the target foil and on the energy of the incident electrons. As shown in § 8.3, under certain simplifying assumptions, the range R of electrons of initial voltage V in a material of density ρ, atomic number Z and atomic weight A, is given by the expression (8.17):

$$R = (A/a_2\rho Z)V^2, \tag{3.25}$$

where a_2 is a constant for most materials at a given voltage and varies only slowly with voltage. By range is meant the average distance travelled by electrons before their velocity becomes zero. The constant a_2 is of order 10^{12} for elements of high Z, so that for $V = 10$ kV and $\rho = 10$ g .cm.$^{-3}$, the range is about 0.2μ (since $A/Z \doteqdot 2$). The sidewards spread of the entrant electrons is more difficult to determine, but it will be of similar order since an electron will suffer several hundred collisions before coming to rest. The small deflexions suffered in each act will sometimes combine to give a total deviation of $90°$ or more by the end of the range, but the average deviation will be much less. The theory of electron diffusion shows that the major part of the beam remains within a cone of semi-angle $30-35°$. The calculations of Langner (1957) reach a similar conclusion: the radius of the disk of diffusion was found to be approximately equal to half the target thickness, for targets not much thinner than the range (cf. intercepts on the μ-axis in Fig. 3.8). So we may take the effective width of the X-ray source, created by an incident pencil thin compared with R, to be equal to R; it is likely to be less than this, since the X-rays emitted towards the end of the range will be too soft to contribute to image formation.

It follows that when a source smaller in diameter than R is to be formed, the target thickness must be reduced to the point where the lateral spread is less than R and the incident electron spot must be reduced to a similar diameter. Fortunately the volume into which scattering occurs is pear-shaped rather than spherical (Ehrenberg & Franks, 1953), so that there is not much broadening in the early part of the range where electron velocity is still high and most of the harder X-rays will be produced. Nevertheless, a considerable loss in X-ray output must result as the target thickness is reduced. It is shown later (§ 8.3) that X-ray output is directly proportional to target thickness in targets thin compared with the range (cf. 8.23). For an incident electron spot of diameter equal to $1/n$ of the range, the target thickness must also be approximately R/n if undue spread is to be avoided. The X-ray intensity would then be $1/n^3$ of what it was for a spot of diameter R with a target of thickness R, if the incident current density were constant; in the presence of spherical aberration it will be $1/n^{\frac{11}{3}}$. If a gold target at 10 kV is used ($R = 0.17\mu$), improvement of resolution from $0\cdot17\mu$ to 500 Å would involve a reduction in intensity of about 50 times.

Assuming an electron spot that would be of negligible size in absence of aberrations, Langner (1957) has calculated the effect on

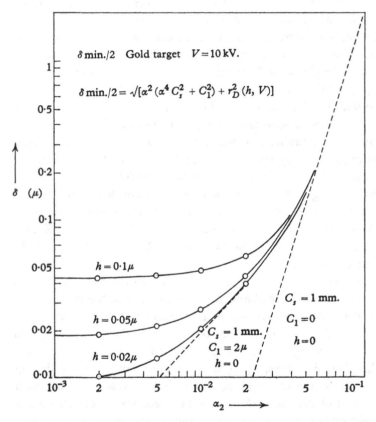

Fig. 3.8. Resolution δ in the X-ray projection microscope in function of angular aperture α_2 of the final electron lens, for different values of target thickness, h, spherical aberration coefficient C_s and chromatic (plus astigmatic) coefficient C_1 (gold target, at 10 kV). (Langner, 1957.)

X-ray source size, and hence on resolution, of spherical aberration and astigmatism in the electron lens and of electron diffusion in the target (Fig. 3.8). Even for very thin targets diffusion is the chief limiting factor. For a gold target at 10 kV, an ultimate resolution of about $0\cdot04\mu$ should be obtained with a thickness of $0\cdot1\mu$ and of

0.02μ (200 Å) with a thickness of 0.05μ. These limits are not appreciably altered by correction of the residual astigmatism of the lens. They would represent an improvement of almost an order of magnitude over the performance of the optical microscope.

The broadening of the X-ray source by electron scattering thus limits both the resolving-power and the intensity attainable in the projection method. The loss in intensity can be made good, in principle, by correcting the spherical aberration which now limits the permissible aperture of the lens, and by developing new types of cathode with higher specific emission. The limitation of source size by scattering, however, can only be countered by means which themselves lead to further fall in intensity: reducing target thickness or using lower anode voltage, since range is approximately proportional to V^2. The achievement of a resolving-power appreciably better than 0.1μ (1000 Å), with a reasonable exposure-time, will require both these latter steps. A resolution as high as 100 Å is unlikely to be reached until new cathodes and corrected lenses become available.

3.6.4. *Choice of target material.* For many purposes the nature of the target material will be fixed by the necessity of using a particular X-ray line or a corresponding waveband at the peak of the continuous spectrum. When it is not, the choice may be made from the metals having the best combination of thermal properties, especially if intensity is a prime consideration: tungsten, copper, silver, gold, aluminium. All these are available in the form of thin foil, suitable for window targets. If the electron range is also a matter of importance, then the order of preference would be tungsten, gold, silver, copper, aluminium, which fortunately corresponds broadly with the order of merit for intensity. Consideration of the efficiency of X-ray production and absorption, as distinct from that of thermal dissipation, may partially reverse this order, however. As the efficiency of emission is proportional to Z (cf. § 8.5), tungsten might be thought preferable to copper. But at 10–15 kV, the absorption of the peak of the continuous spectrum in tungsten is appreciably greater than in copper, whilst the K lines of copper make a contribution but those of tungsten are not excited. For these reasons the total output from tungsten at this voltage is less than that from copper (Dyson, 1956). The absorption coefficients, however, vary considerably through the spectrum, so that the relative merits of any two metals must be assessed in the light of the particular anode voltage used (cf. Long, 1959) and the band of X-ray wavelengths which will give greatest contrast in a particular type of specimen.

3.7. Photographing the image

The photographic requirements in projection microscopy are much less stringent than in the contact method, because the image is now magnified. Indeed an emulsion of almost any grain size may be used, provided that the direct magnification at the plate is high enough to ensure that two resolved points of the image do not fall on the same grain. If the plate is placed nearer to the object than this requires, resolution will be limited by grain size and not by source size. Once the position of the object and the source diameter s are given, the width of the penumbra p in any subsequent plane (cf. Fig. 3.1) is given by $(M-1)s$, by (3.1). The axial distance at which p equals the grain size g can at once be determined, giving the optimum position of the recording emulsion and the corresponding magnification. The value of g taken, however, must be that for the average separation of grains after development and not the size of the individual silver-halide crystallites; except in some special heavily loaded emulsions the grains are very far from being in contact.

It is then a question of secondary importance whether to use a fast emulsion of large grain far from the source, or a slow, fine-grain emulsion closer to it. The speed is directly related to the square of the grain size, in first approximation (§ 2.5), so that it is always compensated by the inverse square law. A finer grain can be used closer to the source, but will be slower in roughly the same proportion as the intensity of illumination is greater at this distance. Since Lippmann and similar emulsions require special handling, it is more convenient in projection microscopy to use emulsions of moderate speed and grain, such as a lantern plate; the photographic resolution, g, is then about 20–30μ for X-rays. If the source size is 1μ, a direct magnification of at least 25 is required, which can be obtained in the compact camera length of

25 mm. if the object is 1 mm. from the source; this is already closer than is needed to suppress the first Fresnel fringe (3.10), even with soft X-ray illumination.

If, on the other hand, the object position is not given, but is variable within a camera of fixed length and with a source of given size, then the exposure-time will be proportional to $1/M^2$ so long as grain size is always made equal to penumbra width. The resolution, as determined by the reduced penumbra width p/M is very little affected, so long as M is always greater than 10 (cf. Table 3.1, p. 83). It is thus advantageous to use the smallest practicable object distance, and largest available grain size, in a camera of given length if the source size is fixed, but these are not necessarily the best working conditions when the size of the source can be varied (cf. § 3.9).

It must be noted, however, that it is desirable to shorten the camera length, or to use a coarser emulsion at fixed length, when resolution is improved by making the spot size s smaller, as otherwise the exposure-time will become unduly long. The current into the spot depends on $s^{\frac{8}{3}}$, by (3.24), so that X-ray output falls rapidly; it may fall even faster if the target has to be made thinner to limit widening of the source by electron scattering. At the same time, to minimize Fresnel diffraction, the object must be brought closer to the target in proportion to s^2 (cf. 3.10). The penumbra width p at given image distance will fall in direct proportion to s, owing to reduction in source size, and will increase in proportion to s^2, owing to reduction in object distance. Thus, although the resolved distance at the object is proportional to s, the magnified distance between resolved points in the image (fixed by p) will vary inversely with s, at fixed camera length. Hence, halving the source size will allow an emulsion with grains twice as big to be used, having four times the speed; alternatively, the same emulsion can be used at half the camera length, giving the same saving in exposure. In this way the rapid fall in X-ray intensity with spot size can be largely offset by shortening the object distance and changing the photographic emulsion. The procedure has a practical limit, in that a camera length of less than a few millimetres is inconvenient and also because the field of view in the image becomes unduly small (cf. § 3.8.2).

3.8. Comparison of the contact and projection methods

The factors governing resolving-power and exposure-time, as well as experimental convenience, differ appreciably as between contact microradiography and projection X-ray microscopy. In this section we compare the two methods as they have been understood and practised up to the present time: that is, with very small image and object distances respectively. In the following section we discuss conditions of intermediate magnification and in particular the 'times-two' method.

3.8.1. *Resolution.* The contact method is limited in resolving-power by the grain of the photographic emulsion and, more immediately, by the performance of the optical system used to enlarge the original one-to-one image recorded on it. The effective grain size in the best emulsions, after processing, may be as small as 0.1μ. Even with ultra-violet illumination it is difficult to resolve detail of this order with an optical system; in practice visible light is normally used and the best resolution obtained is about 0.25μ.

The resolving-power of the projection method is already somewhat better than 0.2μ; in Pl. I A it approaches 0.1μ, to judge from the width of the Fresnel fringe (Nixon, 1955b). In principle there is no insurmountable obstacle to it being considerably improved. Restrictions on the useful field (see below) and on image contrast (§ 2.6), however, make a resolving-power better than 100 Å of doubtful value, even if an X-ray source is obtainable intense enough to keep the exposure-time within reasonable bounds. It would appear that the contact method will be most useful in the range of resolution usually covered in optical microscopy, the projection method being used for the range between 200 Å and 0.2μ. An appreciable advantage in exposure-time and experimental convenience may also make the projection method preferable even when the required resolution is in the optical range.

3.8.2. *Field of view.* The main advantage of the contact over the projection method is in its large field of view. As is evident from the geometrical relations (Fig. 3.9), the field in the contact method will normally cover the whole of the specimen whereas in projection it becomes progressively smaller as the object is moved towards the source, to increase magnification. The width of the object included in an image field of given size decreases proportionately with reduction in b. Since the emitted X-ray intensity from a window target does not vary greatly over a semi-angle of 45° from the forward direction (§ 8.6), the width of the useful field may be taken as equal to $2b$. It is useful to divide the field by the resolution δ, giving the number of resolved lines in the field (N) or

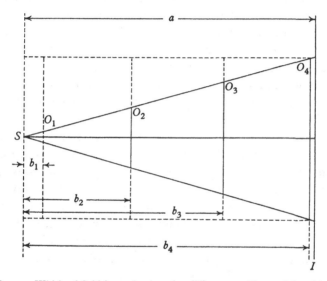

Fig. 3.9. Width of field in projection, for different positions of the object. with fixed source S and image plane I.

the number of picture points (N^2). The former criterion is used in television display, the latter in electron microscopy, as a measure of picture quality.

With $N = F/\delta$ and $F = 2b$, the object distance corresponding to a given resolution and number of lines per picture is given by

$$b = N\delta/2. \qquad (3.26)$$

In projection microscopy, resolution is in first approximation independent of b and is equal to the source size. The number of resolved lines per picture, at given resolution, cannot be increased at will, however, since an upper limit is fixed to b by Fresnel diffraction. The value of b, for which the first fringe width would equal the resolution, is given by (3.10):

$$\delta = (b\lambda)^{\frac{1}{2}}. \tag{3.27}$$

Substitution in (3.26) gives the upper limit for N at given resolution and wavelength:

$$N = 2\delta/\lambda. \tag{3.28}$$

If we accept 100 lines per picture as the smallest useful field, then 100 Å will be the practical limit of resolution at a wavelength of 2 Å; the width of field would be 1μ and the object distance 0.5μ— almost too small to be controllable. For biological specimens, adequate contrast calls for a longer wavelength: at 20 Å, a picture of 100 lines would be obtainable only with the poor resolution of 1000 Å, the width of field now being 10μ and the object distance 5μ, which is still inconveniently small.*

Within the above assumptions, therefore, a resolution of 100 Å would be difficult to achieve by projection, except at short wavelengths, which would entail serious loss of contrast. Paradoxically it may be possible to get better resolution by making positive use of the Fresnel effect, since it is in principle possible to 'reconstruct' a true image from one containing well-defined Fresnel fringes (cf. § 1.8). Considerable difficulties stand in the way of perfecting this method of imaging (cf. Baez & El-Sum, 1957), but if it should prove to be practicable, it would allow the object distance to be increased beyond the above limit, thus giving a larger field. At present, however, only a strictly limited field is available in projection microscopy at high resolution. At the same time, it is not difficult with many specimens to select by optical microscopy the region of particular interest for X-ray microscopy (cf. Long, 1958, 1959), so that smallness of field is not always a serious limitation in projection work.

Although the contact method has the advantage of a large field, this is offset by the fact that it is recorded as a micronegative. To evaluate this record, it has to be searched under a high-powered optical microscope, which has a field of view only a few microns in

---

* It is interesting to note that for light, at a wavelength of 5000 Å, a resolution of 25μ is all that could be attained with a field of 100 lines, or 250μ with 1000 lines. For this reason point projection is not a practicable method of optical microscopy.

width at the magnification of 1000 that is needed to reveal detail of order 0.25μ. Selected areas are then photographed on plates of normal grain size. It is thus a nicely balanced question whether the time spent in the projection method, on initial optical searching followed by X-ray microscopy of a small field, is in fact longer than that spent in photo-micrography of the contact image of a large field. The answer will depend to some extent on the nature of the specimen and especially on whether the features of interest are best recognized in the optical or the X-ray image. Considerations of exposure-time and of experimental convenience also play a part, in both of which the projection method proves to have an advantage.

3.8.3. *Exposure-time.* The comparison of exposure-times in the contact and projection methods is best made by considering the transition from the former to the latter, in given camera length. For recording the contact negative, ultra-fine-grained emulsions of very low sensitivity must be used. If the specimen is transferred from the neighbourhood of the emulsion to that of the source, the penumbra will increase in width and an emulsion of correspondingly larger grain can be used. If the magnification is now M, the grain size can be M times larger so that the exposure-time will be approximately $1/M^2$ what it was (cf. § 2.5). At the same time the source width must be reduced to $1/M$ of the size permissible in the contact method, in first approximation. The emitting area is thus reduced by $1/M^2$, but the permissible current loading on the target is M times greater with the smaller spot (§ 3.6.2.), so that the X-ray intensity is reduced only by $1/M$ whilst the photographic sensitivity has increased by M^2. The exposure-time of the projection method should thus be $1/M$ that of the contact method.

In practice the full prospective advantage may not be attained, because the grain size cannot be increased indefinitely and because the current intensity is finally limited by the imperfections of the electron optical system. The largest usable grain size g may be taken as 0.1 mm. (cf. Table 2.1), so that the necessary minimum direct magnification in projection would be 100 at a resolution of 1μ and 1000 at 0.1μ. If object distance is limited by Fresnel diffraction, this corresponds to a camera length L of 10 cm. for $\delta = 1\mu$ at $\lambda = 10$ Å, or for $\delta = 0.1\mu$ at $\lambda = 1$ Å. In general,

$L = \delta g/\lambda$, so that for a resolution of 100 Å at 10 Å wavelength, a grain size of 0·1 mm. would require the very short camera length of 1 mm. With the more usually used lantern plates of intermediate grain (20–30μ) the camera would have to be shorter still. Optimum use of the photographic material will thus not be possible at the high resolution ultimately to be expected of the projection method.

Limitations in the electron optical system of the projection microscope also prevent indefinite reduction in the exposure-time. The assumption made above, that the current density at the target can be increased in proportion as the spot size is diminished, owing to improved heat dissipation, relies on the ability of the cathode and electron lens system to deliver the required intensity. As shown above (§ 3.6), this is no longer possible for spots of less than a certain size, owing to the limited emission of the cathode and the severe spherical aberration of the final lens. With the best lens available, this critical size lies between 0·5 and 1μ, depending on the other operating conditions (cf. Fig. 8.3). For smaller spot sizes the focused current will vary with $r^{\frac{8}{3}}$, by (3.24), instead of directly with r as assumed in the above comparison.

The conclusion that the exposure-time in projection is $1/M$ that in the contact method, at given resolution and camera length, is thus valid only so long as the emulsion of optimum grain size can be employed and if the resolution is of order 1μ or worse. At higher resolution the advantage of projection becomes rapidly less, and the two methods would have the same exposure-time at a resolution of about 0·1μ—if indeed the contact method can reach this resolution. The advantage would be restored to projection even here, of course, if the prospective development of brighter cathodes and corrected electron lenses leads to higher efficiency in the electron optical system.

3.8.4. *Experimental convenience.* A comparison of the contact and projection methods must also involve questions of the relative difficulty of the various stages: specimen mounting, X-ray tube adjustment, photography and evaluation of the image. The experimental details are described in chapters 10 and 11, respectively, and we discuss here the essential factors only.

In contact microradiography the specimen must be placed very close to the emulsion, either directly upon it or with an intervening film of nitrocellulose or plastic. Careful manipulation in the dark-room is required, and the specimen must be removed again before photographic processing. In the projection method the specimen is mounted in a stage, so that it can be searched, and therefore is clear of the target. Some heating and desiccation may occur during long exposure, but if a low magnification is used the method is more suitable for living material. The projection procedure is also more suitable for stereophotography (§ 11.11), since the specimen is easily moved between the two exposures, whereas in the contact method it must be transferred from one emulsion to another and remain in known orientation.

A great advantage of contact microradiography, on the other hand, is that results of value can be obtained with almost any X-ray tube. If the focal spot is too large, it can be apertured or placed at a sufficiently great distance from the emulsion to reduce the penumbra to acceptable size, though at the cost of increased exposure. For projection a special tube is needed, with precise control of the focusing elements, so that more attention has to be given to setting up the apparatus. This difference, however, is becoming less marked, since the design of fine-focus tubes has been improved so that the electron gun and lenses are largely pre-alined and the focusing conditions are repeatable. At the same time, the contact method has to turn to the use of specially designed tubes for high-resolution work. But it remains true that more thought and care is needed with the X-ray apparatus in the projection than in the contact method.

The photographic procedure is similar in both methods, in that a light-tight camera is needed which may have to be evacuated if very soft radiation is employed. The camera mounting must be stable, since exposure is of the order of minutes in the projection and tens of minutes in the contact method. The main difference lies in the great care that must be observed at all stages in handling the high-resolution emulsions for contact work, since the minutest specks of dust, pinholes or scratches will mar the final image. The initial perfection of the emulsion is also less certain, whereas in projection only gross imperfections cause trouble and these are rare.

Finally, the contact image demands more time and effort for its evaluation, as it must be searched under an optical microscope; when the highest resolution is sought, an immersion objective must be used. The projection image needs at most a hand lens for its appreciation, and that only when the initial magnification has been small. Preliminary searching of the specimen can be carried out with a dry objective at moderate magnification, to demarcate roughly the areas of interest for X-ray microscopy. A contact print, or at most a low optical enlargement, suffices as a record of the micrograph. On the other hand, the contact method demands the very highest skill in photomicrography if all the resolved detail is to be finally rendered visible (Henke, 1959).

In sum, it may be said that the contact method has the advantage of simpler apparatus, the projection method that of simpler photographic procedure. Where only low or moderate resolution is required, contact microradiography may be preferred, but even in this range the cost and relative complexity of a projection microscope would be justified if a great deal of routine work in standard conditions had to be done. For high resolution the projection method has increasing advantage, and for X-ray observations beyond the limit of the optical microscope it is the only promising method. For stereophotography and even more for time-lapse studies it is much the more suitable. In short, projection X-ray microscopy gains in versatility more than it yields in simplicity to the contact method. It will appear later (chapter 6) that it provides the only practicable experimental arrangement for microanalysis by fluorescence or direct excitation spectrometry, besides having some advantage over contact microradiography for absorption microanalysis. The chief disadvantage of projection microscopy is economic: the high cost of the special X-ray equipment required.

3.9. Intermediate magnifications

So far we have discussed only the two extreme conditions, geometrically speaking, of small image throw with effectively unit magnification and of small object distance, with high direct magnification. Until recently the possible advantages of an intermediate magnification had not been considered. The proposal and experimental investigation by Le Poole & Ong (1956) of the 'times-two' method, in which the object is midway between source and emulsion, has now brought them under discussion. We shall treat the problem in general terms before coming to the particular system of Le Poole and Ong.

3.9.1. *Arbitrary magnification.* The object is now assumed to be situated between source S and image plane I (cf. Fig. 3.1), so that object distance b and image distance a are of the same order. The width of the source s and that of the penumbra p are both small; that of the object is w. Both p and the Fresnel fringe width x vary with object position, and it is convenient to discuss them separately.

The value of p in terms of s, b and c ($= a - b$), is given by (3.1), which may be rewritten

$$p = s(a/b - 1) = s(M - 1), \tag{3.29}$$

where M is the magnification. If resolution δ is determined only by penumbra width, i.e. $\delta = p_0 = p/M$, and if the grain size g is always chosen to be equal to penumbra width ($p = g$), we have

$$M = g/s + 1;$$

$$1/\delta = 1/s + 1/g, \tag{3.30}$$

and $$\delta = s(M - 1)/M. \tag{3.31}$$

The resolution can thus be expressed either in terms of M and s, or of s and g, as may be more convenient. The variation of δ with b is shown in Table 3.1 and Fig. 3.10. When M is very large and s small, we have projection conditions and $\delta \doteq s$; with M very small we have contact conditions and $\delta \doteq p$, in agreement with conclusions already reached above. When $M = 2$, so that $p = s$, we have $\delta = p/2 = s/2$. There is no optimum magnification, the resolution improving continuously as M approaches unity. Assuming that the source size is given and the grain size is adjusted *pari passu* to penumbra size, the exposure-time will be proportional to $1/M^2$ as the object position is varied in a camera of fixed length, but it will remain constant if magnification is varied by changing image distance, with object position fixed (cf. (3.34) and (3.35) below).

C & N

The effect of Fresnel diffraction on resolution varies symmetrically about the half-way position. The width of the first fringe, x_0, referred to the object plane, is given by (3.8):

$$x_0 = (bc\lambda/a)^{\frac{1}{2}} = (c\lambda/M)^{\frac{1}{2}}. \qquad (3.32)$$

Its variation with magnification is also shown in Table 3.1 and Fig. 3.10, with a maximum for $b = c$. At high magnification it is small compared with p_0, but at low magnification it exceeds it by

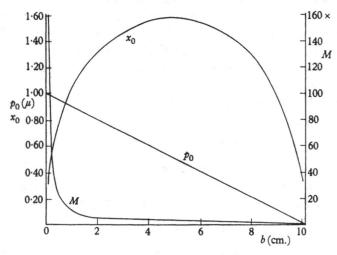

Fig. 3.10. Intermediate magnification: variation with object distance b, in fixed camera length $(a+b)$, of magnification M, reduced penumbra width p_0, and reduced Fresnel fringe width, x_0 (cf. Table 3.1).

an increasing margin, since in first approximation $p_0 \propto c$ whilst $x_0 \propto \sqrt{c}$. Hence, in contact conditions, the fringe width is much larger than the penumbra width, as formed by a source of the small size assumed. In practice, this would allow the source to be increased in size until the penumbra equalled the reduced fringe width.

In order to compare the exposure needed in different conditions of microscopy, we now allow s to vary whilst resolution is held constant by appropriate choice of M in (3.31). The exposure-time will be determined by X-ray output, which depends on source size s, by the camera length L and by the sensitivity of the recording emulsion, which depends on grain size g and hence on penumbra width p; all these quantities can be expressed in terms of magnification M. From (3.31), with δ constant,

$$s \propto M/(M-1). \tag{3.33}$$

If grain size is always arranged to be equal to penumbra width, (3.29) gives

$$g \propto s(M-1)$$

$$\propto M. \tag{3.34}$$

TABLE 3.1. *Penumbra width p, reduced penumbra width p_0 ($= p/M$) and reduced Fresnel fringe width x_0, at varying object distance b in camera length $a = 10$ cm. at wavelength 1 Å and with source width $s = 1\mu$*

b	M	p (μ)	p_0 (μ)	x_0 (μ)
100μ	10^3	999	0·999	0·1
1 mm.	10^2	99	0·99	0·32
2 mm.	50	49	0·98	0·44
5 mm.	20	19	0·95	0·68
1 cm.	10	9	0·90	0·95
2 cm.	5	4	0·8	1·27
3 cm.	$3\frac{1}{3}$	2·5	0·7	1·45
4 cm.	$2\frac{1}{2}$	1·5	0·6	1·55
5 cm.	2	1	0·5	1·6
6 cm.	$1\frac{2}{3}$	0·68	0·40	1·55
7 cm.	1·4	0·43	0·30	1·45
8 cm.	1·25	0·25	0·20	1·27
9 cm.	1·11	0·11	0·10	0·95
9·5 cm.	1·05	0·05	0·05	0·68
9·8 cm.	1·02	0·02	0·02	0·44
9·9 cm.	1·01	0·01	0·01	0·32
9·99 cm.	1·001	0·001	0·001	0·1

The camera length, from the definition of M, is given by $L = Ma$, so that

$$L \propto M. \tag{3.35}$$

Assuming the object position to be fixed, the X-ray output to be proportional to s^2, emulsion speed proportional to g^2 and the inverse square law to hold, the exposure time T will, therefore, vary with M according to

$$T \propto \{(M-1)/M\}^2. \tag{3.36}$$

The rapid increase in exposure-time with magnification is clear from Table 3.2, being due primarily to the rapid increase in image distance.

It is more practical to vary magnification by moving the object between source and emulsion, in fixed positions, i.e. with given camera length. From (3.33) and (3.34), the exposure-time will now vary according to

$$T \propto \{(M-1)/M^2\}^2. \qquad (3.37)$$

This relation has a maximum for $M = 2$, so that exposure-time now decreases rapidly towards high magnification as well as low magnification (Table 3.2).

TABLE 3.2. *Exposure-time T (in arbitrary units) in function of magnification M, at given resolution, when the X-ray image is formed by projection, (i) with source–object distance constant (T_1), and (ii) with camera length constant (T_2), according to (3.36) and (3.37) respectively*

M	100	10	5	2	1·5	1·2	1·1	1·01
T_1	98	83	67	25	11	2·8	0·83	0·01
T_2	0·01	0·83	2·5	6·3	5·0	2·0	0·69	0·01

3.9.2. *The times-two method.* The use of a magnification of two was advocated by Le Poole & Ong (1956) primarily in order to obtain better focusing of the electron optical system in a projection microscope. As the source size is reduced, to improve resolution, the X-ray image on a fluorescent screen becomes too weak for visual focusing. The brightness on the screen can be increased, by the inverse square law, if the camera length is reduced. With a source of given size, the exposure-time would remain unchanged, as noted above, so long as the emulsion was always chosen to have grain size equal to the penumbra width. The latter gets smaller and, as contact conditions are approached, it may become less than the finest grain size available, especially when the source is small, as in the example of Table 3.1.

At a magnification of two, however, Le Poole and Ong pointed out that both source size and grain size can be doubled without loss of resolution (cf. 3.30). Assuming fixed source–object distance, the intensity on the viewing screen (or emulsion) will be increased four times as compared with high magnification projection (cf. 3.36, and Table 3.2). The increase will be even more if the target current is being limited by spherical aberration in the electron lens; by (3.24), the increase could be as much as 6.3. The times-two method thus gains visual intensity both by the inverse square law and by increased X-ray output, as compared with projection at the same source–object distance. The exposure time is reduced only in proportion to the increase in X-ray output. At the same time, a much larger field is obtained and the inconvenience is avoided of having the object too close to either the source or the emulsion.

Fresnel diffraction, however, sets a serious limitation on resolution in the times-two condition, since the fringe width is a maximum for $M = 2$. It follows from (3.32) that when $b = c = a/2$, the width $x_0 = (a\lambda/4)^{\frac{1}{2}}$, as compared with the much smaller values of $(b\lambda)^{\frac{1}{2}}$ and $(c\lambda)^{\frac{1}{2}}$ in projection and contact conditions respectively. In the example of Table 3.1, with typical experimental values of a, s and λ, the fringe width in the times-two condition is several times greater than the resolution as determined by geometry, $s/2$. Le Poole & Ong (1956) in fact used $\lambda = 8\cdot3$ Å and $a = 5$ mm., so that the fringe width was $1\cdot1\mu$ as against a geometrical resolution of $0\cdot5\mu$ with a 1μ source (or 1μ with a source of double the size). A resolution of this order is shown by their micrographs. The fringe width can be reduced by shortening the camera length still further, but a limit of convenience is soon reached. Since the reduced fringe width is proportional to the square root of camera length, a fringe of $0\cdot1\mu$ at $8\cdot3$ Å wavelength could only be obtained with the impracticably small source–image distance of $0\cdot05$ mm.

At given camera length, higher resolution is obtainable by the projection than by the times-two method (Cosslett, 1957). On moving the object closer to the source, the magnification is increased from 2 to M and the reduced fringe width is diminished in the ratio $2/\sqrt{M}$ (cf. (3.32)). An improvement in resolution will result so long as the fringe in the times-two condition was wider than the source width, which now sets the resolution limit.

The main gain from using the times-two method is, therefore, in brightness of the viewing screen, and to a lesser extent in exposure-time. There is little or no advantage in field of view, and so in speed of working, as claimed by Ong & Le Poole (1958b). It is certainly true that the times-two method shares with contact microradiography the ability to image a large field in a single micrograph, whereas only a small field is available in projection.

But, as discussed in § 3.8, a proper comparison of speed of working should include the final evaluation of the record (Cosslett, 1959). When the process as a whole is considered, the time spent in enlarging the contact (or × 2) negative, piece by piece, may be found to exceed the time required to take a set of projection micrographs of the same areas, especially as the exposure-time for the contact picture is in the first place very long.

Experimentally the times-two method combines the disadvantages of both the contact and the projection methods, since it requires a special point focus tube as well as a fine-grain emulsion. Its use, to increase screen brightness, will thus only be tolerable so long as no alternative means are available for fine focusing of the X-ray image. It now seems probable that a more practical solution will be provided either by the method of imaging the focal spot by back-scattered electrons (§ 15.4) devised also by Ong & Le Poole (1957, 1958a), or by using a very short camera length and a viewing screen of ultra-fine grain (§ 10.5), as developed by Pattee (1958). There may well remain applications for the times-two method, however, in investigations in which the specimen must be kept well clear of both target and emulsion—for instance, for treatment at high or low temperature, and by abrasion or etching.

Additional references added in proof

To §3.6.3: Langner, 1960*b*.
To §3.7: Ong, 1959.
To §3.8: Langner, 1960*a*; Pattee, 1960*b*.
To §3.8.2: Cosslett, 1959; Ong & Le Poole, 1959.

CHAPTER 4

REFLEXION X-RAY MICROSCOPY: MIRROR-SYSTEMS

Two types of reflexion-systems have been developed for X-ray microscopy: in one the X-rays are totally reflected, at glancing-incidence, from polished concave mirrors, and in the other they are focused by the curved surfaces of single crystals, by Bragg reflexion (see chapter 5). The method of total reflexion relies on the fact that the refractive index of matter for X-ray wavelengths is less than one, so that a critical angle i_c exists for rays incident from air (or vacuum) on a clean glass or metal surface. The refractive index n being now equal to $\sin i_c$, and less than unity by an amount of order 10^{-4}, the critical angle is close to $90°$ and total reflexion can occur only over a narrow range of glancing-angle $(90° - i_c \doteqdot 30')$. Conditions of grazing-incidence are not met with in light optical instruments, so that the imaging requirements and particularly the aberrations have had to be investigated from the beginning. The practical difficulties are also being overcome, and the total reflexion method can now give a useful direct magnification of one hundred times with exposures of the order of one hour. The ultimate limits of resolution and intensity are bound up with the possibility of aberration correction, as discussed below.

4.1. Total reflexion of X-rays

The refractive index of all materials for visible radiations is greater than one, and for some it is above two. In the far ultra-violet, however, its value decreases rapidly to below unity, as is confirmed by measurement of the critical angle (Compton, 1923). It is found that total reflexion takes place at a surface for rays arriving from the less dense medium, and not from the denser medium as with light. For rays incident at very small glancing-angles θ, a surface behaves like a mirror (Fig. 4.1) and reflexion is almost 100% efficient for very short wavelengths, such as X-rays. But if, on the other hand, they are incident near-normally on a mirror, in the conditions used for forming images with light, X-rays pass into the surface and are transmitted and absorbed in proportions determined by their wavelength and the nature of the material.

The theoretical treatment of total reflexion is based on the optical theory of metals, which in its quantum formulation gives close agreement with experiment. The classical theory of dispersion shows the dependence of refractive index n on wavelength and the physical constants of the material. It assumes that there is one free electron per atom, so that the number of resonating electrons

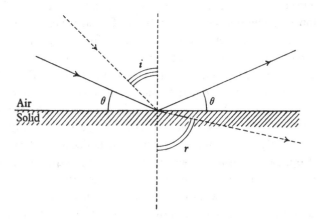

Fig. 4.1. Total reflexion of X-rays.

per cubic centimetre (N) can be taken as equal to the number of atoms per cubic centimetre. It is convenient to introduce a quantity δ, the unit decrement of refractive index, defined as the difference of n from unity:

$$\delta = 1 - n. \tag{4.1}$$

The theoretical expression for δ is then

$$\delta = \{Ne^2/2\pi mc^2\} Z\lambda^2 \tag{4.2}$$

$$= 2.70 \times 10^{10} Z\rho\lambda^2/A, \tag{4.3}$$

where ρ is the density, Z the atomic number and A the atomic weight of the element concerned; e and m are the charge and mass of the electron respectively. The value of δ calculated from (4.3) agrees well with experimental results at frequencies not too near

an absorption edge (Nähring, 1930; Bearden, 1938). Close to such
an edge there are small departures from the calculated value, but
they are less than would be expected from the use of classical
(Lorentz) dispersion theory. The discrepancy is partly explained
by the quantum treatment (see Compton & Allison (1935),
chapter 4), but difficulties still remain in the long-wave region
near the L and M absorption edges (Parratt & Hempstead, 1954).

The critical angle of incidence i_c for total reflexion is immediately
given when n is known:

$$\sin i_c = n, \qquad (4.4)$$

since with X-rays it is measured in the reference medium, not in
the unknown. In terms of the glancing-angle $\theta_c (= 90° - i_c)$, we
have $\cos\theta_c = n$, and

$$\sin\theta_c \doteq \theta_c \doteq (2\delta)^{\frac{1}{2}}. \qquad (4.5)$$

From (4.3) it follows that, for elements of medium Z and wavelengths of order 1 Å, the value of δ is of order 10^{-5} to 10^{-6}, so that
θ_c is of order 10^{-2} to 10^{-3} radian, or 3–30 min. of arc. The values
of δ and θ_c depend strongly on λ, and are given in Table 4.1 at
three wavelengths for three metals of low, medium and high atomic
number. The highest value of θ_c, for gold at 8·3 Å, corresponds to
an angle of 3° 4′ only.

TABLE 4.1

		$\lambda = 0·708$ Å		$\lambda = 2·28$ Å		$\lambda = 8·3$ Å	
	Z	$\delta \times 10^6$	$\theta_c \times 10^3$	$\delta \times 10^5$	$\theta_c \times 10^3$	$\delta \times 10^4$	$\theta_c \times 10^2$
Al	13	1·75	1·87	1·82	6·05	2·41	2·19
Ag	47	6·35	3·55	6·60	11·5	8·75	4·18
Au	79	10·5	4·55	10·9	14·7	14·4	5·35

The above treatment has neglected absorption of the incident
X-rays in the reflecting medium, which becomes appreciable for
wavelengths longer than 1 Å. Its effect can be investigated (Dershem, 1929) by introducing a complex index of refraction, as in
the optics of metals. The ratio R of the reflected to the incident
intensity is given by a modified form of the Fresnel equations (cf.
Parratt & Hempstead, 1954). Reflexion is never 'total', being
already below 100 % at $\theta \doteq 0$, and falling more or less rapidly
with glancing-angle, depending on λ. Typical curves for a gold

surface at three wavelengths are seen in Fig. 4.2 (Hildenbrand, 1956); the shorter the wavelength the more nearly 'critical' is the change in the reflexion coefficient R in the region of the value of θ_c given by (4.5). Conversely, an appreciable amount of the longer wavelength radiation is reflected at grazing-angles much greater than the critical value given by elementary theory; for this reason Farrant (1950) has advocated the use of very soft radiation for microscopy. As a corollary to the lack of a critical limit, there is a

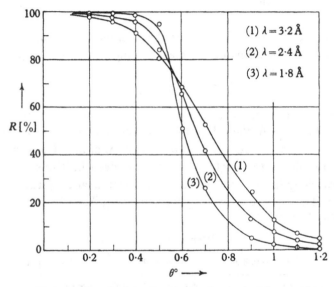

Fig. 4.2. Variation in reflectivity R with glancing-angle θ for a gold layer 600 Å thick, at three different wavelengths λ. (Hildenbrand, 1956.)

refracted wave even at small glancing-angles, the effect of which on the reflected intensity depends on the thickness of the reflecting layer. For very thin layers the effect is small and the transition sharp, as shown in Fig. 4.3 for various thicknesses of gold (Hildenbrand, 1956).

The influence of the nature of the reflecting material (cf. (4.3)) makes itself felt in an approximate proportionality of θ_c to the square root of the density. But, since the absorption coefficient μ depends strongly on Z, the actual variation of R is more complex. Fig. 4.4 shows the comparative behaviour of four metals of widely differing Z. The critical angle is larger, as predicted, for the heavier metals but the transition is sharper for the lighter ones. Experimental results for the reflexion of X-rays in the wavelength range from 1·54 Å to 44 Å, from glass as well as from several different metal surfaces, have been given by Rieser (1957a, b). He concludes that there is little advantage, and some disadvantage, in using a gold instead of a glass surface for 44·5 Å radiation. (See also Hendrick, 1957; Henke, 1960.)

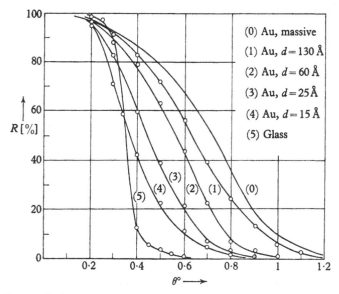

Fig. 4.3. Variation in reflectivity R with glancing-angle θ for gold layers of different thicknesses and for glass, at a wavelength of 2·4 Å. (Hildenbrand, 1956.)

4.2. Dioptric focusing

The focusing of X-rays with a lens is possible in principle. As in optics the focal length of a refracting element having faces curved in opposite senses with radii R_1 and R_2 will be given by

$$1/f = (n-1)(1/R_1 + 1/R_2)$$
$$= -\delta(1/R_1 + 1/R_2). \qquad (4.6)$$

The negative sign indicates that the focusing action is in the opposite sense to that familiar with light: a biconvex lens will diverge, a biconcave lens converge, an initially parallel beam of X-rays (Fig. 4.5 a). The realization in practice of any such lens is beset by the difficulty that a lens of high power must be of small radius of curvature, so that for any appreciable physical aperture it must be thick and will then absorb most of the X-ray beam. A thin lens,

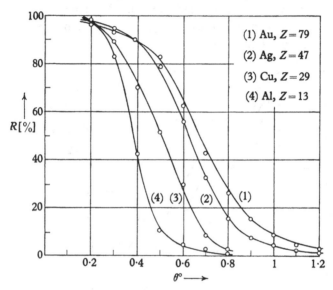

Fig. 4.4. Variation in reflectivity R with glancing-angle θ for thick layers of four metals, at a wavelength of 2·75 Å. (Hildenbrand, 1956.)

of large R, may absorb little, but will have a very long focal length. With $R_1 = R_2 = R$, we have from (4.6):

$$f = R/2\delta. \qquad (4.7)$$

With $R = 1$ cm. and $\delta = 5 . 10^{-6}$, the value of f is 100 m. The value of δ increases in proportion to λ^2, but the absorption coefficient is proportional to λ^3, so that lengthening wavelength to give shorter focal length is a self-defeating procedure. It may prove that such a lens, used at a wavelength close to an absorption edge of the material composing it (where the absorption is low), will be useful for forming a parallel beam of X-rays for transmission over a long distance without appreciable loss in intensity. Any application for image production appears unlikely, since a useful degree of magnification would entail an enormously long optical path.

An alternative method of focusing is to arrange for the radiation from the object to be refracted near the edge of a prism block having adjacent faces concave and plane (see Fig. 4.5 b). In order to obtain maximum deviation the angle of incidence must be large, so that the grazing-angle will be only slightly greater than the

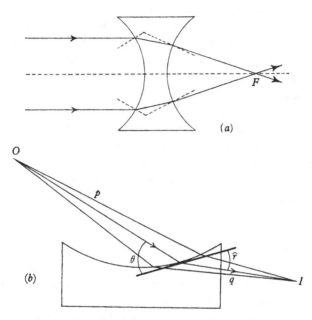

Fig. 4.5. Focusing of X-rays by refraction at a curved surface; (a) at near-normal incidence; (b) at glancing incidence near a prism edge.

critical value for total reflexion, that is, of the order of 1°. An object point O would be imaged in a line normal to the plane of the diagram at I. If p and q are the object and image distances respectively, R the radius of curvature of the lens and c the ratio of the internal to the external glancing-angle (r/θ), then the first order focusing relation is

$$1/p^2 + c^2/q = (1-c)/R\theta \qquad (4.8)$$

and the magnification is q/pc. In the most favourable circumstances a lens of $R = 1$ m. would have focal lengths of 2 cm. and 5 cm. for a wavelength of 2 Å, depending on the direction of propagation. The image will be completely astigmatic, but point to point imaging could be attained with a refracting element formed by rotating the cylindrical lens of Fig. 4.5*b* about the axis *OI*. Alternatively, a stigmatic image would be produced by crossing two such cylindrical lenses, thereby avoiding some of the difficulties of working a surface of revolution. The angular aperture of the beam cannot be larger than θ_c and the path of its central ray through the prism will be of the same order as the focal length, so that the intensity of the imaging beam will be very low.

The deviation of X-rays by such a prism was first demonstrated by Larsson, Siegbahn & Waller (1924) and its converging properties by Stauss (1930*a*, *b*). Hink & Petzold (1958) have recently used it for accurate measurement of the refractive index of glass for X-rays. Its use as a lens was proposed by Kirkpatrick (1949), but no attempt to test it in practice has been reported. The chromatic aberration would in any case be severe, because of the dependence of δ on λ^2. In this respect, as well as in its essential avoidance of absorption, the use of mirrors lends itself much more readily than that of lenses to the focusing of X-rays.

4.3. Focusing by total reflexion

The possibility of obtaining X-ray images by total reflexion from the surface of curved mirrors was first suggested by Jentzsch (1929). He established the essential geometrical relations for focusing by a spherical surface, discussed the most suitable material for use as a mirror and pointed out the need for highly finished surfaces. He concluded that mirrors which were satisfactory for imaging the ultra-violet, at normal incidence, would be adequately smooth for X-ray imaging down to a glancing-angle of 2°. However, he dismissed the practicability of focusing with a spherical surface at glancing incidence, on account of its severe astigmatism, and attempted without success to form an image by reflecting X-rays from the inside surface of a glass tube of 2·7 mm. internal diameter. Although experiments in progress were mentioned, and work on the total reflexion of X-rays was published (Ehrenberg & Jentzsch, 1929; Nähring, 1930), there is no record of X-ray images having been formed. The conditions for imaging by reflexion from the convex surface of a glass rod were discussed by Kellermann (1943), but again no experimental results were reported.

The idea of using total reflexion from concave surfaces for X-ray imaging was revived independently by Ehrenberg (1947, 1949 a), and by Kirkpatrick & Baez (1948), and the experimental exploration of its practicability has been conducted mainly in the laboratory of these latter authors. The small value of the glancing-angle of total reflexion θ_c requires the use of mirrors of large radius of curvature R if the whole of the available aperture is to be used.

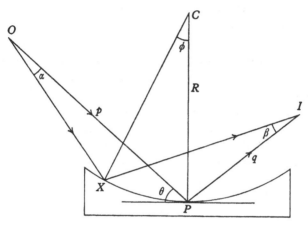

Fig. 4.6. Focusing of X-rays by total reflexion at a curved surface (meridian section).

The focusing conditions are then as in Fig. 4.6, in which, however, the angles are much exaggerated in order to make clear the geometrical relations. An object point O at distance p from the pole of the mirror P is imaged at I, a distance q from P. OXI is any other ray reflected from a point X to the image point; C is the centre of curvature of the mirror. θ is the glancing-angle at P, and α and β the angles contained by the rays at object and image, as shown; the angle PCX is ϕ. It can be shown that

$$\frac{1}{q+2af}+\frac{1}{p-2af}\frac{(2-a)}{(2+a)} = \frac{1}{f(1+a/2)}, \qquad (4.9)$$

where $a = \phi/\theta$ and $f = R\theta/2$. For small aperture (ϕ small), af can be neglected and so

$$1/p+1/q = 1/f = 2/R\theta \qquad (4.10)$$

which is similar to the relation for focusing by spherical mirrors at normal incidence, save that now $f = R\theta/2$ instead of $R/2$. The focal length thus changes with the direction of the incident beam, and, in first approximation, is proportional to the glancing-angle θ. The magnification, as in optics, is given by

$$M = q/p.$$

In the special case of unit magnification, (4.10) reduces to $p = q = R\theta$. In second approximation, with θ taking any value, but α, β and ϕ remaining small, this becomes $p = R\sin\theta$, which is the equation of a circle in polar co-ordinates with p as radial vector and $R/2$ as radius. Hence the points O, P, and I lie on a circle of this radius passing through C and tangent to the mirror at P, a condition met with in the concave grating of Rowland. The circle of radius $R/2$ is the locus of object and image points for paraxial imaging at unit magnification and is known as the Rowland circle of the corresponding concave mirror of radius R. In polar co-ordinates r, θ (Fig. 4.7) the circles are described by the equations

$$r_1 = 2R\sin\theta, \qquad (4.11a)$$

and $$r_2 = R\sin\theta. \qquad (4.11b)$$

For small glancing-angles (4.11b) gives

$$r_2 = p = q = R\theta = 2f$$

which follows also from (4.10) on setting $p = q$. For an image at infinity, on the other hand, $q \to \infty$ and the object is at the focus:

$$p = R\theta/2 = f$$

or more generally $$r_3 = f = R/2\sin\theta. \qquad (4.11c)$$

This is the equation of a circle of radius $R/4$, tangent to the mirror and to the Rowland circle at P, and known as the focal circle (Fig. 4.7). Rays from a point source situated on the focal circle will form a parallel beam after reflexion from the small region of the mirror around its pole P. With $R = 100$ cm. and $\theta = 10^{-2}$, the focal length will be 0·5 cm.

The imaging beams for points O, O' on and near the Rowland circle are shown in Fig. 4.8. A magnification greater than one demands that p be less than q, so that the object must be within

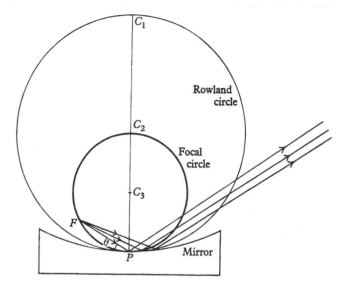

Fig. 4.7. Rowland circle (centre C_2) and focal circle (centre C_3) for a mirror of centre C_1.

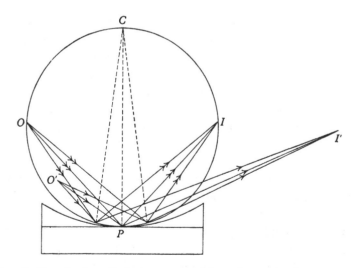

Fig. 4.8. Focusing by reflexion at a curved mirror, for a point O on the Rowland circle and a point O' near to it.

the circle (O') and the image outside it (I'). Owing to the smallness of the glancing-angle, the aberrations of the system are severe, even at unit magnification. An extended object tangent to the circle at O will be imaged in a plane tangent to it at I. If the object were normal to the principal ray from O the inclination of the image would be even greater. For magnifications much greater than one the aberrations are still more severe, the most serious now being spherical aberration, which is absent for $M = 1$. However, it is possible to find conditions for partial correction of the aberrations, even without departing from the use of spherical surfaces, which are the least difficult form of mirror to produce. We discuss first of all the aberrations of a single mirror of cylindrical form, that is to say, of image formation in a meridian plane, and then proceed to the problem of stigmatic imaging and the properties of combinations of mirrors. Finally the ultimate limit to resolving-power imposed by diffraction is discussed (§ 4.10).

4.4. Aberrations of a cylindrical mirror of circular section

In contrast to the imaging systems of light optics, there is no axial symmetry to the ray pencils in conditions of glancing reflexion. The normal optical treatment cannot be taken over intact to the X-ray microscope and the classification of aberrations will be very different. The problem has been discussed in detail by Dyson (1952), Pattee (1953 a) and Montel (1953, 1954), so that the second-order image defects have now been elucidated. A more general approach to ideal imaging requirements, in conditions of total reflexion, has been investigated by Wolter (1952 a, b) and by Herrnring & Weidner (1952).

The image of a point object O formed by a non-ideal reflector will be a disk of confusion, the size and shape of which will depend on the position of O and on the angular width of the focused beam. Rays reflected from different parts of the mirror will intersect the image plane in different points. The paraxial image position serves as point of reference. An infinitely narrow pencil of rays diverging from a given object point O will be focused to I (Fig. 4.9). A pencil making an angle α with this direction will in general be focused to a different point I'. The displacement II' is termed the total aberration for this pencil. Its magnitude and direction will depend on α and on the position of O, defined in terms of the field co-ordinate y measured at right angles to the imaging pencil from some arbitrarily chosen origin. These quantities are sufficient to determine the properties of the image formed by a cylindrical mirror of circular section, since we here confine attention to focusing in a meridian plane. As in normal optics, the total aberration is broken down into a number of component aberration terms, which differ in the way in which they depend on α and on y.

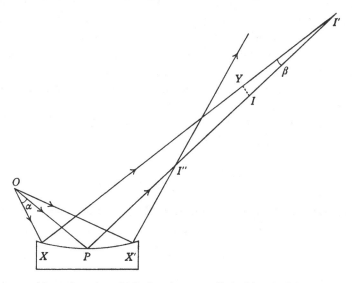

Fig. 4.9. Aberration of a spherical surface: pencils incident at points X, X' on either side of the pole P are focused at I' and I'' instead of at I.

The most important aberration is the spherical error, which persists even for an object point on the reference axis. It can be reduced only by reduction in angular aperture, with consequent loss in image brightness. Coma, on the other hand, does not limit resolution at paraxial focus but leads to a rapid deterioration in image quality for off-axis points. It limits the useful field and cannot be corrected in a simple system. The remaining aberrations of importance, curvature and obliquity of the field, can be corrected by a suitably placed aperture stop (Dyson, 1952), if more than one mirror is used. The effect on resolving-power is small, however, owing to the predominating influence of coma, which can only be eliminated by using aspherical surfaces, such as crossed ellipsoids (Pattee, 1953 a).

4.4.1. *Spherical aberration.* In Fig. 4.9 let I be the position of the image of O formed by a very narrow pencil about the principal ray, incident at P on a mirror of radius of curvature R. Another ray from O, incident at X a distance x from P, will intersect the principal ray at I' and the paraxial image plane at Y. Then II' is the longitudinal and IY the transverse spherical aberration. In centred optical systems the aberrant rays form a disk of confusion symmetrical around I. In conditions of grazing incidence it is found that all rays about the principal ray are deviated to the same side of I; the ray OX', for instance, is focused at I'' short of I and then passes above it. The image spot will be sharp on the side nearer the mirror and fade off on the far side.

Writing $\Delta y'$ for the transverse aberration IY and M for the magnification, it follows from (4.9) that

$$\Delta y' = \left(\frac{3Mpa\phi}{M+1}\right)\left(\frac{M^2+2aM-1}{M+1-2aM}\right).$$

In conditions of high magnification ($M \gg 1$) and small aperture ($a \ll 1$) this reduces to the simple form

$$\delta_s = \Delta y'/M = 3R\phi^2/2, \qquad (4.12)$$

where ϕ is the angular aperture of the mirror from which reflexion takes place, so that $\phi = x/R$. The quotient $\Delta y'/M$ gives the value of the transverse aberration δ_s, referred to the object, which would be the resolving-power of the mirror if spherical aberration alone afflicted the image. Obviously it can be reduced by limiting the aperture of the mirror. At high magnification $\alpha = 2\phi$ (see below), so that

$$\delta_s = 3R\alpha^2/8. \qquad (4.13)$$

For a beam of given aperture, and so long as spherical aberration alone determines resolution, a mirror of smallest possible radius should thus be used.

In fact *diffraction* at the aperture will always set a limit to resolution, even in an aberration-free system (cf. § 4.10). As in optics the diameter of the disk of confusion due to the diffraction of a beam of angular aperture α and wavelength λ may be taken as

$$\delta_D = 2\lambda/\alpha. \tag{4.14}$$

Since (4.13) and (4.14) vary in opposite senses with α, an optimum value for the aperture should exist (Prince, 1950). As in electron lenses, this can be found by equating the expressions for δ_s and δ_D, or by combining them in some other way. As a first approximation, to give orders of magnitude, we equate (4.13) and (4.14) to obtain the 'optimum' aperture (α_0); substitution in (4.13) then gives the resolution (d), defined as in optics as the radius of the corresponding diffraction disc:

$$\alpha_0 = \left(\frac{16\lambda}{3R}\right)^{\frac{1}{3}}, \tag{4.15a}$$

$$d = \frac{\delta_D}{2} = \lambda^{\frac{2}{3}}\left(\frac{3R}{16}\right)^{\frac{1}{3}}. \tag{4.15b}$$

With the typical values $R = 100$ cm. and $\lambda = 2.28$ Å, we have $\alpha_0 = 1.06 \times 10^{-3}$ and $d = 2100$ Å. The optimum value of α_0 is well below the maximum permissible aperture, given by the critical angle for total reflexion, which is 6.10^{-3} for aluminium at this wavelength. Even at a mirror radius of 1 cm., the optimum aperture is only 4.8×10^{-3} and the resolution would be 480 Å.

A more exact treatment of the combined effect of diffraction and spherical aberration requires wave optics. Dyson (1952) has calculated the change in size and shape of the intensity distribution in the focal plane with increasing amounts of spherical aberration. However, his final expression for the resolving-power differs from (4.15b) only by the omission of the factor $3^{\frac{1}{3}}$, or 1·44.

Thus the resolving-power of a mirror of circular section is limited to the order of 1000 Å for a wavelength of 1 Å and a mirror radius of 1 m., by this combination of diffraction and spherical aberration. It is clearly desirable to use the shortest focal length and the largest glancing-angle practically possible, but the latter cannot exceed the critical value, θ_c. Since θ_c is directly proportional to λ, the resolving-power can be expressed as

$$d = k(f\lambda)^{\frac{1}{2}}, \tag{4.16}$$

where the value of the constant k depends on the material of the mirror surface, by (4.3) and (4.5). Some improvement may be obtained by reducing the focal length or the wavelength, but it is evident that the variation is slow. Substantially higher resolving-power can be achieved only by correcting the spherical error. This may be done either by figuring the mirror to some special form, such as an elliptical section (§ 4.7), or by using a combination of two or more mirrors (§ 4.6). With such systems it is also possible to devise means of correcting obliquity and curvature of the field.

4.4.2. *Coma.* The most serious of the field aberrations is coma, which is due to different magnifications being produced by different regions of the mirror. A pencil OA to the nearer part of the reflecting surface (Fig. 4.10) will have a shorter object distance and so a longer image distance than the principal ray, whilst a pencil OB to the further edge will be conversely affected. The magnification for OA will be greater and for OB less than that for the principal ray, as $M = q/p$. The sharpness of the image therefore deteriorates rapidly away from the optimum focus at I.

The value of the comatic aberration depends on the distance y of the field point from O and on the square of the aperture, so that it increases rapidly with aperture. In order to be free of coma all paths through an optical system must give the same magnification, in other words the Abbe sine condition must be satisfied:

$$\sin\alpha/\sin\alpha' = \text{constant}, \qquad (4.17)$$

where α and α' are the aperture angles of an image-forming pencil in object and image space respectively. Spherical aberration is then also zero, that is, the system is aplanatic. It can be seen at once from Fig. 4.10 that whilst the value of α is greater for Y than for O, that of α' is smaller for Y_A than for I, so that (4.17) cannot be satisfied by such a mirror. Moreover, this result is independent of the shape of the mirror surface so that coma cannot be corrected by figuring it to some other shape than a circular section. Indeed the comatic defect is even more severe for an elliptical surface, which is the ideal form for correcting spherical aberration.

It is also impossible to correct coma by using compound systems of mirrors of circular section, which eliminate spherical aberration and obliquity and curvature of the field. Coma remains the limiting factor and Fig. 4.16 shows how rapidly its value increases with width of field in these conditions of partial correction. With combinations of aspherical surfaces, however, it would be possible to satisfy the aplanatic condition by arranging that the variation of magnification across the mirror at the second reflexion is the converse of that at the first reflexion (Pattee, 1953a). Such comatically corrected systems have not yet been figured and tested.

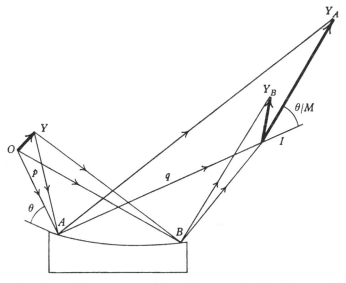

Fig. 4.10. Comatic aberration: an object OY is imaged with different magnification by different parts of the mirror surface (after Kirkpatrick & Pattee).

4.4.3. *Obliquity of the field.* It is clear from Fig. 4.10 that the image of an extended object will be inclined along the direction of the imaging beam at a very small angle. For an object normal to the principal ray, it can be shown that the angle of inclination of the image will be θ/M, where M is the magnification and θ the glancing-angle for the principal ray. It would hardly be practicable to place a viewing screen or photographic plate at this shallow angle to the reflected beam. Alternatively the object could be inclined to the principal ray at the angle θ, so that the image would be normal to the reflected beam; but this solution is also somewhat impracticable, since θ is of the order of 1° at maximum. If, however, an aperture stop is introduced at a distance from the mirror of $\frac{2}{3}f$ on the image side, an object normal to the beam will give an image that is also normal to it (Dyson, 1952). At the same time the effective angular aperture of the beam is reduced and the intensity at the image suffers correspondingly. McGee (1957) points out that the position of the stop must be less than $\frac{2}{3}f$, by a calculable amount, on account of the finite length of the mirror.

4.4.4. *Curvature of the field.* The image obtained through an aperture stop will be normal to the principal ray only at the centre of the field, so that the image formed on a flat plate will deteriorate

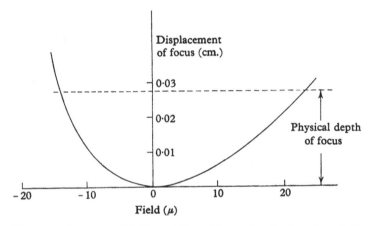

Fig. 4.11. Curvature of field: variation of focus of a circular mirror (with aperture stop), with distance of object point from axis. (Dyson, 1952.)

with distance from the centre. The resulting aberration has been computed by Dyson for a wavelength of 1·6 Å and aperture angle $1·1 \times 10^{-3}$ (Fig. 4.11). For these values of λ and α the resolution d, as limited by spherical aberration and diffraction, would be 1500 Å. The corresponding depth of focus, defined as in optics by d/α, is 275 μ, and is indicated by a broken line at this height in Fig. 4.11. It is seen that only outside a field of total width 37·5 μ does the error due to the curvature exceed this level; that is, 37·5 μ represents the useful field (at the object) in these imaging conditions. Without an aperture stop the useful field would be only 5·5 μ, but the image intensity would be greater.

4.5. Astigmatism and distortion

Astigmatism arises inevitably if a single cylindrical reflector is used, and unfortunately the simplest means of correcting it introduces an appreciable distortion. These defects do not appear in the analysis of the previous section, since they involve rays in the sagittal plane, which is normal to the meridian plane containing the principal curvature of the surface to which that analysis applied. The conditions in a cylindrical system viewed as a whole are shown in Fig. 4.12. The bundle of rays in the meridian plane

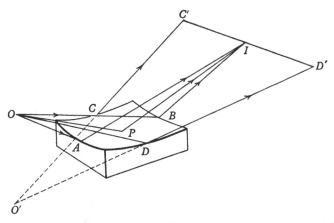

Fig. 4.12. Image formation at grazing incidence on a cylindrical mirror, by meridian rays and sagittal rays, giving a line focus $C'ID'$. (Kirkpatrick & Pattee, 1957.)

form an image of O at I after reflexion along a line APB in the mirror surface, as discussed above. Bundles of rays (OC, OD) diverging from O to either side of this plane will be focused each in their own plane of incidence, but will continue to diverge after reflexion. In other words, the surface acts as a cylindrical mirror for the component of a ray parallel to the meridian plane, but as a plane mirror for the component normal to it. In first approximation the image points for successive bundles fall on a line $C'ID'$ through I normal to the meridian plane; this is the meridian focus, formed at a distance q satisfying the relation (4.10):

$$1/p + 1/q = 2/R\sin\theta = 1/f_m, \qquad (4.18)$$

where R is the radius of curvature of the mirror and f_m the meridian focal length.

The sagittal rays reflected from the zone CPD appear to diverge from the point O', the virtual image of O as it would be formed in a plane mirror tangential to the reflecting surface along the line CD, so that $p = q'$, if q' is the distance of this second 'image' from P. The sagittal bundles reflected from zones of the mirror before and after CD will also pass through the line focus $C'ID'$, but will appear to diverge from points above and below O' respectively. The 'secondary' or sagittal image of O is thus also a line, now normal to the sagittal plane. The cylindrical mirror is, therefore, completely astigmatic, as is a cylindrical lens for light, producing primary and secondary foci in the form of line images that are mutually at right angles and separated by a distance equal to the sum of object and primary image distances.

When the mirror has spherical instead of cylindrical form, the effect of the curvature in the direction normal to the meridian plane is to lessen the divergence of the ray bundles after reflexion, so that the secondary image O' is formed further away from the mirror than is O. It can be shown that the new image distance q'' satisfies the relation

$$1/p + 1/q'' = 2\sin\theta/R = 1/f_s, \qquad (4.19)$$

where f_s is the sagittal focal length. The ratio of the two focal lengths is then $f_m/f_s = \theta^2$, and with θ in the range 10^{-2} to 10^{-3}, it will be of order 10^{-5}. With $R = 100$ cm. and $\theta = 10^{-2}$, the second focal length will be 50 m. Since the focusing action of a spherical mirror in the transverse (or equatorial) direction is so weak, it is more convenient in practice to use cylindrical rather than spherical surfaces for X-ray microscopy, so that each mirror unit performs one focusing action only.

To correct astigmatism the primary and secondary focal lengths must be made equal. There are three ways in which this may be done. In principle it is sufficient to figure a surface such that the radii of curvature in the meridian plane (R_m) and in the equatorial plane (R_e) satisfy the condition

$$R_m \sin\theta = R_e/\sin\theta,$$

or
$$R_e = R_m \sin^2\theta, \qquad (4.20)$$

which follows from (4.18) and (4.19). However, this would require extremely strong equatorial curvature: with $R_m = 100$ cm. and $\theta = 10^{-2}$, then $R_e = 0\cdot1$ mm. The mirror would have to be in the form of a capillary tube of this internal radius, with a curvature of radius 1 m. in the longitudinal section of the bore; that is, the bore must be slightly larger at the centre than at the ends of the tube. Although no such mirror has been made for imaging extended objects with X-rays, because the figuring would need a perfection beyond present technical skill, it has been utilised by

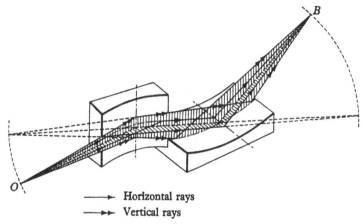

Horizontal rays
Vertical rays

Fig. 4.13. Correction of astigmatism by successive reflexions at two mutually perpendicular cylindrical mirrors (after Kirkpatrick & Baez). (Hildenbrand, 1956.)

Henke & DuMond (1955) for obtaining a point focus of X-rays for small angle diffraction investigations (cf. § 14.1). They use a tube of initial bore 38 mm. and length 11 cm. to focus copper $L\alpha$ radiation of wavelength 13·3 Å. The finished reflector is 0·1 mm. greater in diameter at the centre than at the ends of the tube. The aberrations of such 'internal mirrors' have been discussed in detail by Wolter (1952 a, b).

A more practicable method of correcting astigmatism is simply to cross two identical cylindrical reflectors so that their meridian planes are at right angles to each other (Fig. 4.13). A ray bundle reflected in the meridian plane of the first is then reflected in the sagittal plane of the second mirror, and vice versa. If the curvatures are the same, the focal lengths will be equal and the astigmatism of first order fully corrected. This arrangement has been used in the reflecting X-ray microscope of Kirkpatrick & Baez (1948), with which the first images of extended objects were obtained, and in the more recent work of McGee (1957) and Hink (1957) (see § 4.8).

Inevitably distortion is introduced by such an arrangement. Although the focal lengths are identical, the object distance for

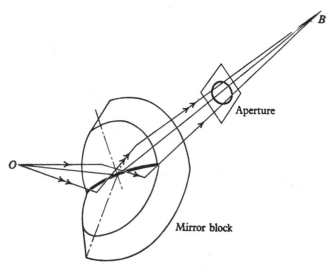

Fig. 4.14. Correction of astigmatism by successive reflexions at two curved mirrors ('roof-prism') at the same distance from the object O (after Montel). (Hildenbrand, 1956.)

meridian reflexion is less than that for sagittal reflexion, for a given ray, so that the magnification will be different in the two directions. This anamorphotism can be eliminated by tilting the recording emulsion by an equivalent amount. Alternatively the meridian reflexion could be shared between two mirrors in the same plane, symmetrically placed on either side of that giving sagittal reflexion.

Such a three-mirror system would require very careful adjustment, especially if a second triple unit had to be combined with it for removing other aberrations. A more elegant means of correcting distortion has been designed by Montel (1954). The reflecting surfaces are brought to the same distance from the object by cutting segments from two blocks, in which spherical surfaces have been worked, and uniting them as shown in Fig. 4.14. Rays incident first in one plane on the lower mirror are then reflected in the second plane by the upper mirror, and vice versa. By suitably positioning an aperture, the image is corrected for obliquity as well as for astigmatism and distortion. The cutting, joining and mounting of the mirror block calls for great technical skill, but Montel (1957) has already obtained micrographs showing encouraging resolution with his 'catamegonic' objective.

The astigmatism of a single cylindrical mirror of circular section can thus be corrected by employing a pair of mirrors in mutually perpendicular planes. The minimizing of other aberrations requires the use of more than one mirror in a given plane, and the complete correction of spherical aberration and coma is only possible if aspherical surfaces are figured. The essential features of such complicated systems will now be discussed; their practical realization is not yet at hand.

4.6. Compound systems of circular mirrors

Accepting the use of a pair of mirrors as essential for correcting astigmatism, we use the term 'compound' to describe a system employing more than one such pair. By 'circular' is meant the possession of circular form in at least the meridian section; in the equatorial plane the curvature will normally be of infinite radius.

It can be shown that, by employing two mirrors in each plane and an aperture stop suitably placed between them, the primary spherical aberration as well as curvature and obliquity of the field may be corrected (Dyson, 1952). The ray paths in one plane are shown in Fig. 4.15, where O is the object and I_1 the image formed by the first mirror; the final image is effectively at infinity. With given values for the glancing-angle θ_1 and radius of curvature R_1 of the first mirror M_1, with the aperture stop at M_1 and the first image at known distance q_1, Dyson obtained the required radius of curvature of the second mirror R_2 by equating first the expressions for the spherical aberration at the two mirrors and then those for the field obliquities. With $\theta_1 = 10^{-2}$, $R_1 = 100$ cm., and

$q_1 = 0.75$ cm., he found $R_2 = 112.5$ cm., and corresponding values for the separation of the poles of the mirrors $L = 1.5$ cm., object distance $p = 1.5$ cm. and glancing-angle at the second mirror $\theta_2 = 1.33 \times 10^{-2}$. The focal lengths of the mirrors will be $f_1 = 0.5$ cm. and $f_2 = 0.75$ cm., the first stage magnification $\frac{1}{2}$ and the second stage magnification infinite, since I_1 falls at the focus of the second mirror.

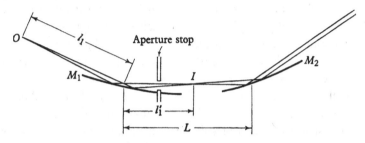

Fig. 4.15. Correction of spherical aberration (in one plane) by two mirrors and an aperture stop. (Dyson, 1952.)

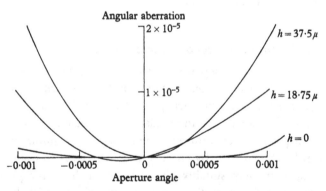

Fig. 4.16. Residual aberration in function of aperture angle for the mirror system of Fig. 4.15 for three values of the field width h. (Dyson, 1952.)

The residual spherical aberration is extremely small, but, with the aperture at the pole of M_1, the curvature of the field remains severe and coma limits the useful field, as always with circular mirrors. Dyson showed that by placing the stop at a distance of 0.4 cm. from the pole of M_1 the field curvature can be eliminated at the cost of introducing some third order distortion, which is not of serious consequence. The angular aberration in such a corrected system, mainly now due to coma, is plotted against aperture angle α in Fig. 4.16 for different widths of field. The maximum value of the glancing-angle at either mirror is 0·0143 radian, requiring an X-ray wavelength of 2·06 Å or longer. With the criterion that the longitudinal spherical aberration should not exceed one-quarter of this wavelength, the permissible aperture α is $1·78 \times 10^{-3}$ radian for a field of 37·5 μ and the resolving limit is 1150 Å. If the spherical aberration is allowed to be $\lambda/4$ at half-field, the useful aperture becomes $2·6 \times 10^{-3}$ and the resolving-limit 800 Å.

It is seen that the addition of a second mirror to correct spherical aberration improves the resolution only by a factor of less than two, with an increase of like order in the useful aperture, as compared with the performance of a single mirror and stop (§ 4.4). Taking into account the added difficulties of mutually alining the additional mirrors (two are in fact needed, one in each plane) and the loss of intensity incurred by doubling the number of reflexions, it is very doubtful if the setting up of a system of four circular mirrors would be sufficiently rewarding. It might be worth while if the mirrors could be so figured that coma were also corrected.

4.7. Compound systems of figured mirrors

The possible advantages of using aspherical surfaces have been explored by Pattee (1953a, 1957a), Baez & Weissbluth (1954), Baez (1957) and, in a more general way, by Wolter (1952a, b) and by Herrnring & Weidner (1952). Compound systems again have to be invoked if any major improvement in imaging is to be obtained, since a single pair of aspherical mirrors cannot be completely corrected. With a single elliptical mirror in each plane, for instance, the spherical error is zero, but the field obliquity very great and coma still predominant. Starting from a parabolic surface, Dyson (1952) has shown that it can be perturbed to a form such that obliquity is eliminated, and spherical aberration reduced in comparison with a circular mirror. The residual aberrations, however, are such that the resolving limit is no better than 900 Å.

The general principle of correction by a system of four mirrors, arranged as in Fig. 4.17, is that the points of reflexion at the second pair are inversely correlated with those at the first pair. An object normal to the axis will be reflected in the latter mirrors so as to give a first image lying almost along the axis, as in Fig. 4.10. This highly comatic image acts as object for the second mirror pair, in which the ray paths are approximately the reverse of those in Fig. 4.10, so that the final image is almost normal to the axis. The rays reflected at the near edge of the first mirror will be reflected from the far edge of the second mirror in this plane, and vice versa, so that the overall magnification is the same for all paths, that is, coma is corrected. For this to be so, it is necessary for the mirror to be figured so that the local radius of curvature changes rapidly across the meridian section. Correction is impossible with the uniform curvature of a circular mirror.

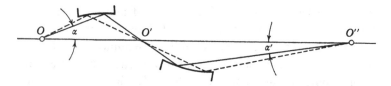

Fig. 4.17. Aplanatic system of mirrors, with real intermediate image
(after Pattee). (Kirkpatrick & Pattee, 1957.)

The properties of four-mirror systems of various sections have been computed in some detail by Pattee. He showed that coma can be corrected by combining a pair of crossed circular mirrors with a pair of crossed ellipsoids, the ray paths being as in Fig. 4.17 (Pattee, 1953 a). At an X-ray wavelength of 2 Å the resolution would be 500 Å within a field of 10 μ. Some improvement is obtained by using a pair of crossed ellipsoids, at unit magnification, followed by a pair of modified parabolas (Pattee, 1957 a). Such a system would be expected to have a resolution of 500 Å over a field of at least 20 μ at a wavelength of 4 Å. Baez (1957) has investigated, by ray tracing with an electronic computer, the effect on image quality of the continuous deformation of a circular mirror. He came to the conclusion that resolution can be improved, whilst preserving usable field, by deforming the circle into a cubic curve.

A more fundamental approach to the problem has been made by Wolter (1952 a, b), and independently by Herrnring & Weidner

(1952), starting from the mathematical formulation by Schwarz-schild (1905) of the conditions necessary for perfect imaging in an optical system. The mirror forms obtained are of high geometrical order, but can be approximated by conic sections, and are then very similar to those used by Baez and by Pattee, except that they are now developed into surfaces of revolution. One of the arrangements proposed by Herrnring and Weidner is shown in Fig. 4.18. The imaging properties are 'perfect' only in second order approximation; higher order aberrations are sufficiently severe for the useful field to be still restricted to a region close to the axis.

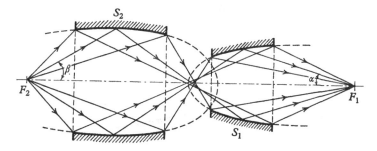

Fig. 4.18. Combination of aspherical surfaces of revolution, with perfect imaging properties (after Herrnring & Weidner). (Hildenbrand, 1956.)

Wolter investigated the approximations to these ideal shapes which are possible by combining paraboloids, hyperboloids and ellipsoids; with such conic sections it is possible to satisfy the sine condition in first approximation. The most promising form, from the point of view of construction and testing, is shown in Fig. 4.19; it is formed by joining a paraboloid to a coaxial hyperboloid so that the back focus of the latter coincides with the focus of the former. Its optical properties are only a little inferior to those of the 'ideal' system, since the higher order aberrations affect both equally, and it has the advantage of being composed of curves of simple analytical form. The calculated resolution is 250 Å over a field of 40μ at an X-ray wavelength of 20 Å. However, the difficulties of figuring even such a comparatively simple aspherical surface to the required degree of accuracy are considerable, and Wolter estimates that the dimensions of the mirrors could not be reduced below a size corresponding to a focal length of 2 cm.; for a direct magnification of ×150 this would demand an image throw of 3 m. The correct positioning of the one mirror with respect to the other will be equally critical, and Wolter puts forward a step-wise testing procedure. No reports of the successful construction of an X-ray microscope of this type have so far appeared.

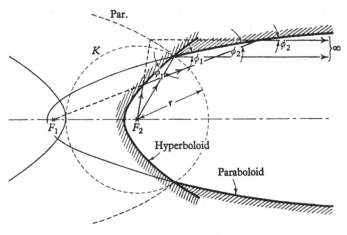

Fig. 4.19. Hyperboloid and paraboloid, combined to give highly corrected imaging properties (after Wolter). (Hildenbrand, 1956.)

4.8. Practical design of the reflexion X-ray microscope

The main lines of the design of a practical mirror microscope for X-rays follow from the considerations discussed above. The physical properties of reflexion at grazing incidence determine, within more or less narrow limits, the illuminating wavelength, the maximum glancing-angle, the focal length, curvature and nature of the reflecting surfaces, and the tolerance for finishing and alining them. The relative difficulty of correcting the various aberrations requires a balance to be struck between resolving-power and width of field, thus determining the usable angular aperture and the intensity at the image. Very high resolution over a minute field at extremely long exposure could doubtless be obtained with a very complicated system of mirrors, but a practicable microscope must have a more realistic performance. The width of field should be at least a few microns and exposure times not longer than a few hours.

The *glancing-angle* cannot be much less than 1°, for reasons both of experimental convenience and of intensity, since it determines the maximum possible aperture. As Table 4.1 shows (p. 89), this requires a wavelength of at least 2 Å and a metal reflecting surface of high atomic number, such as gold or platinum. Aluminium or glass, the properties of which are well known from optical technology, would require a wavelength of at least 8 Å. In designing a practical system it is essential to use experimentally determined values of the critical glancing-angle (cf. Rieser, 1957a, b) as these may differ appreciably from the theoretical values as given in Table 4.1.

The *illuminating wavelength* is further restricted by considerations of intensity and contrast. The efficiency of X-ray production is approximately proportional to the exciting voltage and hence inversely proportional to the wavelength. On the other hand, absorption increases approximately with the third power of the wavelength. Image intensity, therefore, demands a short wavelength, but image contrast, which is especially a limiting factor at high resolution, a long wavelength. The present tendency is to move towards longer values; 24 Å has the advantage of low absorption in water, 43 Å of high absorption in carbon, both important considerations with biological specimens.

The *focal length* and the *radius of curvature* are mutually dependent, since $f_1 = R\theta/2$. Since $\theta \sim 1°$ and, at high magnification $p \sim f_1$, the object has to be very close to the tangential plane of the leading edge of the mirror. It is then convenient to employ a focal length of one or two centimetres, requiring $R = 100\text{–}200$ cm. The production of mirror surfaces of large R is a familiar task in optical technology.

The *optical system* is then also determined by practical convenience. An image resolution of 0.2μ, for example, would need a total magnification of at least × 100 to render all detail directly visible to the eye. To obtain this by two stages of magnification, as in the light microscope, would call for a set of mirrors acting as projector as well as a set acting as objective. Apart from the loss of intensity in so many reflexions and the great optical path length, the complication of setting up such a system strains our present technical ability. A one stage system has been used in most experiments so far, the image thus formed being recorded on fine-grain emulsion and subsequently enlarged optically. An initial magnification of × 100 would bring resolved detail up to 20μ. At a focal length of 2 cm. the image distance would be 200 cm., an X-ray path already long enough for absorption loss in air to be appreciable.

The *intensity of illumination* at the image, formed with a source of given brightness, depends on the square of the angular aperture and the magnification. The aperture is limited to the order of 10^{-3} by the severity of the aberrations, even in a partially corrected system, so that less than 10^{-6} of the output of the source can be utilized. Since the intensity at the image is $1/M^2$ that at the object, this forms an additional reason for not using a high direct magnification. It becomes of first importance to use an X-ray source of highest possible brightness, placed as close as possible to the specimen. The first experiments were carried out with the type of X-ray tube used in crystallographic analysis, in which the target loading was of the order of 100 W/mm.² In none of the early publications is the exposure time given for the micrographs reproduced, but it would appear to have been several hours. More recently use has been made of the type of microfocus tube developed for projection X-ray microscopy (§ 3.4), in which the target loading is several orders of magnitude greater and the brilliance of the source correspondingly high. Optimally the size of the focal spot should equal the useful width of field of the reflexion microscope and should be brought as close as possible to the specimen. In this way Hink (1957) has obtained exposure times of order one hour at direct magnifications of × 100 and more. Montel (1957), using a rather larger aperture, required exposures of 10–30 min. at this magnification. McGee (1957) prefers a low direct magnification (× 7) for high resolution work, to keep the exposure-time in the range 5–30 min.

4.8.1. *Experimental arrangements.* The first reflexion microscope of Kirkpatrick & Baez (1948) employed a single pair of crossed mirrors of large radius of curvature, focusing Cu $K\alpha$ radiation (1·5 Å). The experimental arrangement is shown in Pl. III (see also Baez, 1949). The X-ray tube is in the foreground, set across the optical bench on which is mounted a helium-filled tube for minimizing absorption in the image throw. The specimen and mirror holders are set up just in front of the helium tube and the viewing-screen and plate-holder at the far end. The mirrors are in the form of glass disks, 1 cm. in diameter, cut from a larger mirror having a spherical surface of radius 10 m. The maximum glancing-angle was 0·3°, corresponding to the condition for total reflexion of 1·5 Å X-rays from glass. Photographs of metal grids and foraminifera at initial magnifications of ×50 were obtained with this microscope, showing a resolution of about 1 μ at the centre of the field; outside this the quality of the image deteriorated rapidly on account of coma.

On the basis of this experience, experiments were begun with a four-mirror system laid out on essentially the same lines (Kirkpatrick & Pattee, 1953; Kirkpatrick, 1957). The method of mounting the mirrors and specimen is shown in Pl. II. Each mirror is clamped on a small plate, seen at bottom left. The specimen is mounted at the top of the block and is adjusted in position with the top two micrometers. The four mirrors follow, in planes successively at right angles as in Fig. 4.17, and a micrometer is provided for adjusting the angle of inclination of each to the axis. Experiments on imaging with aspheric surfaces are in progress with this microscope (Pattee, 1957 a).

In place of mirrors of circular section, Lucht & Harker (1951) explored the possibility of using optical flats, supported along two sides and bent by application of a couple as in the original experiments of Ehrenberg (1947, 1949 a). The surface takes the form of a parabolic cylinder. Using a stigmatic pair of such mirrors, at a radius of curvature of 20 m., images at direct magnifications of ×10 and ×15 were obtained in Cu $K\alpha$ radiation. The resolving-power was 'about 1μ', at best, in agreement with theoretical estimates by Prince (1950). The arrangement has the advantage that the focal length of the mirrors can be varied by adjusting the applied couple, which is a considerable aid to focusing. On the other hand, it appears doubtful whether a material can be found that will both take a high optical finish and remain stable under the high pressures that must be applied.

Hink (1957) has used a pair of crossed silvered mirrors with a point focus X-ray tube of high intensity producing Cu $K\alpha$ radiation. Appreciating the reduction in aberration that should follow from using a smaller radius of curvature, as discussed above (§ 4.4), he employed mirrors of $R = 4\cdot89$ m. At a direct magnification of × 120 and an exposure of 90 min. he obtained photographs showing a resolution of about 1μ, estimated from photometer traces. He discusses, in some detail, the reliability of experimental tests of resolving-power. Montel (1957) used a much larger radius of curvature ($R = 19\cdot62$ m.) for the 'catamegonic' objective in which the two mirrors are cut so that two half-mirrors may be rejoined. They were coated with gold by evaporation and an adjustable aperture was placed beyond them to correct obliquity of the field. Micrographs taken on Kodak Definix film required only 10–30 min. exposure at a magnification of × 100. The 'limit of perception' was estimated at 1μ; application of the classical definition of resolution would probably give a value of 3–5μ.

McGee (1955, 1957) has developed further the original Kirkpatrick and Baez two-mirror system by computing the best relative positions of aperture stop and mirror. He used two uncoated spherical mirrors of silica and two aperture stops to limit field obliquity and spherical aberration. The X-ray source was a continuously pumped tube with an electrostatic focusing system that produced a focal spot of about 1 mm., some ten times larger than the field of view of the microscope. Aluminium $K\alpha$ radiation was generated ($8\cdot3$ Å), absorption losses being reduced by using an aluminium exit window and a helium-filled tube. The exposure-time was kept below 30 min. by working at a small direct magnification (× 7). Micrographs were taken with the aperture stop at different positions, to investigate the variation of image quality and width of field, the image being too faint for direct observation. The best photographs of a silver grid test specimen show a resolution approaching $0\cdot25\mu$ (Pl. I B).

4.9. Practical limitations

Apart from the limitations set by the aberrations of glancing reflexion, discussed earlier, there are at present more serious practical difficulties in the way of realizing the predicted resolving-power. These arise from the high precision with which a mirror must be figured and the high degree of finish necessary if the mirror surface itself is not to introduce confusing detail into the image.

In optical technology a surface finished true to one-tenth wave-length ($\sim$ 500 Å) is regarded as sufficiently perfect for the most exacting requirements. In conditions of grazing incidence, however, the equivalent tolerance will be $\lambda/10\theta$, where θ is the glancing-angle. For $\theta = 10^{-2}$, this becomes 10λ or from 10 to 100 Å for the range of X-ray wavelengths at present used for microscopy. Optical figuring to a tolerance some ten times finer than previously demanded must now be aimed at, and appropriate testing procedures devised to control its course. Multiple beam interferometry, as developed by Tolansky (1948), already provides a tool accurate enough for testing surface contour down to 10–20 Å (cf. Koehler, 1953). By this means Hink (1957) found that his metal-coated glass mirrors were true to ± 20 Å, over most of their surface, although a few faults of depth 60–100 Å were also observed.

In addition to this fine tolerance for the truth of the surface to the designed form, whether circular, elliptical or paraboloidal, the surface must be smooth to molecular dimensions (Kirkpatrick, 1957). Such a degree of polish is again likely to be a severe tax on technical skill. Even if it is attainable by lapping, it may then be desirable to evaporate or otherwise deposit a layer of metal to improve reflectivity. However, the investigation of such layers by electron microscopy has shown how difficult it is to obtain a continuous layer; local aggregation usually occurs, either on deposition or by subsequent recrystallization (see Faust, 1950; Sennett & Scott, 1950; Blois, 1951). It was already remarked by Ehrenberg (1947) that his first experiments with bent gold-coated optical flats showed 'wings' to the line image of the source on prolonged exposure. In a more detailed study, Ehrenberg (1949b) found that highly finished surfaces of both glass and quartz, with and without metal coatings, produced marked striations in the reflected X-ray image. He deduced that the surfaces were in fact corrugated in form with slow variations in height of the order of 10 Å; in fact, he obtained similar striations from a pattern of gold bars, 5–10 Å thick and 0·7 mm. apart, formed by evaporation. Butcux (1953) reached the same conclusion after studying a wide variety of glass, crystal, metal and plastic surfaces. The further complication arises that a metal-coated surface is liable to deteriorate under X-ray illumination, as noted by Ehrenberg (1947, 1949b). The probable cause is the deposition of dust, and particularly carbonaceous material, from the surrounding atmosphere, either from direct ionization by X-rays or from the effect of secondary electrons.

The tendency at present is to avoid such complications by using uncoated glass or silica mirrors (Kirkpatrick, 1957; McGee, 1957), which do not appear to contaminate so rapidly. For the soft X-rays now used such surfaces have a large enough critical angle for convenient setting up of the object, so that a heavy metal coating is not called for. It is also the case that the cut-off at critical reflexion is much sharper for light than for heavy elements (Rieser, 1951, 1957a, b); the former also appear to have a higher reflectance than the latter for long wavelengths (Farrant, 1950). In these circumstances it may prove useful to employ a coating of carbon, evaporated from an incipient arc (Bradley, 1954), which is practically structureless and smooth, certainly down to 10 Å; it has a high optical reflectance (Cosslett & Cosslett, 1957), which would be an aid in testing and preliminary alinement by optical means. It is to be noted that reflexion systems, being naturally achromatic, have the same focusing properties whatever the illuminating wavelength. It is, therefore, possible to set up a mirror system for X-rays with the aid of a beam of visible light, to the point where apertures have to be inserted; these are at present so small that diffraction effects then prevent further optical observation.

4.10. The ultimate limit set by diffraction

Analogously to the light microscope, the resolving-power of the X-ray reflexion microscope will be ultimately limited by diffraction effects if and when the geometrical aberrations are fully corrected. The Abbe diffraction theory of image formation leads to the conclusion that two object points will be resolved if they are separated by a distance greater than d, where

$$d = k\lambda/\sin A,$$

λ is the illuminating wavelength, A the angular semi-aperture of the beam leaving the object and k is a constant, the value of which depends on the structural pattern of the object; for present purposes its minimum value may be taken as 0·5. At grazing reflexion the aperture is so small that $\sin A \doteqdot A$, and so

$$d = \lambda/2A. \qquad (4.21)$$

The aberrations of existing X-ray microscopes limit the permissible semi-aperture to the order of 10^{-3} radian. The resolving-power, on diffraction grounds alone, therefore cannot be better than about 500λ; that is, 1000 Å at a wavelength of 2 Å. In a fully corrected system a larger value of A can be used, but it can never exceed θ_c, the critical glancing-angle for total reflexion, as follows from the geometry of grazing reflexion. The maximum value of glancing-angle is θ_c and so the maximum value of the total aperture is $2\theta_c$. The semi-aperture A is thus limited to a maximum of θ_c and on substitution in (4.21) we have:

$$d = \lambda/2\theta_c.$$

From (4.3) and (4.5), θ_c is directly proportional to λ, giving

$$d = 215(A/Z\rho)^{\frac{1}{2}}, \qquad (4.22)$$

where Z, ρ and A are the atomic number, density and atomic weight respectively of the element composing the reflecting surface, and d is in Ångström units.

The ultimate resolving-power of a mirror X-ray microscope is thus independent of the illuminating wavelength, in first approximation, and does not vary greatly with the nature of the reflecting surface.* For the three metals listed in Table 4.1 (Al, Ag and Au) the respective resolving limits would be 190 Å, 100 Å and 80 Å. Such a resolution is roughly midway between the limits for visible light (2000 Å) and for electron microscopy (10 Å). It is this ultimate prospect which stimulates the continuing effort to improve the mirror method. However, there remain very severe

---

* The comparatively slow decrease in reflectivity with angle which occurs for soft radiation, instead of a sharp cut-off at θ_c, leads to an optimum value for the aperture (Rieser, 1957a, b), depending on the nature of the surface.

technical problems in the way, first, of figuring reflecting surfaces to the proper aspherical form for correcting the field aberrations, and secondly, of giving them the requisite finish, some ten times better than the best achieved in present optical working. The attainment of a resolution of 1000 Å will be the next target; at present the best results are two or three times poorer than this.

Additional references added in proof

To §4.4.3: McGee & Milton (1960).
To §4.4.2 and 4.9: Montel (1960).

CHAPTER 5

REFLEXION X-RAY MICROSCOPY: CURVED CRYSTALS

The reflexion of X-rays at a smooth surface, discussed above, is essentially similar to specular reflexion in optics and is governed by the same laws. Unlike visible radiations, however, X-rays are reflected also from crystals, which, appropriately shaped, can be used as mirrors. The 'reflexion' is due to diffraction from the regular lattice formed by the constituent atoms, and will thus occur only in certain well-defined directions. As this is a three-dimensional structure the 'reflecting' surface is not necessarily the physical surface of the crystal and the diffracted beams may emerge through any of its faces. In general, reflexion and transmission go on together, and the refractive index and absorption coefficient of the material at the wavelength used will enter into a full discussion of the direction and intensity of the emergent ray.

A number of systems of curved crystals have been proposed for X-ray microscopy, but experimental investigation of their image-forming properties has been confined to two or three of them, and this mostly at low (or even 1 to 1) magnification. Much greater use has been made of them for monochromatizing and focusing X-ray beams in spectrometry and structure analysis, where requirements are less stringent. Attention will be confined here to features which are of interest for image formation. The fundamental relations are identical with those for specular reflexion, with the added requirement that the Bragg diffraction expression

$$n\lambda = 2d\sin\theta$$

must be satisfied (Fig. 5.1); at other angles of incidence the intensity reflected is negligible. The first order theory is straightforward and the evaluation of aberrations has been attempted in a few special cases, but little information either theoretical or experimental is available on the intensity in the image.

5.1. Efficiency of Bragg reflexion

The ideal case of a narrow beam of monochromatic X-rays reflected by a perfect single crystal can be treated theoretically. The coefficient of reflexion R is defined as the ratio of the reflected to the incident energy flux, at given glancing-angle θ and at given wavelength (Compton & Allison (1935), § 6.12). It is proportional to λ^3/μ and inversely proportional to $\sin 2\theta$, where μ is the absorption coefficient, which also depends on λ (cf. § 2.2). In positions where the Bragg condition is satisfied for planes of given spacing d, reflexion should be almost 100 % efficient, but only over a very

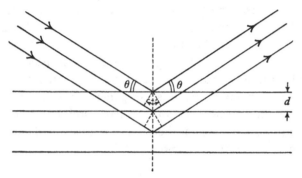

Fig. 5.1. Diffraction of X-rays at lattice planes of spacing d
(Bragg spectrometer).

small angular range (a few seconds of arc). In practice it is extremely difficult to test the theoretical relationship, since highly perfect crystals are hard to come by, but its validity has been established by indirect experiments.

Measurements can more easily be made in the conditions of X-ray spectrometry, where a beam of appreciable divergence is reflected from a nearly-perfect single crystal. With the source and detector placed symmetrically with respect to it, the crystal is rotated at steady angular velocity ω through the position of reflexion. If the energy collected in this time is E, the energy flux in the direct beam P_0 and in the reflected beam P, then the 'integrated coefficient of reflexion' R_i is given by

$$R_i = E\omega/P_0 = \int P/P_0 \, d\theta = \int R(\theta) \, d\theta. \qquad (5.1)$$

Thus, R_i is equal to the angular interval which would contain the reflected intensity E if reflexion were 100% efficient, that is, it is the width of a rectangle of height P_0 equal in area to that under the 'rocking-curve' $\int P\,d\theta$.

Since the reflected intensity is high only over an angular range small with respect to the angle of rotation, the value of R_i as measured is usually very low. In first approximation it is equal to the width of the relevant line in the diffraction pattern, expressed as a fraction of a radian. Values given by Compton & Allison (1935, p. 401) for calcite lie between 10^{-4} and 10^{-5}, depending on the degree of perfection of the crystal. With freshly cleaved surfaces Allison (1932) obtained values of 3·80 and 3·46 × 10^{-5} for the reflexion of Cu $K\alpha$ radiation from the (100) face, in good agreement with the calculated value of 3·82 × 10^{-5}.

The efficiency of reflexion is roughly proportional to the wave-length and depends strongly on the lattice structure. Recent calculations of Renninger (1954) for Cu $K\alpha$ radiation gave values (× 10^5) of 29·5 for Al (200), 71·5 for Cu (200), 86 for diamond (111), 110 for LiF (200) and 500 for graphite (002). The variation with wavelength is also appreciable. For calcite (100) the calculated values (× 10^5) ranged from 0·582 for $\lambda = 0·21$ Å to 16·4 for $\lambda = 4·94$ Å; the measured values agreed well for the softer wave-lengths, but were appreciably higher at $\lambda = 0·21$ Å (Allison, 1932).

The value of the integrated reflexion coefficient is a useful guide in the conditions of X-ray spectrometry, showing that only a few parts in ten thousand of the incident intensity are received in a particular diffracted beam. Hence it determines the time needed to obtain an analysis in emission microspectrometry (§ 7.1) (Long, 1959). For X-ray microscopy, however, it is somewhat mis-leading because of the very different experimental arrangement. As will be shown below, it is possible to design the crystal geo-metry in such a way that an appreciable angular width of the incident beam is all reflected in the same direction, that is, a degree of focusing is obtained. Ideally a crystal may be so curved that every ray of a divergent monochromatic beam meets a facet of the surface at the correct angle for Bragg diffraction. The efficiency of reflexion would then be 100%, apart from absorption loss. In special conditions also there may be a common focus for rays of different wavelengths, so that monochromatic illumination is not demanded.

In crystal reflexion an important practical factor is the range of glancing-angle, about the Bragg position, over which appreciable reflexion takes place from a given surface of a single crystal. The 'rocking curve' can be measured in a double crystal spectrometer in which the X-ray beam, monochromatized by the first crystal (A), is diffracted by a second crystal (B), usually arranged parallel to it (Davis & Stempel, 1921, 1922). The energy collected from a particular reflexion is plotted against the setting of B as it is rotated through the Bragg position. The 'width' of the rocking curve, $d\theta$, measured at half the peak intensity, is closely related to the integrated reflexion R_i. Its value is found to depend on the degree of perfection of the crystal, which may be disturbed by internal stresses as well as by mosaic structure. The smallest values recorded by Compton & Allison (p. 729) are about $2''$ of arc ($\sim$ 10 microradians), being thus of the same order as the 'natural' width of a diffraction line. Reflecting crystals for use in an X-ray microscope should be as perfect as possible: the narrower the rocking curve the sharper will be the image detail. In effect, imperfections act in the same way as departures of the shape of the surface from ideal curvature, in producing an aberrated image: an object point will be reproduced as a disk of confusion of size dependent on the width of the rocking curve and on the magnification. This will set a limit to the attainable resolving-power, over and above the diffraction effect due to use of a finite aperture, even if a curved crystal microscope can be constructed with minimum spherical aberration as proposed by Thathachari (1953), (§ 5.3).

5.2. Focusing at glancing incidence

A flat crystal surface behaves like a plane mirror for X-rays incident at a particular angle, which is determined by the wavelength and by the lattice spacing in a direction normal to the surface. As the optical laws of reflexion are observed for such rays, they will be brought to a focus if the crystal surface is given a suitable curvature, in the same way as concave and convex mirrors act on light. Two limiting cases arise, according as the curvature is in the plane of incidence or normal to it. The former case is directly analogous to the reflexion of X-rays from polished surfaces at glancing incidence, treated in chapter 4, and has been much used for monochromators and focusing spectrometers. Crystals curved at right angles to the plane of incidence have been used as spectrometers as well as for forming images at unit magnification. For higher magnifications doubly curved crystals must be used, on account of intensity limitations as well as of the aberrations of systems of single curvature.

5.2.1. *Crystals curved at right angles to the plane of incidence.* The idea of focusing a cone of X-rays by diffraction from a cylindrical shell, on the axis of which the source and its image are located (Fig. 5.2), was first proposed and investigated theoretically by

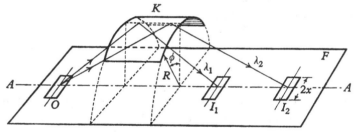

Fig. 5.2. Focusing of X-rays by a cylindrical shell, curved normally to the plane of incidence; different wavelengths are focused at different axial distances. (von Hámos, 1938*b*.)

Gouy (1916). He realized that true focusing would require an elliptical shell, with the source and image at the two foci, but that it would be impossible to bend a thin crystalline lamella into such a shape. Only developable surfaces come into consideration and Gouy discussed the imaging properties of a cylindrical shell (for example, of mica), primarily with its use in a spectrograph in mind. He showed that not only an axial point, but a small object on the axis would be truly imaged in all three dimensions, at unit magnification, by a monochromatic beam of X-rays. With inhomogeneous radiation a series of images would be obtained along the axis at distances (x) from the source related to the wavelength by

$$1/\lambda = n/2d\{1 + (x/2R)^2\}^{\frac{1}{2}}, \qquad (5.2)$$

where n is the order of diffraction resulting from a crystalline shell of radius R and lattice spacing d in the radial direction. In order to record the spectra given by a slit source, the photographic plate would thus have to lie along the axis. He appreciated that only a segment of the cylinder could be used in practice, owing to aberrations, but he does not appear to have constructed such a spectrograph himself.

The first practical investigation of the 'image spectrograph' was made by von Hámos (1934, 1936), using the highly monochromatic fluorescence X-radiation from metals. He developed it into an 'X-ray microanalyser' in which the individual components of an alloy or mixture can be distinguished (Fig. 5.2): the characteristic radiation of each element is reflected to a different axial position, giving a series of 'spectral images' of its distribution in the specimen (von Hámos, 1938a, b; von Hámos & Engström, 1944). Subsequently he examined in more detail the imaging conditions and aberrations (von Hámos 1939, 1953).

As skew rays must be considered as well as those in the plane of symmetry, it is convenient to represent the ray paths in transverse and longitudinal projection (Figs. 5.3, a and b), with XYZ co-ordinates as shown. The Z-axis lies in the shell and parallel to the axis about which it is bent, at radius R. The XZ-plane is tangential to the inner surface of the crystal and the YZ-plane is the plane of symmetry, containing the axis of the cylinder and the object O. The azimuthal angle ϕ is measured from the YZ-plane, and the position of object and image points are specified by their distances from the Z-axis, a and A, as shown. It may be shown that for paraxial imaging the co-ordinates of the image point are

$$x = 0; \quad y = Ra/(2a - R); \quad Z = (a + A)\cot\theta.$$

When the object point is on the axis ($a = R$), then also $y = R$ and $Z = 2R\cot\theta$. All rays are focused to I whatever their azimuth, but the system has no real focal length according to the optical definition, since a beam initially parallel to the axis would suffer no reflexion. A crystalline lamella bent to the shape of a segment of a cone would, however, form a true focus and have a finite focal length.

Detailed investigation of the imaging of points in the neighbourhood of O shows that they form a focal surface, indicated in longitudinal section by the line II' in Fig. 5.3b. For an object small in extension compared with R, this is approximately a plane making an angle β with the XZ-plane. In general the magnification is constant in the direction of the YZ-plane, but normal to this plane its value varies with a, giving rise to trapezoidal distortion. In the neighbourhood of the centre of the image it is possible to obtain the same magnification in both directions by appropriate choice of the inclination of the object plane α.

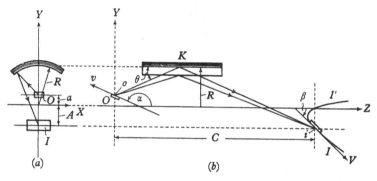

Fig. 5.3. Ray paths (a) in transverse, and (b) in longitudinal projection, for reflexion from a cylindrical shell. (von Hámos, 1938b.)

As in light optics, the geometrical aberration is found by calculating the position of an image point for an off-axis point near to O. For such a point the rays reflected from different azimuthal zones of the crystal do not unite in a single image point but form a disk of confusion, not necessarily circular. Expressions for the value of the geometrical aberration in and normal to the YZ-plane are derived by von Hámos (1939), and contain terms in ϕ as well as in higher powers of it. In general the longitudinal is greater than the transverse aberration. Both are a minimum for $a = R$, that is for an object at the axis, when also $\alpha = -\beta$. The object and image planes are normal to the incident and reflected rays respectively (Fig. 5.4) and the magnification is unity.

The high dispersive power of the crystal, which is its essential feature as a spectrograph, entails an appreciable chromatic aberration. The axial separation of object and image is proportional to

9

$\cot\theta$ and the glancing-angle θ depends on the wavelength λ. The finite spread in wavelength $d\lambda$, which occurs even in fluorescent line radiation, thus entails a broadening of the image point by an amount δ_c which, for the special case of Fig. 5.4, is given by

$$\delta_c = 2R/\cos\theta \cdot d\lambda/\lambda. \qquad (5.3)$$

For first order reflexion of wavelengths of order 1 Å, the chromatic aberration is of order $10^{-3}R$; like the geometrical aberration it is minimal when object and image are normal to the X-ray beam.

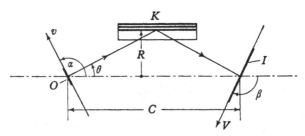

Fig. 5.4. Position of object and image for minimum aberration in reflexion from a cylindrical shell. (von Hámos, 1938b.)

This is a natural limitation, which in practice will be added to by the angular spread due to the 'rocking curve' of the crystal. The aberration δ_r produced by a rocking curve amplitude $\Delta\theta$ is given by

$$\delta_r = 2R\Delta\theta/\sin\theta. \qquad (5.4)$$

As $\Delta\theta$ can be less than 10^{-4} microradians in a good single crystal, the aberration due to lack of perfection should be less than the chromatic aberration by an order of magnitude. It is thus necessary to select crystals that are as nearly perfect as possible and to bend them to the smallest practicable radius. At a radius of 100 cm., usual in curved crystal monochromators, the chromatic aberration would be 1 mm. For the smallest radius used by von Hámos ($R = 5$ mm. for a rocksalt crystal), it is 5μ.

The geometrical aberration can be reduced below this chromatic limit, at cost of lower intensity, by reducing the aperture ϕ; as in electron lenses, the ultimate limit set by the Abbe diffraction conditions is negligible by comparison. For the Cr $K\alpha$ line (2.28 Å) and an aperture of 0·1 radian, the geometrical aberration would be $10^{-4}R$. The aperture involved here is the azimuthal width of beam that is focused, determining the size of crystal segment that can be used. The effective aperture in the longitudinal plane is very small, being no more than that allowed by the rocking curve of the crystal. The fraction of radiation from the specimen actually focused to the image is therefore extremely small and the exposure-time correspondingly long, even if primary rather than secondary X-rays are used.

The arrangement of Fig. 5.4, in which aberrations are minimal, has been used by von Hámos (1938a, b, 1953) to obtain images of a number of specimens. The images are recorded on fine-grained emulsions and subsequently enlarged. A rocksalt crystal was used, bent to a radius of 17·2 mm. Molybdenum (or zinc) primary radiation excited fluorescent emission in the metals in the specimen. Enlarged X-ray images of test objects, containing iron and copper, with an optical micrograph at the same magnification, are shown in Pl. IV. The resolution might be estimated at 50μ, as against the 17μ expected from chromatic aberration alone, which would indicate an appreciable contribution from the rocking curve of the crystal. The exposure-time is in the range 10–30 min. for thick layers of pure elements; it becomes 1 hr. or more when a small amount is to be detected.

The 'microanalyser', interesting as it is as an application of focusing by singly curved crystals, lacks both resolving-power and speed as compared with other types of X-ray microscope. It is unlikely to exceed 10μ in resolution, owing to the difficulty of curving crystals to small radius without impairing the perfection of the lattice, and exposure must remain long owing to the poor efficiency of fluorescent excitation and the small beam aperture. The particular feature of the method, the ability to distinguish the elementary constituents in a sample, is now shared by the scanning microscope, which can work at higher speed and greater resolution (§ 7.3). Some improvement can be obtained in image intensity, however, by using doubly curved crystals in the manner discussed below.

5.3. Focusing at near normal incidence

The possibility of forming images with doubly curved crystals follows logically from the focusing properties of crystals curved in the one or other dimension with respect to the plane of incidence. It offers a larger aperture of illumination, with correspondingly shorter exposures, and also the advantage of stigmatic imaging, as compared with these simple systems. It can employ very much larger glancing-angles than the total reflexion method, and therefore the optical problems are more tractable. In particular the chromatic aberration is zero, in first approximation, so that a band of wavelengths can be focused, with consequent gain in image intensity. Nevertheless the limitation of poor efficiency of reflexion remains serious, and is likely to prevent the realization of some of the attractive properties of combinations of curved crystals, involving multiple reflexions, which would give high magnification and low spherical aberration in a comparatively small optical path.

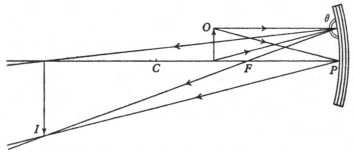

Fig. 5.5. Focusing of X-rays by diffraction from a curved crystalline lamella, when the reflecting planes are parallel to the curved surface (after Cauchois).

The imaging properties of doubly curved crystals were first investigated, experimentally and theoretically, by Cauchois (1946, 1950). Taking as starting-point the focusing spectrometer, which employs a cylindrically curved mica flake, she indicated how it could be modified to form real images and obtained pictures of metal grids and other objects at direct magnifications of between 1·5 and 2; the exposure-time is not stated.

Since focusing from curved crystals depends on Bragg reflexion from lattice planes, they may be utilized either in transmission or reflexion. The latter arrangement (Fig. 5.5) is similar to that used in spectrometry, but is here used at high instead of low glancing-angles; we return to it later. In the former case (Fig. 5.6), the effective reflecting planes are *normal* to the crystal face before it is bent, instead of being those parallel to it as in the imaging systems so far discussed. Rays parallel to the axis from different parts of the object meet the surface at different heights and so make glancing-angles with its radii that increase with height. The wavelength that satisfies the Bragg condition in the first order

$$\lambda = 2d\sin\theta \qquad (5.5)$$

will vary correspondingly, so that rays of these wavelengths are selectively focused into the image by different zones of the mirror. Conversely, a bundle of rays diverging from a given point in the object will make different angles with the normal at the points of

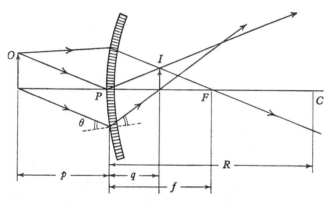

Fig. 5.6. Focusing of X-rays by diffraction from a curved crystalline lamella, when the reflecting planes are normal to the curved surface (after Cauchois).

incidence; the lattice will select and focus the wavelength appropriate to each direction. In principle, therefore, the image will be formed by a wide range of wavelengths from a maximum given by the angular aperture of the mirror, and approaching $2d$, to a minimum set by the short-wave limit of the X-ray source. In practice, absorption in the finite thickness of the crystal will attenuate the softer rays and the image will be formed primarily by the harder component. The paraxial zones of the mirror contribute all wavelengths to the image, the marginal zones only the softer rays. Alternatively seen, of the rays emitted in a given direction from a given object point, only those in the narrow range of wavelengths $d\lambda$, given by the rocking curve width, are focused to the corresponding image point; the remainder form a general background over a much larger area. The image intensity is therefore low.

The transmission-mirror was used by Cauchois (1950, 1952) in her original experiments (see also Despujols, 1953). Focusing in planes through the axis other than that shown in Fig. 5.6 depends on the existence of other networks of planes normal to the surface, and on the practicability of bending a crystal in two directions. Possibly for this reason her best results were obtained with cylindrically instead of spherically curved crystals, of which Fig. 5.6 would then represent a right section. As noted earlier, the image formed by such a lamina is completely astigmatic but stigmatism can be restored by using a line source, a narrow slit or a narrow crystal. Thin sheets of rocksalt and artificially prepared single crystals of aluminium and iron have been used by Cauchois and also by Hulubei (1946). The quality of the images is poor, however, and the technique of preparing and curving the crystals will require developing before a two or three stage system can be usefully built.

The individual reflexions from crystal planes, although subject to the Bragg restriction, must otherwise obey the simple laws of optics. The relations for the focusing properties and aberrations of curved mirrors can therefore be taken over in their entirety to the X-ray case. If, in Figs. 5.5 and 5.6, R is the radius of curvature of the crystal, p the object and q the image distance then they are related in the usual way:

$$1/p + 1/q = 2/R = 1/f \qquad (5.6)$$

$$M = q/p.$$

M is the magnification and f the focal length defined as the image distance when the object is removed to infinity. For paraxial conditions, that is, using a small angular width of the mirror around the pole O, point to point imaging will be obtained. The image will be enlarged or diminished, upright or inverted, depending on its position along the axis and the sense of curvature of the mirror. The main difference from light optics is that transmission and reflexion focusing are equally possible and indeed normally go on together, the relative intensities of the two beams depending on the density of atoms in the planes normal and parallel to the curved surface respectively, apart from absorption loss. In other words there is always a real image formed, which alone can be observed when a single mirror is used; but, as in optics, the virtual image may be of interest in compound systems.

The use of a concave mirror for focusing X-rays (Fig. 5.5) was studied by Thathachari & Ramachandran (1952). It has the advantage over the convex mirror of isolating the image from the incident illumination. The transmitted radiation is lost from the image space and so does not form a general background as in transmission focusing. The only unwanted contribution to the image comes from Compton scattering in the crystal, which is weak and in any case incoherent. A disadvantage, however, is the high value of glancing-angle θ that must be used in anything like paraxial imaging conditions, so that $n\lambda \sim 2d$. The intense first order reflexion can then only be utilized with long wavelength illumination, since d cannot be less than a few Ångström units, and unfortunately soft radiation requires vacuum. On the other hand, use of second or higher order reflexions with harder radiation also meets with difficulty, since intensity falls rapidly with order of reflexion. Use of intermediate values of glancing-angle would necessitate widening the angular aperture with consequent increase in spherical aberration, which depends on the third power of the aperture. However, analysis shows that this aberration can be reduced by multiple reflexion between two concave mirrors (Thathachari, 1953), so that the system warrants further investigation. A compound system also has the advantage of a much shorter image distance for given magnification than a single mirror. Unfortunately, it appears to have serious practical limitations in resolving-power, apart from the long exposure-time to be expected after two Bragg reflexions.

The only details so far published of experimental tests of these ideas relate to the production of images with a single concave mirror (Ramachandran & Thathachari, 1951). The mirror was formed by sealing a strip of mica 0·04 mm. thick to the end of a brass tube of diameter 1 in. (Fig. 5.7) which is then evacuated. At a pressure difference of about half an atmosphere, a radius of curvature as small as 15 cm. could be obtained in the mica. An object O, placed in front of an X-ray tube X, was imaged by reflexion to I; the portion of the mirror utilized was restricted by means of a lead stop L. The position of the image was first found by using the mica as a mirror for visible light. With an X-ray tube operating at 40 kV and a beam current of 15 mA, magnifications of ×4 and ×6 required exposures of 30 min. and 6 hr. respectively. The system is best used, therefore, at low magnification with fine-grained photographic plates, followed by optical enlargement.

5.4. Focusing by ring mirrors

Instead of imaging at high angles of incidence, with the object–image axis normal to the mirror segments as above, it is equally possible to form images at glancing incidence from ring mirrors, formed by rotating a segment about the axis (Fig. 5.8). Such a

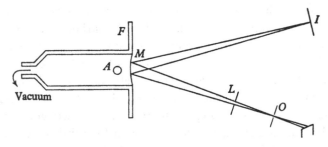

Fig. 5.7. Focusing of X-rays by a concave mirror of mica M, sealed across the end of an evacuated tube. (Ramachandran & Thathachari, 1951.)

system is the three-dimensional form of the singly curved crystal surface used in monochromators and possesses its favourable optical properties.

The use of such an arrangement for forming X-ray images was first proposed by Trurnit & Hoppe (1946), who named them 'zonal lenses', but since they are in function catoptric, not dioptric, it is more appropriate to call them ring mirrors. They obey the simple mirror relation

$$1/u + 1/v = 1/f, \quad \text{with} \quad f = R/2\theta,$$

so long as the object and image distances are large compared with the radius of curvature R, that is, the glancing-angle θ is small. In the simplest form the radii of curvature in and normal to the axial plane are equal, so that the imaging surfaces are segments of spherical mirrors. Hoppe & Trurnit (1947) have shown that, for one particular object position, the geometrical aberrations are then all zero, except for curvature of the field. As in the image spectrograph, however, the chromatic aberration is severe, and they estimate the practical limit of resolving-power at 1μ, even if perfect surfaces can be prepared.

Surfaces of this type are non-developable and cannot be formed by simple bending of a crystalline lamella. Hoppe and Trurnit

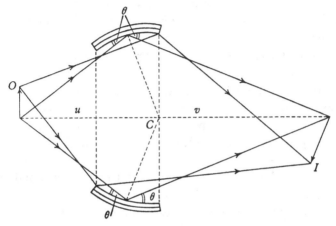

Fig. 5.8. Magnified X-ray image obtained by reflexion from ring mirror (after Hoppe & Trurnit, 1947).

overcame this difficulty by grinding a glass or metal surface to the correct shape and then coating it with layers of barium stearate, by the technique used for forming interference filters (Blodgett & Langmuir, 1937). They obtained X-ray images at low magnification from such 'artificial crystals', but the reflected intensity was low. This is due to the small angular aperture of the imaging beam which is limited, first, by the low glancing-angle imposed by the large lattice constant (~ 50 Å), and secondly, by the narrow tolerance in Bragg reflexion. As in all crystal reflectors the useful beam is practically confined to the surface of a cone of semi-vertical angle equal to the glancing-angle. Hoppe and Trurnit suggested that a finite angular aperture, in longitudinal section, would be obtained if the lattice constant increased around the arc of the mirror segment, so that the required glancing-angle would be smaller for rays incident on it further from the object. In principle this could be brought about by depositing fatty acids of increasing lattice spacing in successive rings around the reflecting surfaces. Further consideration has been given to the optical problems of ring mirrors by Asselmeyer (1954), but their practical value remains questionable. Apart from the difficulty of producing them, it would appear that the smallness of the illuminating aperture and the severe chromatic aberration more than offset the almost complete absence of geometrical aberrations.

In sum, it appears doubtful whether any type of X-ray microscope using doubly curved crystals can approach a resolving power of $1\,\mu$. Those with favourable geometrical properties suffer from chromatic aberration, and must employ monochromatic illumination; those that can use white radiation are afflicted with spherical aberration. All are limited in intensity by the finite rocking curve width of even the most perfect single crystals obtainable, by limited aperture and low efficiency of reflexion. They compare poorly with the contact and projection methods in exposure time as well as resolving-power; with the total reflexion method certainly in resolving-power and probably also in exposure. The value of curved crystal reflectors is thus confined to monochromators and focusing spectrometers, where image quality is of secondary importance.

CHAPTER 6

X-RAY ABSORPTION MICROANALYSIS

The techniques of elementary analysis with X-rays are of two kinds, making use of absorption and characteristic line emission respectively. The latter may be further subdivided, depending on whether emission is produced by direct excitation with electron bombardment or secondarily under irradiation with X-rays (fluorescence analysis). On the macroscopic scale, little use has been made of absorption as an analytical method, whereas spectroscopy, particularly of secondary X-rays, has been widely employed. In microanalysis, the position has been the reverse: absorption methods have been developed very thoroughly, particularly by Engström and his collaborators for biological materials, but emission has been largely neglected until very recently, although it has distinct advantages for inorganic analysis. Both techniques are capable of high sensitivity, amounts of an element of the order of 10^{-12} to 10^{-14} g. being detectable in a volume as small as $1-10\mu^3$.

The absorption technique happens to have been developed primarily in connexion with the contact method of microradiography, the emission technique with the projection and scanning methods. We discuss first the fundamental physical principles of the absorption technique and the experimental details and limits of accuracy of their realization by the contact and projection methods respectively. The emission technique, in which the object field is explored point by point, involves somewhat different considerations and is discussed in chapter 7.

6.1. Microanalysis by differential absorption

The possibility of elementary analysis by absorption rests upon the fact that the mass absorption coefficient varies rapidly with atomic number, being approximately proportional to Z^3 at soft X-ray wavelengths (1–10 Å). The differential effect is further increased by the existence of absorption edges at particular wavelengths; these are sufficiently infrequent to be an aid rather than a complication in element discrimination. When two (known) elements are present, or an element of high Z in a matrix of roughly uniform elementary composition as in biological tissue, their relative concentrations can be found locally by measuring the absorption of X-rays and comparing it with that of the pure element or of a standard reference material, such as nitrocellulose. The measurements may be made photometrically on a negative or directly with Geiger or proportional counters. The former method records a large field in one exposure, but this must then be explored locally; the latter method proceeds point by point in the first place, gives direct quantitative readings and is ultimately capable of higher accuracy. For the most accurate determinations line radiation must be used, so that well-defined absorption coefficients are involved. The appropriate wavelength is isolated by reflexion in a crystal spectrometer or by selective recording with a proportional counter and pulse analyser. For the most accurate work a combination of both methods is necessary (cf. § 6.5); alternatively, fluorescent radiation can sometimes be used. For some purposes, however, and particularly in dry weight determination (see § 6.2), the entire continuous spectrum can be used, with consequent reduction in exposure-time. A mean value for the absorption coefficient must then be used.

The fundamental features of the absorption method, including the choice of optimum wavelength, have already been discussed in chapter 2 in connexion with qualitative microradiography. When two substances are in the path of an X-ray beam (Fig. 6.1a), of initial intensity I_0, the ratio of the respective transmitted intensities (cf. (2.7)) is

$$I_2/I_1 = \exp\left[-\{(\mu/\rho)_2 m_2 - (\mu/\rho)_1 m_1\}\right], \qquad (6.1)$$

where $(\mu/\rho)_1$, $(\mu/\rho)_2$ are the mass absorption coefficients and m_1, m_2 the mass thicknesses in g. cm.$^{-2}$ of the elements respectively. The absorption coefficients may refer to elements or may be integrated values for a compound or mixture, as given by (2.14).

When one substance is present as a local concentration of thickness t in a thickness D of the other (Fig. 6.1b), as it very often will be, (6.1) reduces to the simple form

$$I_2/I_1 = e^{-(\mu_2 - \mu_1)t}, \qquad (6.2)$$

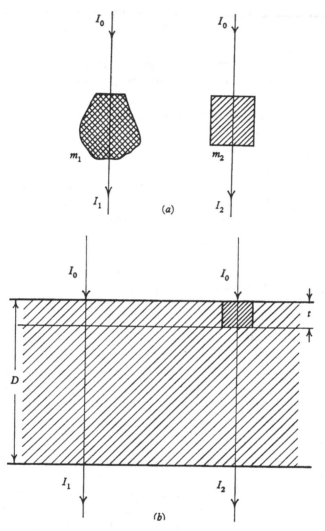

Fig. 6.1. Relative X-ray absorption in (a) two free particles, and (b) in a particle and the matrix in which it is embedded. (I_0 = incident intensity; I_1, I_2 = transmitted intensities.)

so that we are now concerned with the difference in *linear* absorption coefficient. So long as the intensities differ by no more than 20 %, we may use the approximation (2.9) for the effective contrast:

$$1 - I_2/I_1 = \Delta I/I = (\mu_2 - \mu_1)t. \qquad (6.3)$$

As noted earlier (§ 2.3), the intensity ratio of the emergent beams is the same whether the second substance is locally concentrated or is distributed throughout the first in greater or smaller dilution. In the latter case, the experimentally obtained value t is an 'equivalent thickness' of concentrated substance of normal density ρ_2. It is usual to evaluate it as the mass thickness ($m_2 = t\rho_2$) of the second substance in the path of the X-ray beam.

A direct evaluation is only possible if the specimen includes regions free of the second element, so that the value of I_1 is unambiguously known. When this is not so, it can be obtained from a reference measurement on a pure sample of element 1, preferably of the same thickness D; if the thickness is in fact D', giving an intensity I_1', the reference intensity is given by

$$I_1 = I_1' e^{-\mu_1(D-D')}.$$

The methods of accurately determining the thickness of thin sections have been discussed by Lange & Engström (1954).

6.1.1. *Measurements at an absorption edge.* It has been assumed above that the absorption measurements are made at a selected wavelength for which the values of the mass absorption coefficients of the respective elements are known. The considerations governing the choice of wavelength have been set out in § 2.2. The use of a reference absorber can be avoided, however, if advantage is taken of the sudden drop in absorption coefficient that occurs at an absorption edge (Fig. 6.2), by taking measurements at two wavelengths, one above and one below the wavelength of the edge (Glocker & Frohnmayer, 1925). We substitute absorption determinations at two wavelengths, on the same region of the specimen, for those made on two different regions of the specimen, at the same wavelength, in the previous method of analysis.

With primes indicating the absorption coefficients at the two wavelengths employed, λ' and λ'', we have

$$\begin{aligned}
I'/I_0 &= \exp\left[-\{m_1(\mu/\rho)_1' + m_2(\mu/\rho)_2'\}\right], \\
I''/I_0 &= \exp\left[-\{m_1(\mu/\rho)_1'' + m_2(\mu/\rho)_2''\}\right],
\end{aligned} \tag{6.4}$$

where m_1 and m_2 are the mass concentrations (g. cm.$^{-2}$) of the substance to be determined and of its matrix, respectively, so that the specimen as a whole weighs (m_1+m_2) g. per unit area. Then, provided that I' and I'' are referred to a common value of incident intensity I_0,

$$\log(I''/I') = m_1(\mu'/\rho - \mu''/\rho)_1 + m_2(\mu'/\rho - \mu''/\rho)_2. \qquad (6.5)$$

The gradient I/I_0 is often termed the *transmission* T and I''/I' the *transmission ratio*, R. When suitable line radiation is available to allow λ and λ'' to be very close together, we can assume that the change in absorption coefficient of the matrix is so small between

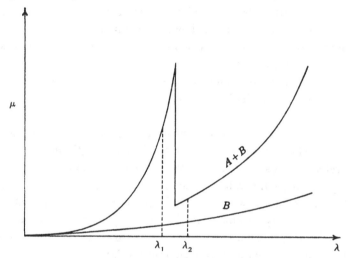

Fig. 6.2. Variation in absorption coefficient for a mixture of two elements A, B in the neighbourhood of an absorption edge of A.

these wavelengths that the term in m_2 can be neglected. The mass thickness of the first element is then given by

$$m_1 = \log(I''/I')/(\mu'/\rho - \mu''/\rho)_1. \qquad (6.6)$$

If the available wavelengths are further apart, we can assume that the absorption coefficient of the matrix depends on the Pth power of λ, so that now

$$m_1 = \frac{\log(I'/I_0) - (\lambda'/\lambda'')^P \log(I''/I_0)}{(\lambda'/\lambda'')^P(\mu''/\rho)_1 - (\mu'/\rho)_1}. \qquad (6.7)$$

For most purposes P can be taken as 3; for more accurate work the values of μ/ρ at the relevant wavelengths may be obtained from data such as that given in the Appendix. Alternatively, measurements may be made at four or more wavelengths and the value of P determined directly from a logarithmic plot (Engström, 1946, p. 25).

The accuracy of the analysis depends in the first place on that with which the transmitted intensities can be measured, but also on the purity of the radiation reaching the recording system and on the accuracy to which the absorption coefficients are known. Except for the last, these factors depend on whether contact or projection methods are employed; for instance, in projection the X-ray intensity is recorded with a counter, but in the contact method it has to be obtained by microphotometry of the photographic negative. It is, therefore, convenient to discuss limits of accuracy later, in connexion with the experimental methods employed for particular purposes.

6.1.2. *Microweighing and microanalysis.* The procedure outlined above, whether carried out at a single wavelength with the aid of a reference absorber or at two wavelengths about an absorption edge, both yield a value for the mass per unit area m at points in the specimen. As m is equal to the product of the local thickness t and the density ρ (which may be an average with respect to thickness), absorption measurements may be used to determine any of the following quantities, on the microscale:

Thickness, when the density is known or can be otherwise determined.

Density, or average density, when thickness is measured.

Mass, included within a measured area, or the mass of a body of known shape, such as a spherical inclusion.

Relative concentration of elements, by comparative absorption measurements.

Absolute concentration of an element in a matrix, if the nature of the matrix and its thickness are known, or if the wavelengths are chosen so as to eliminate its effect.

In practice the microabsorption technique has been mainly used in two directions: for mass thickness determination, especially the dry weight per unit area of biological material, and for microspectrometry, giving quantitative elementary microanalysis of organic and inorganic specimens. The former aim is pursued by use of either a monochromatic beam or the continuous spectrum entire, along with some type of reference system. Quantitative microanalysis requires observations at two or more wavelengths on each side of the absorption edge of the element in question, with or without measurement of the intensity of the incident beam. The techniques and especially the sources of error are sufficiently different to demand separate treatment.

6.2. Dry weight determination

6.2.1. *Principles of the method.* The absorption resulting from the presence of a single substance, of density and thickness t, is given by the simple relation (2.3):

$$I = I_0 e^{-(\mu/\rho)m}, \qquad (6.8)$$

where I_0 is the incident and I the transmitted X-ray intensity, μ/ρ the mass absorption coefficient, and m ($= \rho t$) the mass per unit area of the part of the specimen intercepted by the beam. To find m it is only necessary to measure I and I_0 with radiation of a wavelength at which the absorption coefficient is known for the substance under investigation. In contact microradiography the specimen is almost in contact with the photographic emulsion and it is enough in principle to measure the blackening produced in its image and in an empty space beside it. The non-linear properties of the photographic emulsion, however, and the variation of its response with processing (cf. § 6.4), usually make it desirable to measure the intensity I_r through a reference absorber instead of the initial intensity I_0. In the simplest form, the reference sample consists of a known thickness of substance of known absorption coefficient, preferably giving an extinction approximately equal to that due to the important features in the specimen investigated, so that both produce much the same photographic effect. It may well be of the same composition as the specimen, especially when this is an element or simple compound. For biological investigations a foil of nitrocellulose is most used, prepared as a step-wedge so that a graded series of images is produced on the emulsion (van Huysen, Hodge, Warren & Bishop, 1933; van Huysen, Bale & Hodge, 1934). It is then possible, by interpolation on a calibration plot of optical density against thickness, to determine accurately the mass thickness of reference material m_r which gives the same density as the region of the specimen measured. As the reference foils are exposed simultaneously with the latter on the same emulsion, variations in emulsion response are almost entirely eliminated. I_0 will be the same for both and, if I is made the same by use of the calibration plot, the exponents of e for the two densities must be the same.

$$(\mu/\rho)_s m_s = (\mu/\rho)_r m_r, \qquad (6.9)$$

where the subscripts s and r refer to specimen and reference material respectively. It is thus a straightforward matter to find the mass thickness for any specimen which lends itself to the contact method of microradiography (cf. Fig. 6.3).

The method was first used by Engström & Lindström (1949, 1950) for the determination of the dry weight of biological material. It was further developed by Brattgård (1952) and Brattgård & Hydén (1952), and has since been investigated in detail by Lindström (1955) and by Clemmons (1957). They give an account of the precautions to be observed in preparing and mounting reference systems of known properties (see also Clemmons & Webster, 1953). Some applications are described in § 12.5.

In order to obtain high contrast from thin sections of tissue Lindström used X-rays from a tube operated at 1·5 kV, the construction of which is described in § 10.1. The output of soft X-rays being low, he decided to use the whole of the continuous spectrum, in so far as it penetrated his filters, rather than a single wavelength; the specimen chamber was evacuated and separated from the X-ray tube only by a thin aluminium foil to exclude light. The ratio between the mass absorption coefficients of the elements (C, N, O, H) which compose organic material varies very little with wavelength in the soft X-ray region, from 4 to 24 Å, so that use of the continuous radiation adds little complication. It is fortunately also the case that the elementary composition of a wide range of the compounds that make up such tissues is very similar and their composite absorption coefficient, calculated by (2.14), even more constant.

As a first approximation, a biological specimen may be treated as if it had the absorption coefficient of nitrogen, for finding the value of $(\mu/\rho)_s$ for (6.9) (Engström & Lindström, 1949, 1950). For greater accuracy, a composite value for average animal protein may be used (Lindström, 1955). Lindström evaluated with great care the systematic error introduced by this latter assumption, for specimens of different real composition. He found that they fall into three groups, according as they have positive or negative errors in the main wavelength ranges of interest: glycogen and the nucleic acids, the proteins, and the lipids. He gives tables showing the value of the errors both for specific wavelengths and for the continuous spectrum taken as a whole; for the most part they lie between 0·5 and 2·5 %, except for calcified tissues.

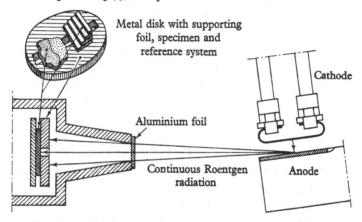

Fig. 6.3. Method of mounting a specimen for contact microradiography. (Lindström, 1955).

6.2.2. Microphotometry. The X-ray intensity is best measured direct from the negative with a microdensitometer. Practical details of the technique are given by Swift & Rasch (1956). Indirect intensity measurement, by photometry of an enlargement made from the micronegative (Engström, Wegstedt & Welin, 1948), is subject to appreciably greater error (Brattgård, Hallén & Hydén, 1953).

For direct microphotometry, a spot of light only a few microns in diameter is required and this is best obtained by forming a reduced image of the light source with an optical condenser system focused on the negative, which is carried on the stage of a high-power microscope. After the area to be measured has been selected by visual inspection, the transmitted light may be measured by means of a photomultiplier and valve voltmeter (Engström & Wegstedt, 1951).

A detailed specification has been given by Bourghardt, Brattgård, Hydén, Kiewertz & Larsson (1953) of a microphotometer for microradiography, in which the light source is reduced to a width of $1-3\mu$ by a condenser of numerical aperture $1\cdot4$. The main body of the microscope contains a prism for use with an eyepiece while searching the negative. A half-silvered mirror is used to project the selected area on to a viewing screen at the same time as the transmitted light is recorded by the photomultiplier. The specimen stage is replaced by a microscope table that can be moved by a synchronous motor in two directions at right angles. A pen recorder operates in synchronism with the negative movement, so that a line across the image may be scanned and the transmitted light intensity recorded automatically. The light beam is chopped by a rotating drum to produce an alternating output in the photomultiplier in order to use a.c. amplification of the signal. The response of the amplifier may be adjusted so that it matches the density curve of the photographic plate. With a test system using cellulose step-wedges, the standard error of the intensity measurements was $\pm 1\cdot5\%$.

This apparatus has since been extended to handle rapidly a large number of negatives (or cells directly) by automatic scanning in both directions, with automatic printing of the results (Bourghardt, Hydén & Nyqvist, 1955; Hallén & Hydén, 1957b). The scanning is accomplished by a perforated steel band moving in front of the light source, the holes being equivalent to an area of $(3\mu)^2$ in the image. Alternatively, the sample may be moved in $1-3\mu$ steps along a line and then down by the same displacement to the next line, in front of a stationary aperture (Larsson, 1957). In either case, 12,000 points can be scanned in a period of 4 min. The transmitted intensity is detected by a phototube, with special circuits to ensure synchronism with the scanning speed. Since the electrical scanning speed is much faster than the available mechanical printing rate, each scan of 12,000 points is stored on magnetic tape and later played back and printed over a period of 80 min. The field of view is separately photographed and the printed data then superimposed on an image of the specimen magnified to 45 × 45 cm. A calculating unit converts the phototube output directly into mass thickness values before the printing stage. The complete automatic analyser is shown in Pl. VA; the microscope and scanning system is on the left, followed by the calculating unit, the memory unit and the printer.

6.2.3. *Experimental errors.* Both Brattgård and Lindström have investigated very fully the experimental errors in microphotometry, using standard specimens (see also Clemmons, 1955). Microradiographs were recorded by Lindström on three different types of fine-grained emulsion: Lippmann, Kodak Maximum Resolution and Kodak High Resolution. Selected areas of the negative were projected through an optical microscope of the highest performance, so that the image fell on a screen behind which was mounted a photomultiplier tube. A pinhole of 2 mm. diameter in the screen made possible the microphotometry of areas as small as $7\mu^2$ in the original negative. From observations on the variation of the photocurrent from different areas and in different conditions the relative importance of various errors was deduced. In decreasing order of importance they were:

Granularity of the emulsion, which varies appreciably with the type of emulsion and method of developing. It was made the subject of a separate investigation by Lindström (cf. § 6.4). In the best circumstances the error from this cause was about 2 %.

Variation in the reference foil, the weight per unit area of which is not constant to better than about 1·5 %. This, the most important source of error under experimental control, is also discussed by Brattgård & Hydén (1952) and Clemmons & Webster (1953).

Uncertainty in *the values of the absorption coefficients* of the collodion foil and of protein. Lindström used primarily the data of Victoreen (1949, 1950), the uncertainty in which was estimated at 1 %. The more recent calculations of Henke, White & Lundberg (1957) are more reliable (see Appendix) and indicate that this was an underestimate.

Thickness *variation in the support film* on which the specimen and reference wedge were carried. With care in preparation this can be reduced well below 1 %.

Variations in *intensity of the light source* used in the photometric procedure. These were reduced to negligible proportions by stabilizing the voltage and by waiting for 30 min. after switching on before commencing readings.

Artefacts in the emulsion, in the form of pinholes, chips of glass formed in cutting the plate, and gross local variations in intensity. Lippmann-type emulsions were found to be far inferior to the Kodak Maximum Resolution and Kodak High Resolution in this respect.

Inhomogeneous illumination of the specimen by the X-ray beam. If the specimen holder is not too close to the target the illumination will be uniform enough over the required area of about 1 cm.^{2} In Lindström's arrangement the target–specimen distance was 185 mm.

The aggregate effect of these experimental errors is estimated at 4·4% by Lindström and at 3·2% by Brattgård and Hydén. The systematic error in the investigation of particular tissues is given as 2–3% and 5–7% respectively, depending on how far the actual composition differs from the average value assumed in calculating the absorption coefficient. Compounding them in the usual way, the overall error in such dry weight determinations will be in the range 5–8%. Clemmons (1955, 1957) points out that the errors increase rapidly for specimens of small absorption equivalent.

6.2.4. *Resolving-power and limit of detection.* The spatial resolution of the method involves both lateral extension and depth. The former is limited more by the difficulty of photometering areas smaller than about 1 μ in diameter than by the resolving-power of the contact process itself, which is better by a factor of 2 or 3. Lindström made measurements on areas down to 1·5 μ^2, Brattgård and Hydén down to 4 μ^2. The limit of thickness is set by the amount of absorption in the tissue, which in turn depends on the wavelength used. For soft radiation (8–20 Å) it is about 2 μ, according to Lindström. The minimum volume of material that can be analysed is therefore about 3–10 μ^3. The limit of detection of dry tissue will then be 10^{-11} to 10^{-12} g. Rather better discrimination has been claimed, down to 0·2 × 10^{-12} g., by Engström & Glick (1950), Hallén & Hydén (1957 a, b) and others. The limit of dry weight determination by optical interference microscopy is approximately an order of magnitude better (Davies, Engström & Lindström, 1953).

The limits of accuracy in determining total mass M, or the mean physical density ρ of the irradiated area, are fixed primarily by the uncertainty in thickness t, since $M = mt$ and $\rho = m/t$. The methods of measuring the thickness of sections of order 1μ have been investigated by Lange & Engström (1954), who concluded that the error is at best 10%. Hallén (1956) has developed a method of fine-focusing an optical microscope, using the shadow of a metal filament, which gives slightly better accuracy. The error in determining the mass of whole cells, or inclusions, must be correspondingly greater than that of thickness, involving as it does assumptions as to the shape (Lindström, 1955, p. 129).

6.2.5. *Improvements in the contact method.* As the preparation of the reference system is a limiting factor, both in the time required and in its probable error, Combée & Engström (1954) have proposed a method of eliminating it. They expose up to six samples of the same material on one piece of emulsion, but for differing times. The X-ray beam intensity is assumed constant, within the remaining experimental error, as it will be for a sealed-off tube such as they used. The density–exposure curve for the emulsion is then plotted from photometer measurements of the several images. In order to obtain absolute values for the mass per unit area of selected details, the effective mass absorption coefficient must be known for the continuous spectrum as incident at the specimen. It is found in terms of the equivalent absorption in nitrogen by a separate measurement with a given air path at the voltage used.

The necessity for using a reference absorber rests upon the non-linearity and limited extent of the response curve of the photographic emulsion. It can be avoided by employing counter recording (Rosengren, 1956), since no saturation effect then enters, at least until very much higher rates of count than usually used. An alternative procedure is to choose an emulsion with a linear response over a major part of its characteristic curve, and to control exposure so that all densities to be measured fall in this range. As described later (§ 6.4), a number of emulsions will give a sufficiently linear characteristic under suitable processing conditions. A standardizing photometer measurement on a single reference foil, instead of several made on a step-wedge, will then suffice. Carlson (1957) has calculated a standard calibration curve of mass per unit area against optical transmission T for these conditions. The middle region of this curve is also almost linear, so that the dry weight can be taken as directly proportional to the transmission, for values of T between 30% and 60%. Essentially the same method was used by Wallgren & Holmstrand (1957) in a quantitative determination of the mineral content of bone.

6.2.6. *Water content.* From microradiographs made of a biological specimen before and after removal of water by freeze-drying, the water content may be found by difference. Photometry of the first image gives the total mass per unit area and that of the second the dry mass per unit area (Engström & Glick, 1956). The recording of the image of the fresh material by the contact procedure is facilitated if it is frozen as soon as prepared, and then sectioned with a freezing microtome. The main error, as before, is in the photometric measurement. The morphological effects of the drying process must also be considered carefully.

It was pointed out by Engström and Glick that a semi-quantitative determination of water content may be made from a micrograph of the dried material only, if one assumes an average value for the density of dried tissue ρ_d. If the mass per unit area of dry matter is m_d, as determined by photometry, then the 'effective dry thickness' is m_d/ρ_d. The mass of water per unit area m_w, is then

$$m_w = T - m_d/\rho_d, \qquad (6.10)$$

where T is the thickness of the wet section, which is separately determined. The main error enters in the measurement of T. Relative values for the water content of different parts of a given specimen can be obtained at once, within the limits of constancy of its thickness. Engström and Glick, for instance, give values of 70, 76 and 85% for the chief cells, epithelial cytoplasm and parietal cells in dog gastric mucosa, respectively, assuming a value of 1·3 for ρ_d. They also point out that an immediate qualitative view of the distribution of water is presented by the negative of a micrograph, whilst the positive indicates the dry weight distribution.

6.3. Absorption microspectrometry by the contact method

6.3.1. *Procedure.* In addition to dry weight determination, the contact method also allows quantitative analysis of the main elements present in a specimen. Each element has a series of characteristic absorption edges (§ 2.2) and by making absorption measurements at wavelengths on either side of (say) the K-edge, the mass per unit area of the corresponding element may be found. As (6.4) shows, the absorption coefficients of the element at the chosen wavelengths must be known and the incident intensity I_0 as well as the transmitted values I' and I'' must be found. Alternatively I_0 need not be measured if the nature and thickness of the matrix is known, or found by separate experiment, when (6.5) applies. If the wavelengths are close enough together to make negligible the difference in the matrix absorption coefficients, then (6.6) can be used.

The method requires transmission measurements to be made successively at two or more wavelengths, and is therefore easier to practise with a projection X-ray microscope, where the specimen is well separated from the recording system, be it photographic or counting tube. It has been successfully practised by the contact procedure, however, by Engström (1946) and his collaborators, who have investigated the optimum conditions for determining elements of biological interest.

The sensitivity attainable is governed by the accuracy with which intensities can be measured and by the height of the absorption edge ($\Delta(\mu/\rho)$) for the element in question. The value of the latter increases steeply with decreasing atomic number (Fig. 6.4), assisting the determination of the biologically important elements, carbon, nitrogen, oxygen, phosphorus, sulphur. For a given element, $\Delta(\mu/\rho)$ is greater for the L than for the K lines and still higher for the M series, but as the wavelength of the edge also increases, the K-edge is experimentally more convenient. Even so, vacuum spectroscopy is required owing to the softness of the radiation. Some experimental estimations of oxygen (at 23·3 Å) have been made by Lindström (1955), but the technical difficulties in regard to nitrogen (31 Å) and carbon (44 Å) remain considerable.

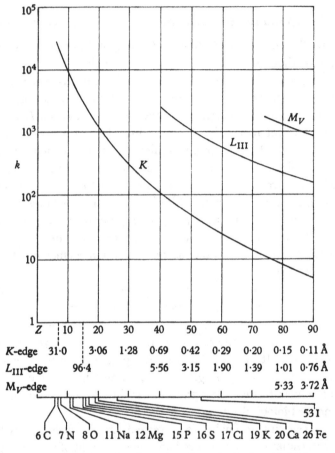

Fig. 6.4. Magnitude of the absorption discontinuity $k = \Delta(\mu/\rho)$ for various elements. (Engström, 1946.)

The possibilities of estimating the latter have recently been discussed by Henke (1957, 1959).

The minimum amount of an element Δm that can be determined depends also on the accuracy in measurement of intensity, $\Delta I/I$:

$$\Delta m = (\Delta I/I)/\Delta(\mu/\rho), \qquad (6.11)$$

where Δm is in g. cm.$^{-2}$ Assuming a value of 0·05 for $\Delta I/I$, Engström (1946) calculated the detection limit for different elements. As Fig. 6.5 shows, it becomes less than 1 μg./mm.^{2} (= 10^{-4} g./cm.^{2}) only for elements below manganese in the periodic table. For the biologist this information is more usefully

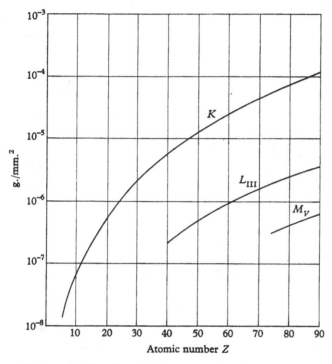

Fig. 6.5. Smallest determinable amount of an element as function of atomic number. (Engström, 1946.)

expressed as the smallest concentration of an element that can be detected in a section of given thickness (Engström, 1951b). In general, the light elements must be present in a concentration of the order of 1 % if they are to be detectable in a section 10μ thick. Apart from the basic constituents of biological material (carbon, nitrogen, oxygen), which cannot yet be determined, only phosphorus, sulphur, and calcium occur in sufficient concentration to be detected in normal tissues; silicon and some of the metals may occur in pathological conditions. Some results are given on p. 320.

6.3.2. *The spectrometer.* A special vacuum spectrometer was built by Lindström (1955) for microanalysis of the three accessible elements (P, S and Ca), in the course of which he made a thorough investigation of the limitations and sources of error of the method. The spectrometer incorporated the same low voltage tube (1·5 kV), described in § 10.1, as was used for the dry weight determinations discussed above. In the plan view (Fig. 6.6), a particular wavelength in the radiation from the tube (B) is selected by the curved

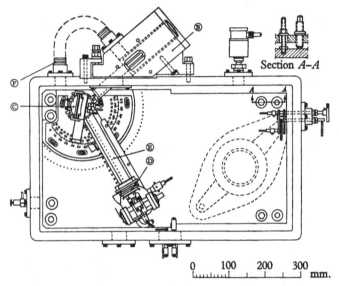

Fig. 6.6. Vacuum X-ray spectrometer of Lindström (1955).

crystal spectrometer (C), and focused through a slit into the camera (D). The spectrometer is of the Johann type and employs a bent crystal of quartz, mica or layered barium-copper stearate, according to the required wavelengths. As the position of the focus on the Rowland circle varies rapidly with reflected wavelength, the camera position is adjustable along the arm (E) and arms of two different lengths are provided. Provision was made for Geiger counter as well as photographic recording. The spectrometer and X-ray tube are evacuated continuously, and are connected to each other by the flexible tube (F).

The procedure is to take a photograph first at a wavelength on one side and then at one on the other side of the absorption edge of the element to be estimated. The specimen must be transferred from one photographic plate to another between exposures, after admitting air to the spectrometer. For the estimation of phosphorus the L-lines of niobium and zirconium were used, on either side of the phosphorus K-edge at 5·79 Å. For sulphur, which has the K-edge at 5·02 Å, the L-lines of ruthenium and molybdenum were used. For calcium no suitable lines occur near its K-edge (3·07 Å) and wavelengths to each side were selected from the continuous spectrum. Lindström used the same photometric procedure as in his dry weight determinations for obtaining relative intensities from the negatives. But, in order to simplify the evaluation, the recorded intensity was kept within the range of exposure for which the response curve was linear for the type of emulsion used (Kodak High Resolution) (cf. § 6.4). If the effect of the matrix is not negligible, the incident intensity must be recorded on the same plate as the micrograph of the specimen, at each wavelength (cf. (6.7)). Lindström shows that the error is about $\sqrt{2}$ greater when (6.7) rather than (6.6) applies.

An optimum thickness of absorber exists (t_x), which gives maximum difference between the intensities transmitted at the two selected wavelengths and therefore results in maximum sensitivity. This thickness is given by the same expression as that which gives maximum contrast between two different absorbers at a given wavelength (2.10):

$$t_x = \log_e(\mu_1/\mu_2)/(\mu_1 - \mu_2).$$

except that μ_1 and μ_2 are now the absorption coefficients of the substance to be analysed, at the two selected wavelengths. The expression holds good only so long as the effect of the matrix is negligible, for instance in the determination of calcium in bone, for which t_x is about 25μ. When the absorption coefficient of the matrix (μ_m) is not negligible compared with μ_1 or μ_2 (whichever is the smaller), then the more accurate expression (2.12) must be used to find t_x.

6.3.3. *Experimental errors.* The experimental error in the method depends on the accuracy with which the relative intensities can be measured and on the difference in mass absorption coefficient $\Delta(\mu/\rho)$ across the absorption edge (cf. (6.11)). For an acceptable limit of 5 % in the accuracy of the final result of the analysis, Lindström calculated the smallest detectable mass per unit area of substance at three values of the error in intensity (0·5, 1·0 and 2·0 %) and for a number of values of $\Delta(\mu/\rho)$, from 100 to 20,000. For an error of 1 % and $\Delta(\mu/\rho) = 10,000$, the detection limit was 0·28 $\mu\mu$g./μ^2 (0·28 × 10^{-4} g./cm.^{2}). As an area of 7μ^2 was covered by the light spot in the photometer, this represents a mass sensitivity of 2 × 10^{-12} g.

The overall experimental error in the analyses, including those of the photometric procedure, are estimated by Lindström at ± 3 %. Allowance must also be made for the systematic error, probably of the same order, due to uncertainty in the mass absorption coefficients: this is more important in the soft than in the hard X-ray region, but improved values have recently been given by Henke, White & Lundberg (1957) (see the Appendix). Line broadening at reflexion in the spectrometer must also be taken into account, but is only appreciable when a flat crystal is used and the required wavelengths are selected from the continuous spectrum; it can be reduced by limiting the width of crystal from which reflexion takes place. Broadening due to crystal imperfections will remain, but this is normally negligible in comparison with the other errors. If the operating voltage is set higher than twice that needed to excite the required wavelength, further error may arise from second-order reflexions (§ 6.5).

Error in interpretation of the results may be introduced by inhomogeneous distribution of the absorbing element within the biological feature examined. This difficulty has been critically examined by Glick, Engström & Malmström (1951), and Engström & Weissbluth (1951) have calculated corrections for specimens in which the unit particles are cubes or spheres. Lindström (1955) regards the effect of inhomogeneity as negligible in his conditions, where an area of 7μ^2 of a section 10μ thick was measured. Averaging by photometry with a line scan has been recommended by Caspersson (1955).

When the element sought is present as a compound of known composition and in a matrix of negligible absorption, an absorption measurement at a single wavelength will suffice (Wallgren, 1957). The mass per unit area of the compound is given by the same expression as is used in dry weight determination:

$$I = I_0 e^{-(\mu/\rho)m}, \qquad (6.12)$$

where (μ/ρ) is the mass absorption coefficient of the compound. Such a situation arises in the case of the mineral content of bone, hydroxyl-apatite embedded in an organic matrix; the absorption of the latter is very small for moderately hard X-rays. Wallgren & Holmstrand (1957) were thus able to determine the calcium content of bone by measurements at one wavelength only, with an estimated accuracy of 1·66% at favourable sites and of 3·25% on the average (cf. Amprino & Engström, 1952).

The main disadvantages of the contact method for quantitative work, as usually practised, are its slowness and technical awkwardness. In Lindström's apparatus the exposure was ½–4 hr. depending on the opacity of the material. The specimen must be transferred from one emulsion to another between exposures, and it is a tedious business locating precisely the same biological features in both negatives for microphotometry. In an attempt to avoid this difficulty, Engström, Lagergren & Lundberg (1957) allowed the narrow beam reflected from the crystal spectrometer to move across the specimen during a single exposure, thus imaging different parts of it with different wavelengths. A very long exposure is needed, and the two wavelength method of evaluation can only be applied if essentially identical features occur in two suitably illuminated parts of the field. The authors determined calcium in bone by this method, but only with a localization of 0·1 mm.² and a limit of detection of 10^{-6} g.

The use of the projection method (§ 6.5), on the other hand, enables measurements to be made at two or more wavelengths without moving the specimen. If counter recording is used, the whole process then becomes quicker and almost automatic. Projection conditions also minimize the possible effect of scattered and fluorescent radiation produced in the specimen, as it will be very weak at the recording system, whereas in the contact method one half of the amount produced is bound to reach the photographic emulsion. Error from this source appears not to have been taken sufficiently into account in contact microanalysis hitherto (Virtama, 1958), but see also Hoh & Lindström (1959).

6.4. Photographic considerations

The properties of the photographic emulsion on which the micrograph is recorded are of essential importance in the contact method. The size and distribution of the silver halide granules, their behaviour in different conditions of development and fixing, the shapes and extent of the curve of density versus exposure, all bear on the accuracy of the final result. The relation between resolving-power and speed is an inverse one, as discussed in § 2.5. The remaining factors, and especially the shape of the response curve, will be treated here, mainly in connexion with the high-resolution emulsions used in contact work. Some aspects of photography important in the projection method will also be noticed.

6.4.1. *The characteristic curve.* The response of an emulsion to radiation is discussed in terms of the optical density produced by a given amount of incident radiation. The latter is usually measured as the exposure E, given by the time t for which it is exposed to a constant incident intensity I. The optical density D is measured by the absorption of light in the image, and by analogy with (2.1) is given by

$$D = \log_{10} I_0/I_t, \qquad (6.13)$$

where I_t is the transmitted and I_0 the incident light intensity in the photometer (densitometer). The blackening produced by visible and ultra-violet light bears no simple relation to the exposure. The plot of D against E shows a more or less extended foot, indicating a lag in production of a visible effect, then rises fairly steeply before levelling off when saturation of the available silver halide is reached. If D is plotted against $\log_{10} E$ a curve of the same general shape is obtained, but it now shows a reasonably uniform rise in the central portion (Fig. 6.7a) the slope of which (γ) is a measure of the contrast in the image:

$$\gamma = dD/d(\log_{10} E). \qquad (6.14)$$

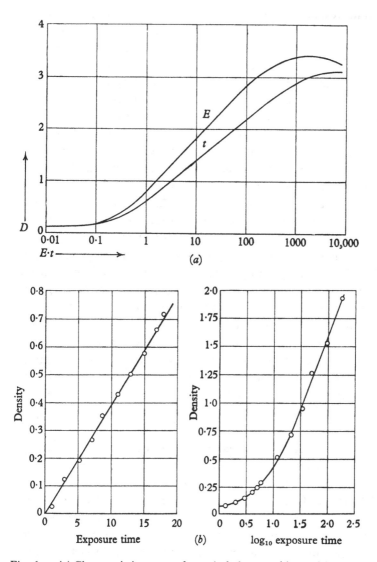

Fig. 6.7. (a) Characteristic curves of a typical photographic emulsion, exposed to light for given time at different intensities (upper curve) and for varying times at constant intensity of illumination (lower curve) (after Michel, *Die Mikrophotographie* (Vienna: Springer), 1957).

(b) Characteristic curve of a high-resolution photographic emulsion (Eastman Kodak 649) exposed to X-rays from copper at 20 kV with exposure plotted (left) linearly and (right) logarithmically against density. (Holmstrand, 1957.)

The response of most emulsions to X-rays is much simpler. They show a straight-line relation between D and E, at least over the initial part of the curve (Fig. 6.7b), so that evaluation of intensity from point to point in the image is greatly simplified. The extent of the linear region varies with wavelength, as well as with type of emulsion. It becomes shorter as the X-ray wavelength is increased and for very soft radiation the curve approaches that for ultra-violet light. For the fine-grained emulsions used in contact microradiography linearity extends to a density approaching 1. For Eastman Kodak 649 plates, Wallgren (1957) found it held up to $D = 0.8-0.9$ for $1.5-1.6$ Å radiation, and Lindström (1955) up to 0.6 for 2.75 Å and to 0.5 for 5.73 Å. For quantitative microanalysis, therefore, a lightly exposed plate is desirable so as to keep photometry in the linear range. On the other hand, the photometric errors increase for both very low and very high optical density. Henke, Lundberg & Engström (1957) give the desirable range of D, from this point of view, as $0.5-1.25$, although optimum *differentiation* of image features would require a value of 1.7 in the experimental conditions,—an Eastman Kodak 649 plate exposed to 13.3 Å radiation. For visual estimation the best value is still greater.

The D–E curve has been so far discussed in terms of variable exposure time t at constant X-ray intensity I_x. When both vary, the reciprocity (or Bunsen–Roscoe) law is found to hold for X-rays, but not for light:

$$E = I_x t. \qquad (6.15)$$

The same density is obtained for a given value of the product $I_x t$, no matter what the individual values of I_x and t, over a range of 100 to 1 in intensity and with fair accuracy up to 10,000 to 1 (Bell, 1936). This result, as well as the linearity of the D–E plot, indicates that the effect of the radiation on the emulsion depends on the number of quanta n. Experiments show (Bromley & Herz, 1950) that for moderately hard X-rays ($0.37-1.2$ Å) each quantum absorbed renders one grain developable. For very hard X-rays more than one grain is affected per quantum (Hoerlin, 1949, 1951), and for very soft X-rays more than one quantum is needed per grain, as reflected in the lack of linearity of the photographic response. For visible light very many quanta are needed to affect one grain: 10–100, depending on the type of emulsion.

The shape of the characteristic curve, and hence the range of linearity, can be markedly influenced by the conditions of development. Fig. 6.8 shows the effect on the Eastman Kodak 649 plate of using a fast, high contrast developer (D 8) or a slow finer-grained type (DK 50), for different times (Bellman, 1953). With both, a

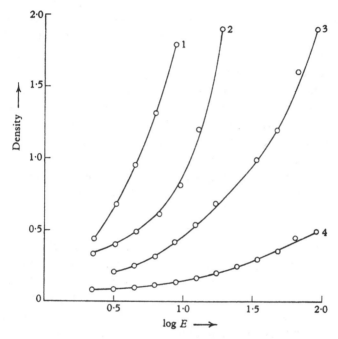

Fig. 6.8. Characteristic curves for Eastman Kodak Spectroscopic Plate 649-GH, for various developing conditions. (1) Kodak D8 for 2 min. (2) Kodak D8 for 1 min. (3) Kodak DK50 for 10 min. (4) Kodak DK50 for 4½ min. (Bellman, 1953.)

longer development time increases both the total density and the contrast, measured as the slope of the curve. The high contrast developer is of value in qualitative work, when the micrograph is visually examined, but has the disadvantage of causing large grain size. It is preferable to use a slow acting, fine-grain developer when high resolution and quantitative evaluation is needed.

6.4.2. Grain size and granularity. The initial size of the grains in an emulsion appears to have little to do with the granularity seen in the final image, which is influenced more by the statistical distribution of the grains and the clumping effect of development. In high-resolution emulsions the individual grains are found to be about 0.05μ in size, but some four times as far apart (Combée & Recourt, 1957). On statistical theory the fluctuation to be expected in the number n present in a given field (in photometry, for instance) will be $\sqrt{n}$, giving a relative error of $1/\sqrt{n}$ in the photometric procedure. Lindström (1955) was able to detect the influence of such fluctuations when photometering areas of $1.5\mu^2$. As the emulsion was of order 10μ in thickness, the number of grains in the volume examined would be 3000 or more, giving a probable fluctuation of about $1/55$ or 1.8%, just within his experimental error. Over an area of $7\mu^2$, with which most of his analyses were made, the expected error would be 0.8% as compared with an observed variation of $\pm 1\%$ in experimental values. Careful comparative tests of different emulsions showed that Kodak High Resolution plates had slightly lower granularity than the Eastman Kodak Maximum Resolution type, especially when processed in D 158 developer. The Lippmann emulsions showed appreciably greater granularity.

The best resolution attainable in practice with such plates is of order 0.25μ (cf. § 2.5), whereas there is evidence that the individual grains may be as small as 0.02μ (Ehrenberg & White, 1957). The larger distance in the observed image must be due to the separation between grains and the effect of development (cf. Combée & Recourt, 1957). At the same time, since the range of soft X-rays in the emulsion is only a few microns, there is a case for using thinner emulsions carrying a higher concentration of grains (cf. § 2.5). However, a reduction from 5μ to 1μ in emulsion thickness, and a hundredfold increase in the number of grains per cubic micron, was found by Combée & Recourt (1957) to give no improvement in granularity or resolution, although sensitivity was increased some 5 times. Aggregation during development thus appears to be the limiting factor.

The relation between sensitivity and size of grain has been discussed in § 2.5, in the light of the experimental investigations of Engström & Lindström (1951). In general the exposure-time needed at given beam intensity is inversely proportional to the square of the grain size, as would be expected if one quantum per grain is required for activation. For quantitative microanalysis, only emulsions of very fine grain and hence of low sensitivity come in question. For qualitative work, and especially for arteriography of living tissues, where kinetic blurring in any case limits resolution (§ 2.4), coarser and faster emulsions are preferred. Saunders (1957a) used ordinary lantern plates, which have a resolution of about 10μ and required an exposure of a few seconds at a distance of 35 cm. from a tube operated at 30 kV and 20 mA. In such cases a contrast developer such as D 8 may be used with advantage, although the capillaries in the subject are in any case normally injected with an opaque medium to enhance their visibility.

6.4.3. *Variation of sensitivity with wavelength.* Two factors influence the response of an emulsion as the wavelength of the X-radiation is changed: the variation in its absorption coefficient and the number of quanta per unit intensity of the beam. The value of (μ/ρ) for silver bromide rises approximately with the third power of the wavelength, apart from the breaks due to the K- and L-absorption edges of silver and bromine (Fig. 6.9). So long as the emulsion is thick enough to absorb almost all the incident quanta, this rise will make no difference to the exposure needed to produce a given photographic density. But the layer of emulsion in which the image is contained will decrease rapidly in thickness with increasing wavelength. For X-rays of 1 Å wavelength the emulsion would have to be at least 500μ thick, for 10 Å wavelength only 10μ thick, to absorb 90% of the intensity, on reasonable assumptions of grain distribution.

The number of quanta in the beam (n) is related to its energy E, and thus to the intensity I, by

$$E = It = nh\nu = nhc/\lambda, \qquad (6.16)$$

where h is Planck's constant ($6 \cdot 625 \times 10^{-27}$ erg-sec.), c the velocity of light, ν the frequency and λ the wavelength of the X-rays. Hence, for a given exposure, the number of incident quanta is directly proportional to the wavelength. Within the range over which a single quantum suffices to render one grain developable, the effectiveness of a given exposure should increase in direct proportion to the wavelength. Seemann (1950) found evidence to this effect in measurements made at 0·5 and 2·5 Å. Qualitative results with softer X-rays also indicate that photographic efficiency is at least not less than for hard radiation. For very soft X-rays a fall-off is to be expected when the range becomes of the same order as the grain dimensions: at 50 Å the intensity is reduced by $1/e$ in 0·05 μ of silver bromide.

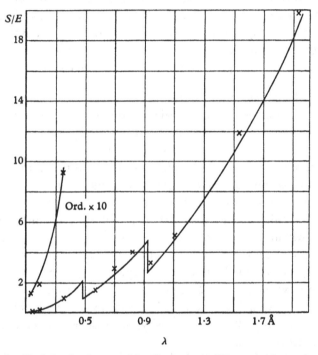

Fig. 6.9. Variation in photographic effectiveness of X-rays, with wavelength, for constant incident energy; the steps in the curve correspond to the K absorption edges of silver and bromine respectively. (Glocker, *Materialprüfung mit Röntgenstrahlen* (Springer, Berlin), 1958.) S = Optical Density; E = Exposure.

6.4.4. *Equivalent effective wavelength.* In discussing the contrast obtained in a micrograph, it is of interest to determine the 'effective' wavelength of the radiation with respect to the emulsion: the wavelength of monochromatic radiation which would produce the equivalent photographic effect of the spectrum actually employed. A detailed evaluation has been made by Dyson (1957) for moderately soft X-rays (1–5 Å).

The usual method of plotting the spectral intensity against wavelength (cf. Fig. 8.4) would suggest that the mean effective wavelength must be in the neighbourhood of the peak, which occurs at about 1·5 times the limiting wavelength. But since the photographic action depends on the energy, the effective quantity is the frequency ν (6.16). A plot of intensity per unit frequency interval $(I_\nu d\nu)$ shows that no peak occurs (cf. Fig. 8.5), the curve rising linearly from near the limiting frequency, to which it falls with a slight curvature. For a very thin target the variation with frequency is still simpler, being a straight line parallel to the abscissa (Fig. 8.5, broken line). As the number of quanta for a given amount of energy is inversely proportional to the frequency (6.16), the number of available quanta is greater at lower frequency (softer radiation) as noted above. The plot of $n d\nu$, the number of quanta per frequency interval, thus rises more rapidly than does that of intensity (Fig. 8.5).

According as the photographic effect is assumed to depend on the number of quanta absorbed or on their total energy, the mean effective frequency can be immediately deduced from these plots. In given experimental conditions, however, the contrast produced on the photographic plate will also be influenced by the nature and thickness of the specimen, which will attenuate some parts of the spectrum more than others, by absorption losses in the window of the X-ray tube and in the camera, and especially by the wavelength sensitivity of the recording emulsion. Dyson (1957) made detailed calculations for a half-tone plate and for cellulose specimens of different thickness, irradiated from a tube operated at 10 kV having a window target of gold and with an air path of 2 cm. in the camera. Two limiting conditions were taken: a 'thin' target, in which ideally only single scattering of electrons takes place and a 'thick' target, of thickness equal to the electron range, so that all electrons are stopped in it. The results are shown in Fig. 6.10, where the full and broken lines relate to thick and thin targets respectively. The upper pair of curves apply if the photographic process is regarded as a quantum effect, with one quantum yielding one developable grain over the energy range from 10 kV down to 2 kV. The lower pair of curves assume the process to depend on the total energy absorbed, that is, on the intensity I not on n. The assumption made leads to little difference in the effective wave-length deduced. It is evident that this is much greater than the value expected ($\sim$ 1·8 Å) from the peak of the wavelength distri-bution at this voltage, being about 4 Å and varying little with specimen thickness. For a heavy element specimen, such as gold, the effective wavelength is shorter ($\sim$ 2·6 Å), owing to greater

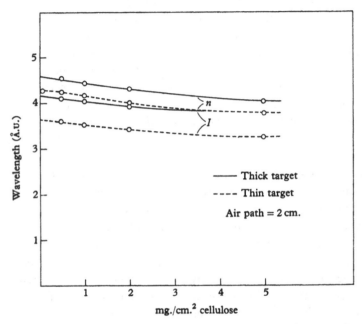

Fig. 6.10. Effective wavelength of X-rays in a half-tone emulsion, as a function of mass-thickness of absorber (cellulose), for thick and thin targets of gold. The upper and lower curves assume the photographic process to be a quantum and an energy effect respectively. (Dyson, 1957.)

absorption of the softer components. With increasing kilovoltage, the effective wavelength falls somewhat (Fig. 6.11), but much less than might have been expected. The fall is greater with a thick than a thin target, since the former absorbs more and more of the soft radiation as it is increased in thickness in step with the electron range.

This type of calculation can be adapted to give the minimum detectable thickness of specimen in given conditions. Assuming that a variation of 3 % in the number of quanta absorbed in the emulsion is required to give a visible contrast difference, Dyson obtained the minimum thickness of cellulose detectable at different voltages from a window target of the thick type (Fig. 6.12). The improvement in sensitivity at low voltage is clear, though it is less pronounced than is sometimes assumed. At 5 kV, a thickness of 1μ could be detected, corresponding to 10^{-6} g. mm.$^{-2}$ if $\rho = 1$ g./ c.c.; with a spot width of 1μ, this means that 10^{-12} g. is detectable, an estimate that agrees well with those made in §§ 6.2 and 6.3. The

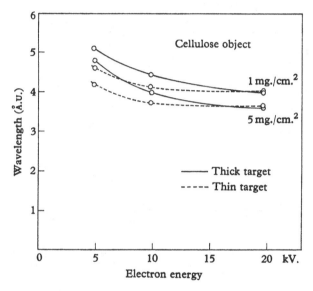

Fig. 6.11. Effective wavelength of X-rays in a half-tone emulsion, as a function of X-ray tube voltage, for two thicknesses of cellulose absorber; air path = 2 cm. (Dyson, 1957.)

sensitivity will increase still further with wavelength, and a similar evaluation of the contrast obtainable with the maximum resolution plates used in contact microradiography has been made by Henke, Lundberg & Engström (1957). They define a 'contrast function' which gives the image contrast in terms of the difference in optical density ΔD produced on the emulsion by a given small change in the mass thickness of the specimen Δm. Their results for the Eastman Kodak 649 plate, in filtered copper-L radiation of 13·3 Å wavelength, are summarized in Fig. 6.13. An optimum average sample transmission exists at about $D = 1.7$, i.e. the exposure-time should be adjusted to give this degree of blackening in the region of interest. The maximum is more pronounced for specimens of low than of high transmission (t).

6.5. Absorption analysis by the projection method

The carrying out of absorption measurements on a projected X-ray image has the advantage that, in certain types of problem,

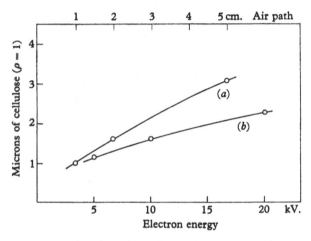

Fig. 6.12. Minimum detectable thickness of cellulose as a function of (a) air path (at 10 kV), and (b) electron energy (for air path of 2 cm.). (Dyson, 1957.)

direct quantitative results may be obtained by means of a proportional counter and pulse analyser. Although the use of a proportional counter is in principle possible in the contact method, it is attended by considerable practical difficulties, particularly in that the area to be analysed must be selected with a very small aperture placed in contact with the specimen (cf. Rosengren, 1956). In the projection method the aperture can be many times larger than the area examined. The possibilities of the method have been discussed by Zeitz & Baez (1957); experimental details are given by Long & Cosslett (1957). Applications to particular problems are described by Long (1957, 1958, 1959); Long & McConnell (1959) and Röckert (1958). (Cf. §12.5.)

6.5.1. *X-ray tube and spectrometer.* The experimental arrangement in the projection method is shown in Fig. 6.14. The specimen is placed immediately in front of the X-ray source formed on the target of a point focus tube. The transmitted radiation is analysed

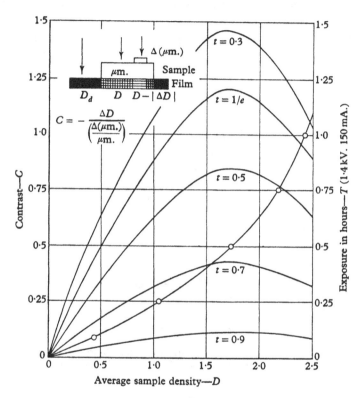

Fig. 6.13. Contrast function C, for 13·3 Å X-rays on Eastman Kodak Spectroscopic Plate 649, in function of average photographic density D in the image of a sample, for five values of the sample transmission t; X-ray tube at 1·4 kV and 150 mA. The characteristic curve of the emulsion (D vs. E) is also shown. (Henke, Lundberg & Engström, 1957.)

by a crystal spectrometer before reaching the proportional counter. Apertures limit the area from which X-rays are accepted, the lower limit to which is set by the size of the focal spot, unless a more complicated collimating system is used. When the specimen shows high X-ray contrast it can be positioned by direct reference to its image on a fluorescent screen. When this is not the case, a

200 mesh per inch copper grid (as used for electron microscopy) is mounted on the specimen, which is then placed over a hole in an aluminium foil at the centre of a stage plate. The latter fits both on to an optical microscope and on to the top of the tube, and is adjustable with high accuracy in each position. The co-ordinates of a selected area are noted optically, by reference to the grid bars, and the stage plate then transferred to the X-ray tube, where it is

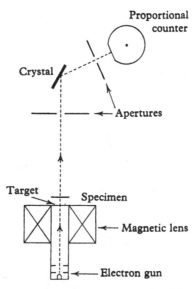

Fig. 6.14. Schematic arrangement of apparatus for absorption microanalysis by the projection method. (Long & Cosslett, 1957.)

adjusted until the shadow image of the grid bars (readily visible) are in the correct position, with respect to cross-wires on the fluorescent screen, to bring the selected area on to the axis. The fluorescent screen is then removed.

The cross-wires on the screen are alined with the spectrometer aperture beforehand with the aid of an aperture 10μ in diameter in the stage of the X-ray microscope. When the aperture is re-placed by the specimen it is thus possible to locate a selected area to an accuracy of $\pm 5\mu$. The screen slides in a plane 2·5 cm. above the target and is located against three fixed points to an accuracy of about $\pm 50\mu$. With a target–specimen distance of 0·5 mm. the magnification of the image on the screen is × 50, so that an accuracy of ± 0·1 mm. in locating the image on the cross wires will correspond to an accuracy of ± 2μ in the specimen plane. The error in locating the selected area could be reduced from ± 5μ to the same accuracy by using a reference grid of finer mesh. The alinement is sensitive to shift in the position of the focal spot, but in practice the latter is so stable that no readjustment of the cross-wires was found necessary even after several hundred hours of operation. Experimental details of the apparatus used by Long (1958, 1959) are shown in Fig. 6.15. To allow the use of soft radiation, provision is made for the X-ray path to be almost entirely in hydrogen. Windows of Mylar, 6μ thick, are sealed to the end of the rubber tube R just above the specimen and to the spectrometer chamber. The air path is thus reduced to about 1 cm. The rubber tube is pushed aside as the fluorescent screen is slid into position. The spectrometer is rigidly mounted above the magnetic lens of the fine-focus tube. The position of the aperture is thus fixed with respect to the focal spot. For given target–specimen distance, its diameter determines the angular width of the beam accepted by the spectrometer and hence the area of specimen examined. The counter window acts as the second aperture and a sealed-off counter is used in place of the continuous flow type used in earlier work (Long & Cosslett, 1957). In order to obtain high reflected intensity, a singly curved crystal is used; the gain over that from a flat crystal is some seven to ten times. A crystal with spacing suitable for reflecting the desired wavelength range must be chosen with regard to the Bragg condition (§ 5.1) and the angular deviation convenient for the recording system. For instance, a lithium fluoride crystal, curved to a radius suitable for a Bragg angle of 30°, will separate the nickel $K\alpha$ line from the copper $K\alpha$ line with a 1·25° aperture. The number of crystals needed to cover a given angular range depends on the aperture and on the resolution desired. For the determination of calcium a pentaerythritol crystal was used, curved by pressing it into a thin layer of wax with a steel plate ground to the appropriate radius.

The projection method, dealing with one point of the specimen at a time, is best suited to the two (or more) wavelength procedure which dispenses with a reference measurement on the matrix or standard sample. The accuracy of the analysis depends essentially on that of the measurement of transmitted intensity. As the amount to be determined (m_1) decreases, the intensity ratio I''/I'

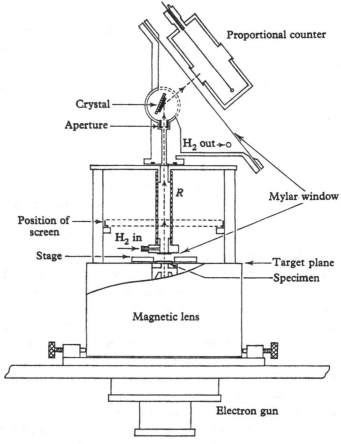

Fig. 6.15. Section through X-ray tube, spectrometer and counter, as used for absorption microanalysis by projection. (Long, 1958.)

tends to unity (cf. (6.5)) and the importance of accurate intensity measurement increases. It is, therefore, necessary to hold constant as far as possible the incident intensity I_0, or otherwise to ensure that the recorded intensities are properly corrected for fluctuations. In X-ray diffraction equipment a monitor counter is employed, but this solution is hardly practicable in the projection arrangement owing to the restricted space available for inserting a counter in the incident beam. The target current in the tube is a sufficient guide for much work, but not when high accuracy is required. However, when the effect of the matrix is negligible or can be calculated (cf. (6.6) and (6.7)), it is only necessary to measure the ratio I''/I'. The absolute value of the transmission I/I_0 is not required, provided that the ratio of the intensities of the two wavelengths is constant in the initial beam and that this ratio can be measured in absence of the specimen. When only moderate accuracy is required, and especially when the recording times are short, it is enough to make a single measurement of the transmitted intensity at each wavelength in turn. For more accurate analysis, possible fluctuations in the primary beam intensity and in the sensitivity of the recording system can be compensated by taking a series of readings at the two wavelengths alternately. In the apparatus of Long (1958) this is achieved by arranging for the spectrometer crystal to alternate between two fixed angular positions corresponding to the two wavelengths at which the transmission is to be measured. The recording time in each position is 7·5 sec. and the total time of recording is usually several minutes, so that the effect of drift in the primary intensity is largely eliminated.

6.5.2. *The recording-system.* If a spectrometer of high enough resolution is available, or other measures are taken to monochromatize the incident beam, a simple Geiger counter can be used (Engström, Lagergren & Lundberg, 1957). A proportional counter alone may be used (Rosengren, 1956), so long as only a rough indication is needed of the amount of element present, but owing to the limited energy resolution of such a counter ($\Delta\lambda/\lambda \sim 10\%$), it is usually necessary to employ it in conjunction with a crystal spectrometer as in the work of Long. It is also then possible to employ a high tube voltage, in the interests of greater X-ray intensity; a crystal spectrometer alone, or with a Geiger counter, would accept second harmonic radiation in the recorded beam which the selectivity of the proportional counter will reject. The proportional counter may be either of the continuous flow type devised by Arndt & Riley (1952) and used by Long & Cosslett (1957), or sealed-off (Long, 1958, 1959). The convenience of the latter has to be set against a tendency to vary in sensitivity with time, whereas the former is very constant in response. Practical details of the construction of counters for X-rays are given by Arndt (1955) and Lang (1956).

When used in conjunction with a spectrometer a mechanism must be provided for moving the counter at twice the angular rate of rotation of the crystal. Alternatively a stationary counter with

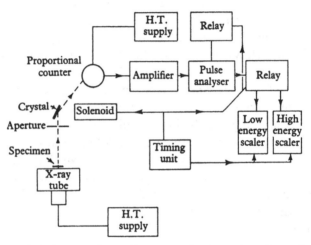

Fig. 6.16. Block diagram of recording-system for absorption microanalysis.
(Long, 1958.)

long window may be used, across which an aperture moves in synchronism with the crystal rotation (Long & Cosslett, 1957).

The pulses from the proportional counter are amplified and passed to an electronic analyser which selects those due to quanta of a predetermined wavelength and largely eliminates the effect of those due to higher order reflexions. The pass-band (or 'gate') of the pulse-analyser is automatically switched in synchronism with the alternation in reflecting position of the crystal (cf. Fig. 6.16), and its output correspondingly routed to the high energy or low energy counting units ('scalers'), which record the intensities of the two selected wavelengths. The ratio of the totals shown by the two scalers gives the ratio I''/I' of the transmissions through the selected area of the specimen at these wavelengths, after correction for the ratio of their intensity in the incident beam, as determined by similar counts in the absence of a specimen. Unless the two wavelengths are very close together, on either side of the absorption edge, correction for the effect of the matrix must be made. For this purpose (cf. (6.5)), its mass thickness m_2 must be known at the point of measurement, that is, its composition and local thickness; for an organic matrix the correction is usually negligible. Other corrections which may be necessary are discussed below.

A quicker but less accurate method of analysis is to feed the counter output to a ratemeter connected to a pen recorder, and to rotate the spectrometer crystal slowly. A continuous record is thus obtained of the variation in transmission with wavelength at a given point in the specimen. In this way the presence of an element may be detected by its absorption edge, a necessary preliminary to more accurate microanalysis when the composition of a specimen is unknown. As the height of the absorption jump is unaffected by the presence of other absorbers, it gives a direct measure of the amount of the corresponding element present. Fig. 6.17, obtained with a focal spot $1\,\mu$ in diameter, shows that 10^{-10} g. of iron is readily detectable. The limit depends upon the background count and in this example would be between 10^{-11} and 10^{-12} g., approximately an order of magnitude less than the limit of detection by the two wavelength method, in

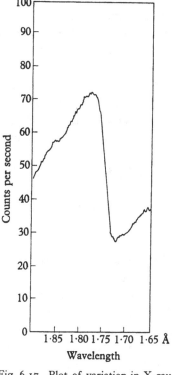

Fig. 6.17. Plot of variation in X-ray transmission across the iron K absorption edge, as wavelength is varied with a crystal spectrometer. (Long, 1957.)

which a much greater total count is obtained. The height of the absorption jump is also independent of the presence of neighbouring elements in the periodic table, so that it is possible to separate iron from manganese, for instance, by this method and obtain their relative concentration.

6.5.3. *Errors in counter recording.* When photographic recording is used, and intensities are subsequently determined by micro-photometry, it was shown above (§ 6.2) that limitations in the response of the emulsion set an unavoidable limit to the accuracy of absorption microanalysis. Although counter recording is free from this saturation effect, it suffers from other limitations which prevent it achieving an appreciably greater accuracy, even when all precautions are taken to minimize their effect. Most of these errors apply whether line or continuous radiation is used, but one or two additional troubles arise with the latter and prevent its application to elements of atomic number below about 15.

The statistical error depends on the total count according to well-established rules, so that natural fluctuations in the number of recorded quanta may be eliminated by counting for a long enough time. In practice some compromise between speed and accuracy of working has to be found, the terms of which will be governed largely by the magnitude of other sources of error. If R is the transmission ratio I''/I', the fractional error in its measurement may be expressed in terms of the standard deviation σ_R:

$$\sigma_R/R = \pm\{(\sigma_s/s)^2 + (\sigma_a/a)^2\}^{\frac{1}{2}}, \qquad (6.17)$$

where σ_s/s is the fractional error for the measurement with the specimen and σ_a/a that for the uninterrupted beam. The standard deviation of any single measurement will be $\pm(Nt)^{\frac{1}{2}}$, from the theory of statistics, where N is the count rate at wavelength λ and t is the time for which it is recorded. Under the conditions obtaining in calcium determination (Long, 1958), $\sigma_a/a = \pm 0\cdot 01$. The fractional error in the final ratio is then

$$\sigma_R/R = \pm\{1/N_1 t + 1/N_2 t + (0\cdot 01)^2\}^{\frac{1}{2}}. \qquad (6.18)$$

N_1 and N_2 are related to the count rate N_0 in the absence of specimen by (6.4), with N replacing I in each case. (6.18) can be evaluated for particular values of the mass thickness m_1, if the absorption coefficients are known for the wavelengths λ_1 and λ_2 at which N_1

and N_2 were taken; correction may also be necessary for the effect of the matrix, m_2. For gypsum, with $N_0 = 2000$ counts per minute and $t = 25$ min., the percentage error in the ratio is about 10% for a mass thickness of calcium of 3·5 mg./cm.², falling to less than 2% at a thickness of 1·5 mg./cm.², and then remaining almost constant down to $m = 0$.

The error in the ratio may be related to the error in m_1 by differentiating (6.5) (with $I''/I' = R$) to obtain dR/R and inserting the respective absorption coefficients for calcium and for the rest of the gypsum molecule. Then

$$dR/R = 765 \, dm_1,$$

so that an error of 1% in the ratio corresponds to a difference of 0·013 mg. in the value of m_1. The variation in the accuracy of determining m_1 with its absolute value is shown in Fig. 6.18. At the lower end of the scale the accuracy is limited by the fact that as m_1 approaches zero the ratio tends to unity, and at the upper end by the low intensity of the transmitted radiation. In both conditions the accuracy could be increased by longer counts. In the experimental circumstances assumed, where an area of $\sim 100\mu^2$ is analysed, an accuracy of $\pm 3\%$ is attained in the estimation of mass thicknesses of calcium between 1 and 3 mg./cm.² corresponding to $1\text{--}3 \cdot 10^{-9}$ g. The limit of detection, with greatly increased error, is in the order of 0·1 mg./cm.², or 10^{-10} g.

To the statistical error must be added that due to unwanted radiation in the beam reaching the counter, which may arise from diffraction and fluorescence in the specimen and from Compton scattering in the crystal of the spectrometer. These effects may be serious if the tube is operated, as is usual, at a voltage well above that needed to excite the wanted line. For specimens which are not too thick, these sources taken together will usually amount to an error of order 1%, on which account the counting conditions were chosen in the example above to give a statistical error of the same magnitude in the measurement of the incident beam.

When the tube voltage is more than twice that needed to excite the desired wavelength, a further source of error is introduced in the form of second-order reflexion from the crystal. If the latter is set to reflect a wavelength λ into the counter, radiation of

wavelength $\lambda/2$ will also be selected from the continuous spectrum
and will reach the counter. As the pulse height distribution of the
counter has an appreciable low energy 'tail', some of the quanta
of this harmonic will be recorded, a process which is assisted by

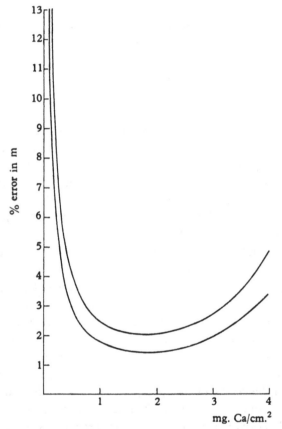

Fig. 6.18. Calculated relation between percentage error and amount analysed,
in mass-thickness determination of calcium in gypsum. Top curve for single
observations, lower curve for double observations. (Long, 1959.)

the presence of the escape peak effect. The second harmonic effect
is particularly important when the required wavelength is not a
line, but is selected from the continuous spectrum. In order to
evaluate the systematic errors in the determination of calcium,
Long (1958) made control experiments on cleavage flakes of

gypsum of varying thickness. He found that the second harmonic comprised 0·24 % and its scattered radiation 1·15 % of the count at 3 Å wavelength, just short of the calcium absorption edge. The impurity radiation appreciably increases the apparent transmission at this wavelength, but has little effect on that at 3·14 Å, just above the edge, where the absorption coefficient for calcium is about ten times less than at 3 Å. The transmission ratio is thus affected by an amount which increases with calcium content, resulting indeed in a fall in the measured ratio beyond a content of 3·5 mg./cm.^{2} (Long, 1958). It is, therefore, necessary to cut or grind the specimens to be examined to a thickness equivalent to a calcium content below this level. Improvement in the range of operation may also be obtained by working at a lower voltage on the X-ray tube and by improved design of the counter, so that it has better energy resolution. These difficulties increase at longer wavelengths, with the result that line radiation is to be preferred to the continuous spectrum for the absorption microanalysis of elements of atomic number less than 15 (phosphorus). Alternatively a totally reflecting surface can be used, since it has no second-order reflexion (cf. p. 238).

6.5.4. *Comparison with the contact method.* The advantage of the contact procedure in absorption spectrometry is that a comparatively large area (several square centimetres) is recorded at one exposure, which can be examined subsequently at selected points by microphotometry. The projection method, although it records only a single point at each 'exposure', gives a direct quantitative measure of the transmission and avoids the errors of the photographic process and the practical difficulties of microphotometry. It can also be adapted to give continuous recording along a line, or other path, across the specimen so that the variation in amount of a given element is immediately displayed. This information may also be obtained by microphotometry of a contact negative, but not the variation in transmission over a range of wavelengths, for a given object point, which projection allows. The contact method also has the disadvantage that the specimen must be transferred from one emulsion to another when the two wavelength method is employed. It may then be a matter of some difficulty to identify corresponding micro-areas on the two negatives. The effect of fluorescent radiation produced in the specimen is a limitation on the accuracy of the contact method, but is negligible in projection. The ultimate areal resolution of the contact method is restricted by the limit set by optical diffraction effects on the size of spot usable in the microphotometer. Standardization of the photographic processes may make it possible to work with better accuracy on enlargements of the contact negative, but statistical limitations set by the grain size and distribution will still remain. The corresponding limitations on the accuracy of counter recording can be overcome by taking longer counts, or using higher initial intensity. These measures become increasingly necessary as the size of focal spot is reduced, but it is only this inconvenience of long exposure that prohibits the use of the projection method at the lowest spot sizes at present attainable, $0 \cdot 1 \mu$ in diameter. Once a fine-focus tube is available, there is little advantage in employing the contact method, especially where an extended series of analyses has to be made. The projection method gives higher speed and somewhat greater ultimate accuracy, especially where matrix absorption is severe. It can also provide a permanent photographic record of the area examined, if required. Its main drawback in the eyes of non-physicists, the use of more complicated recording equipment than a camera, is disappearing as counters, pulse analysers and scalers are rapidly becoming familiar even in biological and metallurgical laboratories.

Additional references added in proof

To §6.2.1: Müller & Sandritter (1960).
To §6.2.2: Hydén & Larsson (1960); Sissons, Jowsey & Stewart (1960a).
To §6.5: Butler, Bahr, Taft & Jennings (1960); Long (1960a).
To §6.5.4: Lindström (1960).

CHAPTER 7

X-RAY EMISSION MICROANALYSIS

7.1. Emission microspectrometry

Emission microanalysis, by direct excitation of X-rays in the specimen with a fine focal spot ('electron probe'), was first investigated by Castaing (1951) and independently by Borovskii (1953) (Borovskii & Il'in, 1956). The natural extension to a flying-spot or scanning system has been made by Cosslett & Duncumb (1956), whereby television technique is employed to display a picture of the specimen in terms of variation in characteristic X-ray emission, that is, the distribution of a particular element is made directly visible. These electron probe methods have rapidly established themselves as valuable research tools, especially in metallurgy. The simplicity of the characteristic spectrum, and especially the great relative strength of the $K\alpha$ line, makes it possible to differentiate between near neighbours in the periodic table with comparatively simple apparatus. When direct excitation is inconvenient, secondary (or fluorescent) radiation may be excited by exposing the specimen to a powerful primary beam of X-rays, as described in § 7.4.

7.1.1. Experimental conditions for direct excitation. Emission microanalysis may be carried out with the type of fine-focus tube developed for projection microscopy (§ 3.4). The essential difference is that the electron spot must now strike the surface of the specimen itself, instead of a target (Fig. 7.1). The X-rays thus produced are analysed with a crystal spectrometer and Geiger counter, or with a proportional counter alone, so that the intensity of the line characteristic of a particular element is recorded. Although photographic recording can in principle be employed, the output of X-rays from an area of a few square microns is so small, and a spectrometer collects so minute a proportion of this output, that exposures would be prohibitively long. Microanalysis by emission has become practicable only because the new point focus tubes produce a very much higher current density than do tubes with macroscopic spots, and because of the high sensitivity of counters. Such an arrangement allows continuous recording, so that the distribution of an element may be explored along a line in the specimen surface, as either the electron spot or the specimen is moved (Plate XXVI). Alternatively, the relative amounts of a number of elements in a given micro-area may be found if the

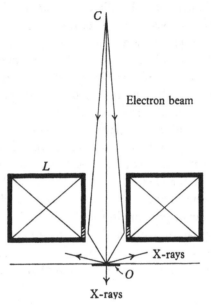

Fig. 7.1. Direct excitation of X-rays for spectral microanalysis. Electrons from a cathode C are focused onto the sample O by a magnetic lens L.

'gate' of the recording system is moved through a range of wavelengths, by rotating the crystal of a spectrometer or varying the acceptance channel of a pulse analyser coupled to a proportional counter (cf. Fig. 7.2).

Experimental details of the micro-focus apparatus and of the recording gear are given in §§ 7.2 and 7.3. In general, the tube may be an adaptation either of the projection X-ray microscope or of an electron microscope. The pole-pieces of the final magnetic lens, or the electrodes of the electrostatic type, are designed so that the specimen can be introduced into the focal plane and so as to allow the X-ray beam to emerge into the analysing system. The simplest solution is to bring the focal plane out clear of the end face of the lens (Mulvey, 1957, 1959*a*; Duncumb, 1957*a*, 1960*a*), allowing specimens to be readily interchanged. It is also necessary to incorporate some means of observing the exact spot on the specimen struck by the electron probe, in order to know which

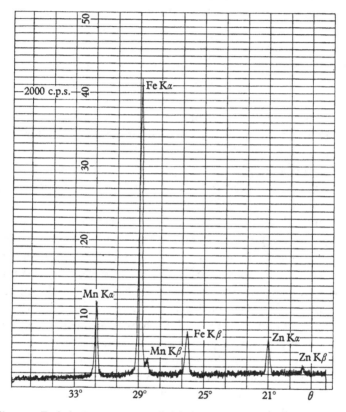

Fig. 7.2. Emission spectrum recorded from an area of $1\,\mu^2$ of ferroxcube— a manganese-zinc ferrite. (Duncumb, unpublished.)

regions are analysed. Castaing (1951) built an optical system into his electrostatic lens, and Mulvey has devised an accurate mechanism for rotating the specimen from beneath an optical microscope, outside the lens, on to the axis of the electron beam. If the focal length of the electron lens is sufficiently long, an optical microscope may be arranged to view the specimen *in situ*, the point struck by the electron beam being lined up with the cross-wires of the eyepiece (Long & Cosslett, 1957; Long, 1959). At the cost of some loss in resolution, the electron beam may be brought out into air, either through a thin collodion foil, as suggested by Zeitz & Baez (1957), or via a minute aperture (Riggs, 1958). In this way it is possible to excite X-rays in large specimens, or even in liquids (Schumacher, 1958), which could not be introduced into vacuum.

The optimum operating conditions will depend on the element to be estimated: self-absorption of the line radiation in the target must be minimized and its intensity relative to the continuous spectrum maximized. Since absorption increases rapidly with wavelength, the hardest convenient characteristic line should be used. The K-series is suitable for almost all elements, but as an unduly high voltage would be needed for those near the top of the periodic table, the L-series is then preferable. For instance, excitation of the K-lines of tungsten requires a minimum voltage of 70 kV, but only 12 kV for its L-lines, which are still hard enough (1·1–1·5 Å) for absorption in the target to be small. Since an element is relatively transparent to its own characteristic radiation (§ 2.2) the absorption is not so great as might at first sight seem probable. For light elements it becomes more serious, and it is then necessary to take off the X-rays at a steep angle to the surface. Collection along the normal would be best, but would interfere with the incident electron beam. An angle of 45° is, therefore, used for light elements, although for heavier elements it may be as small as 10°, to suit experimental convenience, without giving undue self-absorption. Even so, corrections must be applied for such absorption when estimating the amount of an element present from the recorded strength of its line radiation (p. 189).

It is desirable, especially in the scanning method, to ensure the highest possible intensity in the recorded beam. The intensity of X-ray emission depends on a number of factors: on the current in the incident electron beam, on its voltage and its penetration in the target specimen, as well as on the self-absorption of the wanted line in the latter. The electron beam voltage needs to be high, both to increase the current density (cf. 3.20) and the output of line radiation. The dependence of the output of both characteristic and continuous radiation on applied voltage is discussed in detail in the next chapter (§ 8.5). From (8.28) it follows that the energy radiated in (say) the $K\alpha$-line (E_k) is given by

$$E_k \propto (V_1 - V_k)^{1\cdot65} V_k^{0\cdot35}, \qquad (7.1)$$

where V_1 is the voltage across the X-ray tube and V_k that needed to excite the K-series (that is, the voltage corresponding to the wavelength of the K-absorption edge). The *total* energy in the continuous spectrum, E_c, by (8.21), depends on the square of the applied voltage:

$$E_c \propto V_1^2.$$

The variation of intensity through the spectrum has the simple form of Fig. 8.5, when plotted against frequency. The total energy E_c is given by the area under the graph, so that the energy per unit energy interval E_{ck} in the continuous spectrum at the frequency of the $K\alpha$-line, will be related to voltage by

$$E_{ck} \propto (V_1 - V_k) \qquad (7.2)$$

on making the simplifying assumption that the plot in Fig. 8.5 may be extrapolated linearly to zero frequency. The dependence on voltage of the ratio of the strength of the $K\alpha$-line to that of the continuous spectrum at the same wavelength, as selected by a spectrometer of bandwidth dV, is then:

$$E_k/E_{ck} dV \propto (V_1 - V_k)^{0\cdot65} V_k^{0\cdot35}/dV,$$
$$\propto (U-1)^{0\cdot65} V_k/dV, \qquad (7.3)$$

where $U = V_1/V_k$. The constant of proportionality is of order unity (cf. (8.29)), so that the ratio will be of order V_k/dV.

Hence, not only the absolute intensity of the $K\alpha$-line, but also its intensity relative to the continuous background radiation rises with applied voltage, without limit (in this approximation). When the beam intensity is recorded with an energy selective system, as is usual in emission spectrometry, it is thus desirable to use as high an applied voltage as is conveniently possible; but regard must be had to possible second-order effects, when $V_1 > 2V_k$, if crystal reflexion is involved (§ 6.5). For crystallographic analysis it has become usual to employ a value of V_1 no more than about three times that of V_k, for elements in the middle of the periodic table, in order to avoid unduly high voltages on the tube. For lighter elements, as (7.3) shows, the relative yield of K-radiation is less, since $V_k \propto (Z-1)^2$, by Moseley's law. It would then be desirable to use a higher value of V_1/V_k, but a practical limit may be set by the corresponding increase in the depth of electron penetration in the specimen (8.16) and by the heat thereby generated.

Deeper penetration both broadens the effective X-ray source (cf. § 3.8c) and increases the amount of self-absorption, thus worsening the accuracy of localization of the analysis and the available X-ray intensity. Especially when the K-line is in the soft or ultra-soft wavelength range, a limiting voltage is soon reached beyond which the amount of radiation recorded actually begins to decrease (Castaing, 1958). In general it is necessary to keep the voltage below that at which the electron range, given by (8.16), exceeds the desired linear resolving-power, as determined primarily by the size of the electron spot. The similar limit on resolution in point projection imaging has been discussed in § 3.8. It becomes a practical limitation only for spots of less than a few microns in diameter. To ensure that the range is less than the spot size, the voltage must not be greater than 9·5 kV for a carbon, 10 kV for an aluminium or 18 kV for an iron specimen, for a spot of $1\,\mu$ diameter.

For electron spots of this order of size, the heating effect in the specimen is relatively unimportant: the electron optical system is unable to deliver a high enough current density to overheat the metal (§ 8.2). At 10 kV and with a solid copper block as specimen, the beam current would have to be reduced below its maximum possible value only for spots larger than about 10μ (cf. (8.7)). For non-metallic specimens, such as rock sections, the heating should be kept small by using a target current of 10^{-7} to 10^{-8} A instead of 10^{-6} A. To avoid charging effects, it is necessary to coat the specimen surface with a thin evaporated layer of a good electrical conductor such as aluminium. So long as the layer is still slightly transparent to light (100–500 Å) it will be thin enough not to interfere with the analysis; naturally a metal must be chosen that has no characteristic line in the neighbourhood of that to be detected.

7.1.2. *Corrections; limits of accuracy.* In first approximation the amount of the analysed element, in the volume penetrated by the electron beam, is proportional to the intensity of its characteristic X-ray line, in quanta per second, as recorded by the counting and analysing system. The result can be put into absolute terms by a comparison measurement on a pure sample of the element under identical experimental conditions. Corrections must be made, however, for absorption of the line in the specimen itself, for possible fluorescent excitation and for difference in stopping power for electrons as between a light element and a heavy matrix, or vice versa. In some circumstances these corrections can be considerable (cf. Castaing & Descamps, 1955; Duncumb, 1957a). For instance, self-absorption will be high if the line excited in the element is of slightly shorter wavelength than that of an absorption edge of the matrix material. At the same time radiation from the matrix may cause fluorescent excitation of the characteristic line of the element, thus increasing the intensity recorded at this wavelength. For nickel in an iron matrix, for instance, the correction may be as high as 40 %, when the emergence angle is 20°. Sufficient data for such correction is not always available, but the need for it may be avoided if the concentration of the element is approximately known from a preliminary measurement or from other evidence. It may then be possible to prepare a homogeneous mixture of about the same composition for use as comparison specimen in the emission analyser, so that the self-absorption and other effects will be approximately the same as in the sample to be analysed.

Apart from any lack of precision in making such corrections, the accuracy of microanalysis by the emission method has to be discussed from two aspects: the volume of material from which X-rays are excited by the electron probe and the accuracy with which the amount of the wanted element in this volume can be determined. This latter accuracy will depend on the concentration of the element and the nature of the matrix, and will reach a limit when the signal due to the element, over the allowed counting time, is of the same order as the fluctuation in background count. The ultimate sensitivity, or limit of detection, may be defined as that amount of an element for which the signal equals the r.m.s. fluctuation in the background count, that is, for which the probability of detection is e^{-1} (36 %).

The attainable accuracy is determined essentially by the rate at which X-ray quanta reach the counter, that is to say, by the intensity of line emission from the specimen and the proportion of this output collected by the counter. The efficiency of emission depends on the factors discussed in the previous subsection; the efficiency of collection depends primarily on the solid angle subtended at the specimen by the counter window, or by the analysing crystal if a spectrometer is used. As this angle is usually fixed by considerations of experimental convenience, the signal reaching the counter is given by the amount of the wanted element in the volume explored by the electron probe and by the operating conditions of the X-ray tube. Assuming that the window of the counter and the gas filling it have been properly selected so that it records all the quanta entering it (efficiencies of 90–95% are usual), the accuracy attainable is a matter of the statistical error in the recorded count (§ 6.5) and thus depends on the rate of arrival of quanta and on how long a counting time can be allowed. If an accuracy of 1% is required, a total count of at least 10,000 must be obtained, which may take seconds or minutes according to the amount of element present in the region analysed. It is thus important for speed of working, as mentioned above, to ensure the maximum possible X-ray output. In practice, adventitious experimental factors such as imperfect constancy of electron beam current, make it fruitless to seek an accuracy of much better than 1%; with high concentrations of an element, a big enough count to satisfy this limit statistically can be obtained in a few seconds with an electron spot of order 1μ. As the concentration decreases, the accuracy of analysis falls even if the counting time is increased, owing to mechanical and electrical instabilities in the apparatus (of order 1% per hour). The ultimate detection limit, beyond which it is impossible to be sure that the wanted element is present, will depend on the element sought and on the experimental conditions. For elements in the middle of the periodic table it is of order 10^{-14} g., corresponding to 0·1% of a constituent of density 10, in a volume of $1\mu^3$.

It is usually convenient to arrange for the counter output to be integrated in a rate-meter, the signal from which actuates a pen-recorder as the analyser channel sweeps through the spectrum, thus displaying graphically the relative concentration of a number of elements (Fig. 7.2). Alternatively, with a fixed analyser channel, the variation in concentration of a given element across the surface of a specimen can be plotted. In the Castaing type of micro-analyser the specimen is moved mechanically under a stationary electron probe; in the scanning type it is fixed and the probe is scanned across it in two dimensions by deflecting plates or coils (cf. Fig. 7.8). When the signal intensity is sufficiently great, the scanning method can be used to give a visible picture of the distribution of the element, on a cathode-ray tube in which the beam moves in synchronism with the electron probe of the analyser (cf. § 7.3). For flicker not to be visible, sufficient counts must be available within the integration time of the fluorescent screen and this is usually only possible if a long-persistence screen is used. The spot size needed to give the minimum count rate, for an image definition of a given number of lines per picture (cf. § 3.8.2), can be calculated by standard methods and has been evaluated by Duncumb (1957a) for the X-ray scanning microscope (Fig. 7.3). For a spot of diameter 1μ and an element in the middle of the periodic table, the statistical fluctuation for a 200-line picture becomes a serious factor at concentrations below a few parts in a hundred. Unless means can be found to increase X-ray output—by increasing the target current, for instance—it will prevent visible representation of the analysis when the spot size is reduced much below 1μ, in the present type of apparatus, since the beam current depends on the size of focal spot (cf. (3.20)). This is a limitation on convenience rather than of fundamental importance, however, since an image of the surface can nevertheless be displayed by collecting the scattered electrons, which are many times more intense than the line X-radiation. Quantitative analysis can still be carried out, by allowing a longer counting time with the smaller X-ray signal, down to the point at which it merges with the background count.

The degree of localization possible in emission microanalysis is clearly inseparable from the accuracy of analysis. The accuracy depends on the intensity of X-ray emission, and this in turn depends on the target current and on the volume from which emission occurs. In fact the target current is itself dependent on the size of the electron spot, in given experimental circumstances, and on the tube voltage V. The latter is usually chosen, as discussed above, so that the range of the electron beam in the specimen is less than the diameter of the electron spot. Hence the volume analysed is approximately equal to the cube of the spot radius r. The limit of localization will thus be reached when the rate of count from this volume becomes inconveniently small, in

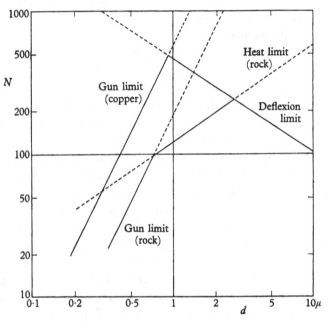

Fig. 7.3. Width of field (number of resolved lines, N) in function of spot diameter d in scanning microanalyser, as limited by errors in beam deflexion system, by electron gun output and by thermal dissipation (in two types of specimen). (Duncumb, 1957a.)

comparison with the background count or the analysing time, whichever is the limiting factor. As the target current is proportional to the voltage V and to $r^{\frac{8}{3}}$, by (3.20), and the electron range depends on the square of the voltage, the X-ray emission falls very rapidly as spot size is reduced. From the electron optical point of view, a spot as small as 10 Å in diameter could be formed, but the X-ray output would then be vanishingly small. At present X-ray emission microanalysis is practicable down to spots of diameter 0.2–0.3μ. Improvement in localization to 0.1μ will depend on obtaining higher target current, by improvements in the electron optical system, and on devising a recording system with much larger collecting efficiency.

7.1.3. *Discrimination between elements.* When a 'proportional' counter is used alone for detecting the line radiation, the element present in a pure sample can be immediately identified, since the voltage of the output pulse is proportional to the energy (and hence the frequency) of the absorbed quantum. If the specimen is composed of two elements well separated in the periodic table, they can be separately identified and their relative concentration will be given by the ratio of the number of pulses recorded per second in each energy channel, within an error of approximately 5 %. If the elements are near neighbours in the table, however, it becomes more difficult to discriminate one from the other: although the X-ray lines are sharp and well separated, the output of the counter system runs them together. This broadening is due partly to the spread in the number of ion pairs per absorbed quantum in the gas filling the counter and partly to statistical variations in the multiplication process that follows. The identity of two X-ray lines may then be completely obscured in the output from the counter. In terms of energy, the half-width of the pulse-height distribution at half its maximum intensity is found to be about 7 % of the mean energy of a pulse ($\Delta E/E_k \doteqdot \pm 0.07$) for elements in the middle of the periodic table (Fig. 7.4*a*), but it increases for elements of lower atomic number (Hendee & Fine, 1954). The energy of a K-quantum is related to the atomic number of the element producing it by Moseley's law:

$$E_k = a(Z-1)^2, \tag{7.4}$$

where a is a constant. Hence $\Delta E/E_k = 2/Z$, so that the K-line of cobalt ($Z = 27$) is separated from that of its neighbours (nickel and iron) by only about 8 %, whereas the counter can separate pulses only if their peaks are separated by more than $2 \times 7 = 14$ %. The pulses from these three elements, when they are present together, will thus be indistinguishable in the counter output—it will fail to 'resolve' them. If nickel and iron are present, but cobalt absent, the respective pulse-height distributions will still

run into each other. The three elements can, of course, be quantitatively resolved by introducing a crystal spectrometer between
specimen and counter. The loss of intensity thus incurred will
require a longer counting time, but this is partly compensated by
a considerable reduction in the background count, which gives an
improved ultimate sensitivity.

In general, therefore, emission microanalysis with simple
counter recording is unable to analyse a mixture of two elements
with high accuracy unless they are separated by at least three places
in the periodic table ($\Delta Z \geqslant 3$) (Fig. 7.4b). Elements separated by
two places may just be resolved by the counter alone, when they
are present in more or less equal proportions, or by applying

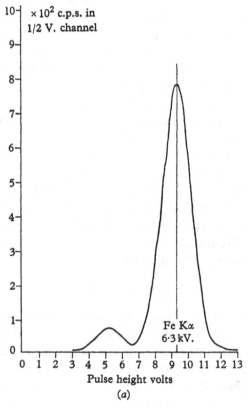

Fig. 7.4a. Form of pulse from proportional counter, compared with natural
width of $K\alpha$ spectral line of iron, showing line broadening due to detection
system. (Duncumb, 1957a.)

refined methods of analysis to the composite pulses when the proportions are very different (Dolby, 1959). For the trio of elements (carbon, nitrogen and oxygen) which are particularly important in ferrous metallurgy, the value of $\Delta E/E_k$ is about 33%, so that they should be easily resolvable by a proportional counter, provided that the response curve does not continue to broaden appreciably for elements of such low atomic number (cf. Bisi & Zappa, 1955; Mulvey & Campbell, 1958). Detailed evidence about the response of a counter to such soft radiation is meagre, but the margin is so

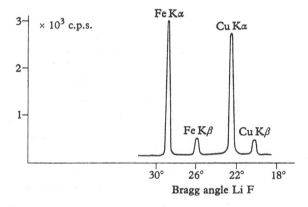

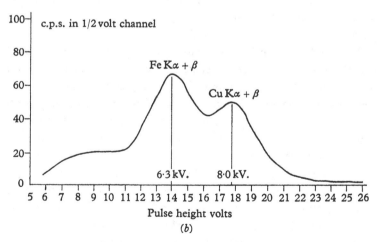

Fig. 7.4b. Overlap of pulse height distributions from proportional counter for two elements differing by 3 in atomic number (copper and iron), compared with resolution obtained with crystal spectrometer (above). (Duncumb, 1957a.)

wide that it appears likely that a simple proportional counter would
be able to separate the pulses from carbon, nitrogen and oxygen,
without need of a dispersive system. The resultant saving in
intensity over a crystal spectrometer, in which there is severe loss
on reflexion, would go far to offset the fall in efficiency of X-ray
emission with atomic number (cf. (8.15)).

7.1.4. *Comparison with other methods*. In comparison with the
micro-absorption method, emission micro-spectrometry has a much
smaller detection limit (10^{-14} to 10^{-15} g.) and accuracy of analysis
(< 1 %). As so far developed, its accuracy of location is about
0.3μ but only intensity limitations stand in the way of its extension
beyond that attainable in the contact-absorption method, 0.25μ,
down to 0.1μ. The main disadvantage is that elements that are near
neighbours in the periodic table cannot be separated by a counter
system alone, except possibly those of very low atomic number.
The need to use dispersion to separate neighbours higher in the
table leads to an appreciable loss in intensity, but the decrease in
working speed is only serious if one of the elements is present in
very small amount, so that a long counting time must be allowed.
It is probable that developments in the technique of analysing the
pulse-height distribution from the counter will make possible the
direct separation even of next neighbours in the middle of the
periodic table (Dolby & Cosslett, 1960).

    The great advantage of the emission over the absorption method
is that it does not require the specimen to be prepared as a section
thin enough to have appreciable X-ray transparency at the
analysing wavelength. Micro-emission can thus be employed on
solid surfaces of any type, so long as they are reasonably smooth.
The drawback of having to introduce the specimen into vacuum is
serious only with volatile materials and with those of unwieldy
size. Even the latter can be attacked if the electron beam is brought
out into the air through a thin window or small hole. The resulting
loss in accuracy of location is small, so long as a voltage of several
tens of kilovolts is permissible. When such a solution is not
practicable, and especially for liquid specimens, the micro-
fluorescent technique may be employed, though with appreciable
loss in localization (§ 7.4).

7.2. The static spot microanalyser

A schematic diagram of the unit constructed by Castaing & Guinier (1950, 1953; Castaing, 1951, 1954) is shown in Figure 7.5. The electron beam forming system is similar to the two lens projection X-ray microscope (chapters 3 and 11). A hot tungsten

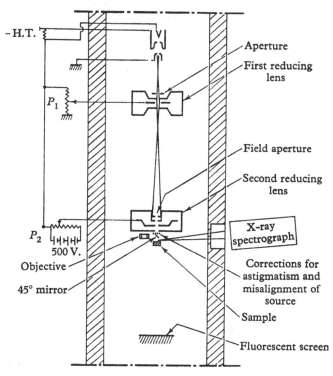

Fig. 7.5. Static spot emission microanalyser in its original form.
(Castaing & Guinier, 1950.)

hairpin filament in a triode electron gun forms the source. The accelerated electrons pass through two electrostatic lenses, which reduce the beam diameter to about 1μ at the final focal plane. The specimen is placed in this position and X-rays are generated where the electron probe strikes it. The X-rays pass out through a thin window in the wall of the instrument into the X-ray spectrometer and Geiger counter for analysis as to wavelength; the chemical composition of the small area of the specimen is given by the position and intensity of the characteristic lines recorded. A 45° mirror and a long working distance optical microscope objective give a view of the surface of the specimen so that different areas of interest may be placed underneath the electron probe in turn. The original instrument used the electrostatic lenses from the C.S.F. electron microscope but with a modified objective lens of 14 mm. focal length and 200 mm. spherical aberration coefficient. The lens was not symmetrical about the central electrode and the specimen was placed 6·5 mm. from the final flat electrode. This system delivered about $1·5 \times 10^{-8}$ A into a 1μ spot; the beam voltage employed depends on the nature of the specimen analysed. Analyses were made of diffusion couples and different phases in metal specimens. An improved version of this instrument, described at the 1954 London conference on electron microscopy (Castaing, 1956), embodied two magnetic lenses and a reflecting optical objective of higher numerical aperture. It is capable of detecting elements down to magnesium (12) (Castaing, Philibert & Crussard, 1957), and is now commercially produced.

The elegance of this method of obtaining exact chemical information about minute areas of metallurgical specimens has led many other investigators to convert or construct similar instruments. These all follow the original conception of two electron lenses and electron gun, X-ray spectrograph and optical viewing system. Counter recording is universally used, with either Geiger or proportional counters, experimental details of which are here omitted (cf. Arndt, 1955; Lang, 1956). The RCA type B electron microscope has been converted to this purpose by Fisher & Knechtel (1955) and by Ogilvie (1957). The smaller and simpler RCA table model forms the basis for the electron probe instrument of Birks & Brooks (1957 a, b). The permanent magnet lenses have been replaced by electromagnetic lenses giving a 1μ spot. This apparatus is designed for simplicity and stability in rapid routine analysis of many samples with different amounts of known elements. The analysing system consists of several crystals set at different angles corresponding to the wavelengths of the known elements. Each crystal is followed by an individual counter, ratemeter or scaler and output recorder. The use of several crystals and detectors allows much faster analysis than when one crystal is swept through the complete spectrum at each point of the specimen. At 25 kV, 0·1 μA and a 1–2μ spot, 3000 c.p.s. were obtained from pure iron using a curved crystal spectrometer. Elements down to 22 (Ti) can be detected.

An electron probe system has been constructed by Wittry (1957) using two magnetic lenses and a hot cathode electron gun, after initial trials with a field emission filament and electrostatic lens.

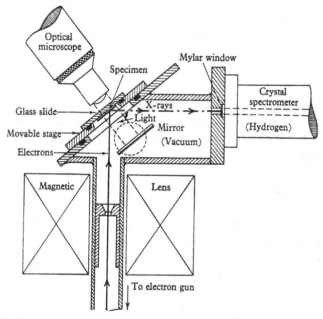

Fig. 7.6. Emission microanalyser (magnetic lens) with long working distance, for examining rock sections and other transparent specimens. (Long & Cosslett, 1957.)

His instrument has an integrating circuit to measure the total charge given to the specimen, as a control on X-ray output instead of the more usual measurement of electron beam current.

A simple single lens emission analysis unit for optically transparent specimens has been constructed by Long and is shown in Fig. 7.6 (Long & Cosslett, 1957; Long, 1957). The lens is operated with a long focal length and an electron spot of 20–30μ. The specimen is mounted on a microscope slide set at 45° to the beam in a movable stage. A small mirror reflects light through the specimen and into the optical microscope used to position the area of interest over the electron beam, by means of the stage controls. The electron beam produces a light blue fluorescence in many materials and the specimen is placed at this point. The X-rays pass out at 90° to the electron beam, though a thin plastic window and hydrogen path to the crystal spectrometer, counter and recording apparatus. At 20 kV and 0·5μA, 500 c.p.s. are recorded from pure iron using a flat crystal in the spectrometer.

Several commercial versions of electron probe analysers have been produced, besides the magnetic lens instrument of Castaing mentioned above and the Soviet model (Borovskii & Il'in, 1957). The Metropolitan-Vickers model, which follows a design due to Mulvey (1959 a, 1960 a), has several new features (Fig. 7.7). The main column comprises the usual hot filament electron gun and two magnetic lenses, L_1L_2, but has the gun G at the bottom. The X-ray spectrometer has a moving crystal and counter completely within the high vacuum. The Rowland circle geometry and the large angular range covered have led to a vacuum enclosure about 3 ft. across. The specimen S is mounted on a rotating drum R so that the area to be analysed may first be selected with an optical microscope M and then rotated (by 180°) on to the axis of the electron beam. The drum must turn with the accuracy expected in locating the electron spot on the chosen part of the specimen, that is, within 1 μ.

Another approach has been followed by Philips Electronics (Norelco) who have converted the Philips EM 75 electron microscope for analysis, in much the same manner as for X-ray microscopy (§ 11.13.2). The recording system relies on the resolution of the proportional counter, without a crystal analyser, in the interests of simplicity and low cost. A 2–10μ probe is formed by the two magnetic lenses and the same voltage range (75 kV) is used as for electron microscopy (Bessen, 1957 b).

A commercial apparatus has also been described by Buschmann & Norton (1957) of the General Electric Company. This instrument uses the electrostatic optical components of the X-ray microscope made by the same firm (§ 11.13.1) with the addition of a third lens at 1 to 1 magnification to bring the electron probe onto the specimen. The optical microscope is focused on the reflected

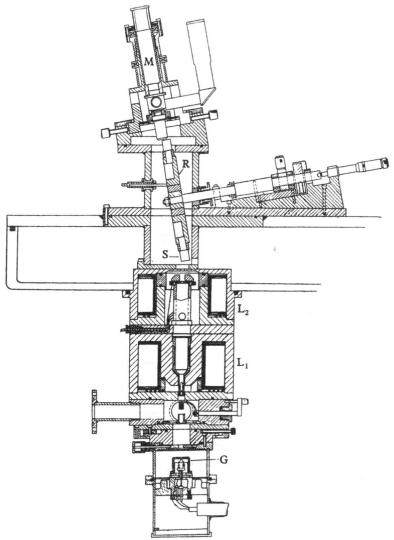

Fig. 7.7. Metropolitan-Vickers magnetic lens microanalyser.
(Mulvey, 1960a.)

image of the specimen as seen on the surface of the third lens. The
aberrations of this lens, used at a long focal length, limit the beam
current to $1\mu A$ at a spot size of 5μ. With a helium X-ray path,
elements down to scandium (21) can be detected. At $0.6\mu A$ and
a 5μ spot some 4000 c.p.s. were recorded from chromium, using a

curved crystal spectrometer. Both lithium fluoride and sodium fluoride curved to 10 cm. radius have been used, followed by Geiger or proportional counters.

7.3. The scanning microanalyser or flying-spot X-ray microscope

7.3.1. *Apparatus.* The scanning X-ray emission microscope (Cosslett & Duncumb, 1956; Duncumb, 1957a, b; Duncumb & Cosslett, 1957) combines the emission method of analysis of

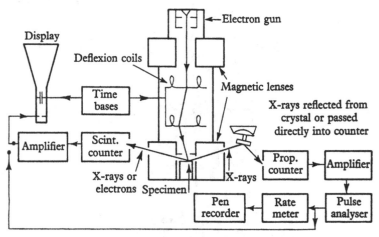

Fig. 7.8. Schematic diagram of the scanning microanalyser. (Duncumb 1957a.)

Castaing (1951) with the techniques of scanning electron microscopy developed by McMullan (1953), Smith & Oatley (1955). The essential features of the method are set out in Fig. 7.8; Fig. 7.9 is a section through the column of the instrument. The electron beam is scanned across the specimen by two pairs of deflexion coils (or electrodes). The double deflexion system ensures that the beam always passes through the final lens near the axis thereby reducing the effect of 'off axis' lens aberrations and also allowing a larger scanning field. Either the excited X-rays or the scattered electrons (at will) can be detected by a scintillation counter (phosphor and photomultiplier), and the output used to modulate the control grid of a cathode ray display tube which is scanned in synchronism with the main electron beam. In this way

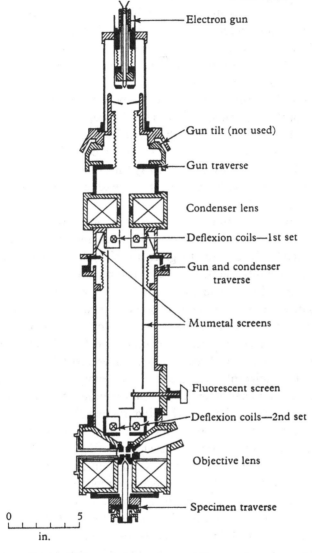

Fig. 7.9. Sectional diagram of the column of the scanning microanalyser.
(Duncumb, 1957a.)

a two-dimensional picture of the specimen surface in terms of the emitted X-rays or scattered electrons is built up on the screen of the display tube and may be photographed if desired. For analysis the X-rays are detected by a proportional counter directly, as shown on the right of Fig. 7.8. As the output pulse is proportional to the photon energy it is possible to add a pulse analyser and determine from the characteristic X-ray lines the elements in the surface of the specimen. The scan can be stopped and the electron probe positioned over any chosen point for analysis by using direct current deflexion in the scanning coils. The fluorescent screen of the display tube has a decay time of about 20 sec., so that the probe may be positioned by reference to the afterglow image of the specimen.

The amplified output from the counter actuates a pen-recorder as the channel of the pulse analyser is swept through the spectrum. Alternatively, the output at a given wavelength can be fed directly to the display tube, so that a picture is formed on the screen of the distribution of a particular element over the area of the specimen scanned. When the variation along a line path is of interest, it can be quantitatively displayed by means of a double beam oscilloscope (Pl. XXVI). The proportional counter can be used alone if the elements present are three or more places apart in the periodic table. For adjacent elements it is necessary to add a curved crystal spectrometer before the counter in order to obtain adequate energy resolution, the crystal being rotated through a range of reflecting angles as the signal operates the pen-recorder.

The final lens in a later form is shown in Fig. 7.10. The electron beam enters from the top and is focused on to the specimen in a spot of about 1μ dia. The specimen is supported from below and can be translated in two directions, moved up and down vertically and rotated. The specimen holder is insulated so that the collected beam current can be monitored for quantitative work. The lens aperture is placed 3 mm. below the top pole face and the specimen just outside the lower surface of the lens. The pole piece is of the asymmetrical type, with a bore of 12 mm. and gap of 6 mm. on the electron beam side and a bore of 3 mm. on the specimen side. At 840 ampere-turns (12,000 turns and 70 mA) the focal length is 12 mm. at 25 kV. The X-rays leave the specimen region below the pole piece gap and then pass through thin beryllium or plastic windows sealing the phosphor and spectrometer tubes. A cone of X-rays of semi-angle 5°, emerging at 20° to the specimen surface, can reach the crystal spectrometer through the hydrogen path. For recording scattered electrons the beryllium window is removed and the scintillation counter is positively biased; the light from the scintillator is conducted by a perspex light pipe to the photomultiplier.

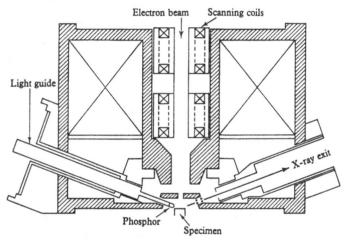

Fig. 7.10. Section through final lens of scanning microanalyser, with medium working distance. (Duncumb, 1960 a.)

7.3.2. *Applications.* Image formation with a selected X-ray wavelength is shown in Pl. VI A. The specimen is formed of two overlapping grids, one of copper with 200 bars/in. and the other silver with 800 bars/in. In the left-hand figure the image has been formed with the pulse analyser channel set astride the copper $K\alpha$ emission line, so that the thick bars of copper appear bright. In the right-hand image, the channel has been set at the silver L line, and the thin bars of silver appear bright. The clear spaces between the bars are dark in both pictures. In this manner the element distribution over approx. 10^5 picture-points is displayed at once instead of each in turn as in the static system.

An example of this method applied to a metallurgical problem is shown in Pl. VII (Melford & Duncumb, 1958). The optical and electron scanning images are obtained first, to locate the area of interest. The specimen was a mild steel known to be subject to cracking during hot working. The three X-ray images show the distribution of iron, nickel and copper respectively and reveal a high degree of segregation around an inclusion. The absolute concentrations were measured by stopping the electron probe on selected areas. Nickel reached 40% compared to an average concentration of about 0·14% throughout the sample, as given by chemical analysis. Other applications of the scanning method have been described by Duncumb (1957b, 1960a). An improved instrument has now been built by Duncumb and Melford (1960) for metallurgical investigations (Pl. VI B).

The method of X-ray emission scanning microscopy is now being developed along several lines. The first is towards the estimation of elements of lower atomic number, since the present limit falls between Al (13) and Cl (17) depending on the crystal and counter used. Eventually it should be possible to detect carbon, nitrogen and oxygen, which are of great importance in both ferrous metallurgy and organic substances. Similarly, the resolution should be capable of improvement from the present limit of 0·3μ by reduction in the beam voltage, use of thinner specimens, and improvements in electron gun and recording efficiency. Of more immediate importance is the extension of the method in its present form to larger samples outside the lens structure. Larger scanning fields may also be possible, giving the opportunity of choosing a smaller area for higher resolution. All of these developments are serving to establish this new method of X-ray microscopy as one of the most fruitful approaches to microchemical analysis. (See also p. 367–8.)

7.4. Fluorescent microspectrometry

X-rays may be excited by irradiation with a primary beam of X-rays, as well as by direct electron impact. The emitted secondary radiation is necessarily of longer wavelength (lower energy) than the primary beam (cf. § 2.2), so that the latter must be appropriately harder than the line to be excited. The fluorescent intensity is much less than that attained by direct excitation of the same line, but, on the other hand, the line is now unaccompanied by a continuous spectrum. The only other radiation in the recorded beam is that due to lines of longer wavelength from the same element and those from other elements present, which will only complicate detection if these latter are near neighbours in the periodic table. There is also a small amount of scattered radiation. The fluorescent line spectrum is thus of greater purity than that obtained by direct

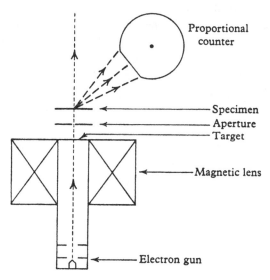

Fig. 7.11. Schematic diagram of apparatus for fluorescence microanalysis. (Long & Cosslett, 1957.)

excitation, for which reason the ultimate detection limit is appreciably smaller, from a given volume of material. It has therefore been increasingly used as a method of analysis since the early work of Glocker & Schreiber (1928).

The experimental arrangement for fluorescent analysis is almost identical with that for micro-absorption (cf. Figs. 6.15 and 7.11). The essential difference is that the detection system is offset from the axis of the incident X-ray beam. Fluorescent radiation is initially isotropic in distribution, so that it can be equally well detected in a direction well away from the incident direction in order to avoid recording the primary beam. Long & Cosslett (1957) have directed attention to the need to suppress secondary radiation from the primary beam passing through the air in front of the counter, by interposing an absorbing screen. The specimen is best prepared as a thin section so that it can be observed in transmission, as in absorption measurements. Solid surfaces may also be examined, but only at greater distance from the primary X-ray source, with resulting loss in intensity (or resolution) by the inverse square law.

The optimum thickness of a section, to give maximum absorption of the primary beam consistent with minimum absorption of the secondary radiation, depends on the direction of observation as well as on the relative values of the absorption coefficients of the specimen for the two radiations. It has been investigated theoretically by Compton & Allison (1935, p. 485) and by Long & Cosslett (1957). Suppose a sample of thickness M to be irradiated with X-rays of a wavelength for which its mass absorption coefficient is μ_1/ρ, producing in it a fluorescent line of absorption coefficient μ_2/ρ. If the latter radiation is collected over a (small) solid angle ω in a direction θ to the primary beam, then the ratio (in terms of quanta) of the fluorescent intensity I_f to that of the incident beam I_0 is given by

$$I_f/I_0 = k\omega/4\pi \cdot \mu_1/(\mu_2 \sec\theta - \mu_1) \exp(-\mu_2/\rho\, M \sec\theta)$$
$$\times \{\exp(M/\rho(\mu_2 \sec\theta - \mu_1)) - 1\}, \qquad (7.5)$$

where k is the fluorescent yield, that is, the number of fluorescent quanta emitted per incident quantum. This expression was evaluated for the cases of calcite and gypsum, excited by copper $K\alpha$ radiation, in the typical experimental conditions $\omega/4\pi = 0.022$ and $\theta = 48°$, with the result shown in Fig. 7.12. The output from calcite shows a maximum for a thickness of about 35μ; the limit of the linear rise is at about 25μ. If a more penetrating primary radiation is used, such as molybdenum $K\alpha$, the linear rise is hardly affected but the maximum is longer and flatter; the sample then behaves, in fact, analogously to an 'infinitely thick' layer of a radioactive substance.

The fluorescent output varies appreciably with the wavelength of the primary radiation, at given incident intensity. The intensity of the primary beam itself will depend on the target element and the voltage used to excite it, so that the choice of primary wavelength will depend largely on factors of experimental convenience and availability of apparatus. The essential condition, inherent in fluorescent conversion, is that the exciting wavelength be shorter than that of the absorption edge corresponding to the characteristic

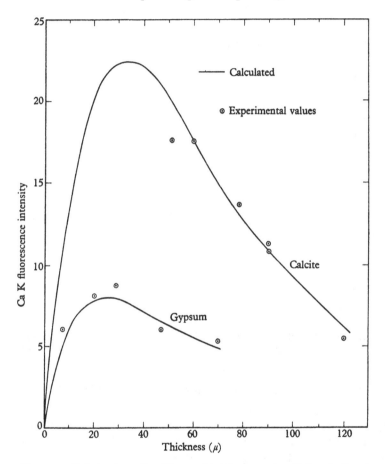

Fig. 7.12. Relative intensity of transmitted fluorescent radiation, in function of thickness of specimen, in calcite and gypsum. (Long, 1959.)

line to be excited. The question of the most suitable target to choose in particular circumstances has been discussed by Hevesy (1932) and in more practical terms by Rogers (1952) in connexion with analysis on the macroscopic scale (cf. § 10.1). The optimum thickness of sample will depend on the primary wavelength employed, since it determines the value of μ_1 in (7.5). The variation with wavelength in the case of an iron sample was calculated by Zeitz & Baez (1957) and shows a sharp rise for harder radiation (Fig. 7.13).

The experimental errors of the fluorescent method are similar to those of direct excitation (§ 7.1), being primarily statistical. Compared with direct excitation, fluorescence analysis suffers from

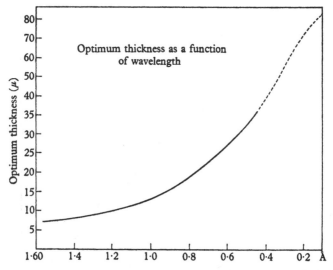

Fig. 7.13. Specimen thickness giving maximum fluorescent intensity in function of primary X-ray wavelength, for an iron sample and an angle of observation of 30°. (Zeitz & Baez, 1957.)

lack of intensity. The sample can intercept only a small part of the primary X-ray beam and, in addition, the fluorescent yield is low, especially for elements of low atomic number. In order to obtain a count of the same statistical significance, therefore, the fluorescent method requires longer recording time than when the same wavelength is excited in the same area of sample by electron bombardment. The same speed and accuracy of analysis can only be obtained by increasing the area of sample inspected. The full advantages of high localization cannot be retained, even by sacrificing speed, because the effect of the background count makes itself increasingy felt as the desired count decreases.

Over such an area, fluorescence scores over direct excitation in ultimate concentration sensitivity because the ratio of wanted line intensity to background is greater, there being no contribution from the continuous spectrum. The remaining background is due to radiation from other elements in the sample and to scattering, which can be largely eliminated if a proportional counter is used. The detection sensitivity thus depends on the level of the background count in given experimental circumstances. Using a subsidiary aperture to prevent fluorescent radiation from the air from reaching the counter, Long & Cosslett (1957) estimated the minimum detectable mass of calcium at 7×10^{-11} g. in an area of the specimen defined by a 50μ aperture at a distance of 0·8 mm. from it. The primary target was copper, at 10 kV and 7μA, and the counting time 10 min. By using a smaller aperture still closer to the specimen, this limit could be brought below 10^{-12} g. Assuming constant background, it could be further improved by operating the primary target at higher voltage and current. Zeitz & Baez (1957) predict a limit of less than 10^{-14} g. of iron with a tube at 40 kV and 25μA, for an irradiated area of sample of $25\mu^2$ and a counting time of 10 min. Such a limit would correspond to an analytical detectability of about ten parts per million. As it is not yet supported by adequate experimental evidence, the more conservative estimate of 10^{-12} to 10^{-13} g. should be taken, corresponding to one part in a thousand and one part in ten thousand.

In comparison with direct excitation, therefore, the fluorescent method offers better concentration sensitivity only for a comparatively large specimen area, and permits distinctly poorer localization. Its great advantage is that it is much simpler to operate. The sample remains under normal atmospheric conditions, instead of having to be inserted into an X-ray tube as target. Any increase in counting time is more than offset by the greater speed and ease of changing specimens. More important, for practical applications, bulky specimens of almost any size and shape can be handled, and liquids as well as solids. For these reasons the fluorescent method is finding increasing application, except where a very high degree of localization is required. When the distribution of an element to a linear accuracy of 1μ or less is necessary, the method of direct excitation has to be used; for speed and convenience, preferably in the scanning system.

CHAPTER 8

PRODUCTION OF X-RAYS

The production of a beam of X-rays requires an electron source, some means of accelerating the issuing electron beam and directing it on to the anticathode (target), a window through which X-rays leave the vacuum and usually some means of cooling the target. For utilizing the output to best advantage, it is important to know something of the spectral and angular distribution of the X-rays and the way in which this depends on voltage and target thickness. The overall efficiency of production is also of interest when quantitative measurements are to be made. Only the factors of practical importance in each of these phenomena will be treated here, references being given to detailed treatments of the fundamental physical processes.

8.1. Electron emission and focusing

8.1.1. *General requirements.* Almost all X-ray tubes have a hot filament, normally of tungsten, as source of electrons and some have a more or less elaborate system of electrodes, or electron lenses, for focusing the emitted beam on to the target. With focal spots of large or moderate size, including the so-called 'fine-focus' tubes used for crystallographic analysis, the cathode can deliver more energy into the beam than can be dissipated even in the most efficient conditions of target cooling. With spots smaller than a few microns, heat dissipation becomes potentially faster than energy supply. Particular attention, therefore, has to be given to the design of the electron gun and lens system in the point projection method, as discussed in § 11.1. When a very small spot is required, of order $0.1\,\mu$ or less, the use of field emission from a point cathode begins to have advantages over the thermionic filament, owing to the very high current density that can be drawn from a metal point of radius less than $1\,\mu$ in a very strong electric field (see below).

The tubes of larger spot size, as normally used for contact microradiography, produce a small amount of focusing through

shaping of a cup electrode around the cathode and usually electrically connected to it (Fig. 8.1). Stronger and variable focusing
may be obtained by applying a bias voltage to this electrode, as in
the tube of Ehrenberg & Spear (1951a) in which spots down to
40μ in diameter can be formed (§ 14.1). All such tubes have a
massive target, cooled with water or some other liquid, and X-rays
from its surface issue through a window close to it, usually of
beryllium. In order to get better cooling and hence higher beam

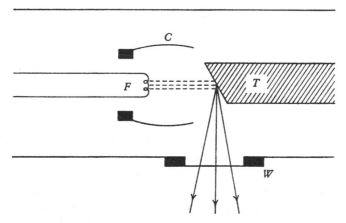

Fig. 8.1. X-ray tube with simple electrostatic focusing. F, filament;
C, focusing cup; T, anode; W, window.

current it is then advantageous, as explained below, to form the spot
into a line rather than a circle by employing a line filament and slit
aperture. The issuing X-ray beam acquires a square cross-section
by foreshortened viewing of this line, the target–window axis being
inclined at a low angle to the target surface in the interests of
maximum intensity (§ 8.6). The chief exceptions to this standard
layout are the low voltage gas-discharge tube of Henke (§ 14.1), in
which the target is flat and the X-ray beam passes back through
holes bored in the cathode, and the point anode tubes described in
§ 3.4.2.

Most fine focus tubes use a thermionic source of electrons, but
as the advantages of field emission from point cathodes have
recently come under discussion, some comparison of these two
types of emitter will be made.

8.1.2. *Thermionic emission.* The emission from the surface of a heated metal depends exponentially on the temperature. If the specific emission is ρ_c A/cm.$^{-2}$ from a cathode at temperature T, it was shown in § 3.5 that the current density ρ_2 in any subsequent image plane, at which the beam angle is α_2, is given by (3.19). Putting $\alpha_c = \frac{1}{2}\pi$ and expressing the emission velocity of the electrons in terms of T instead of an equivalent voltage V_c, we have

$$\rho_2 = 1\cdot16 \times 10^4 \rho_c \alpha_2^2 V_2/T, \qquad (8.1)$$

where V_2 is the final voltage. The current in a spot of radius r_2 is then
$$I_2 = \pi r_2^2 \rho_2. \qquad (8.2)$$

In a perfect imaging system the maximum value of α_2 is $\frac{1}{2}\pi$, corresponding to a minimum magnification M of the cathode (of radius r_c), which by (3.21) is given by

$$M = \frac{r_2}{r_c} = \left(\frac{T}{1\cdot16 \times 10^4 V_2} \right)^{\frac{1}{2}}. \qquad (8.3)$$

Hence the full current would be deliverable into a spot of any size down to the limit given by (8.3). If a smaller spot were required, a cathode of smaller radius than r_c would have to be used.

In actual electron lenses, α_2 is limited by spherical aberration and the target current becomes

$$I_2' = \frac{1\cdot16 \times 10^4}{T} 4\pi r_2^{\frac{8}{3}} \rho_c V_2 (2C_s)^{-\frac{2}{3}}, \qquad (8.4)$$

where C_s is the aberration constant of the lens (cf. (3.24)).

The variation of ρ_c with T was measured by Haine & Einstein (1952) for pure tungsten, repeating the original work of Jones & Langmuir (1927). At 2700° C it is about 1·5 A cm.$^{-2}$ and at 3000° C it is well above 10 A cm.$^{-2}$. The value employed in practice has to be reconciled with a reasonable cathode life, and at 3000° C. this is likely to be in the order of hours only, with filament wire of 0·1 mm. diameter. Such a filament will give a maximum output of about 1 mA, which can be limited at will by varying the voltage on the bias electrode. The value of the target current falls rapidly with spot size, according to (8.4), the absolute value depending on the value of C_s in the focusing lens. This variation was shown in Fig. 3.7 for two values of brightness β; the upper curve ($\beta = 10^5$) corresponds closely to a filament at 3000° C. and $V_2 = 10$ kV, or at 2800° C. with $V_2 = 100$ kV, and the lower curve to a voltage ten times smaller in each case. The target current in a spot of 0·1 μ radius would be approx. 1 μA and 0·2 μA respectively; the value (1 mm.) assumed for C_s is typical of a strong magnetic electron lens. Because of this rapid fall of current with spot size, the possibility of employing field emission from point cathodes has been considered, since they can provide a much higher current density than a hot filament.

8.1.3. *Field emission.* Electrons can be extracted from a metal by establishing a very high field at its surface ($> 10^7$ V/cm.), as well as by heating it. The necessary high field is obtained by making the cathode radius very small ($< 1\mu$). The cold emission properties of such point cathodes have been explored very thoroughly (cf. Good & Müller, 1956; Dyke & Dolan, 1956; Drechsler & Henkel, 1954). Under high vacuum conditions current densities as high as 10^5 A cm.$^{-2}$ can be drawn continuously, and much higher instantaneous values in pulsed operation. If the point cathode is heated to above 1500° C., both thermionic and field emission occur, and it is usual to speak of T-F or Schottky emission (Dolan & Dyke, 1954). In these conditions current densities as high as 10^4 A/cm.$^{-2}$ may be continuously drawn even at moderate vacuum, such as obtains in demountable X-ray tubes.

Since such an emission corresponds to a brightness of $3 \cdot 10^7$ A cm.$^{-2}$ sterad.$^{-1}$ at an anode voltage of 10^4 V, the use of a point cathode in preference to a thermionic emitter was proposed by Marton (1954) and Pattee (1953 b), particularly for X-ray microscopy. Important experimental limitations occur, however, due to the large angle of the emitted beam (~ 1 radian) and the small size of the virtual cathode. A detailed comparison of the thermionic and field emitter has been made by Cosslett & Haine (1956) and by Drechsler, Cosslett & Nixon (1960). With a perfect imaging system, the point cathode would have the advantage for focal spots smaller than about 0·25 μ; at larger spot size, the thermionic emitter can supply more current. When spherical aberration is present the point is limited more than the thermionic cathode. The aperture of the lens must be opened to admit as much as possible of the wide beam emitted, but the disk of confusion grows with the third power of the aperture. An optimum position of the point with respect to the aperture, and thus an optimum magnification, may be shown to exist. Thin lens theory gives it as × 3, but in thick lens theory it is approximately × 2 (Cosslett & Haine, 1956). In these conditions the true image of the cathode is negligibly small, the spot size at the target being given by the disk of confusion. The target current is then given by

$$I'_c = 0.94\pi i_c (r_2/C_s)^{\frac{2}{3}},\tag{8.5}$$

where i_c is the current per unit solid angle in the beam entering the lens, that is, the current emitted by the point cathode divided by $\pi \sin^2 \alpha_1$, where α_1 is the semi-angle of the cone within which it is contained. As this angle is almost one radian in field emission systems of present design, and currents of 10^{-4} to 10^{-3} A may be drawn from points of radius 1μ or so, the value of i_c is 5.10^{-4} A sterad.$^{-1}$ at maximum.

In comparing the field emitter with the thermionic emitter, it must be remembered that the value for C_s in (8.5) will be larger than that in (8.4), owing to the limit placed on cathode–lens distance in the former case by the danger of electrical break-down. The electron lens must thus be of longer focal length than that used for focusing a thermionic cathode, for which a strong lens at some distance from the cathode is needed in order to give the required high demagnification. The value of C_s is about ten times bigger for the field emitter than for the thermionic cathode. It is then found that the two would give the same target current at a spot size of about 0.1μ; only for smaller spots would the field emitter have the advantage. Since the projection microscope now requires spots of this order, for high-resolution work, investigation of the use of a field emitter in continuously evacuated tubes is being carried forward (Hibi, 1956; Drechsler, Cosslett & Nixon, 1960). In view of the sensitivity of field emitters to vacuum conditions, the alternative of increasing the focused current from a thermionic emitter by at least partial correction of spherical aberration also merits attention. At present, thermionic electron sources of the form described in § 11.1 are generally employed.

8.2. Heat dissipation in the target

In all those tubes which form macroscopic or moderately fine focal spots, the main practical problem is to dissipate the heat generated in the target. As shown in § 8.5 less than 1 % of the energy in the electron beam is converted into X-rays, the remainder appearing as heat. A great deal of attention has been paid to this problem, and only the main results will be mentioned here.

8.2.1. *Thick target.* When the target thickness is large compared with the spot size, heat flow is essentially radial into the hemisphere of which the spot is the centre (Fig. 8.2 a). If the focal spot is circular in form, of radius r, and has an average temperature T_r when thermal equilibrium is reached, the rate of heat dissipation E_s is given by

$$E_s = 2\pi k(T_r - T_R)/(1/r - 1/R), \tag{8.6}$$

where k is the thermal conductivity of the target metal, and T_R is the temperature at a convenient boundary at radius R, as determined by the cooling conditions. Rapid dissipation demands a target material of high conductivity, such as copper, and a low boundary temperature—hence the need for forced cooling. So long as R is large compared with the spot size, it has little effect on the rate of dissipation. With $R \gg r$:

$$E_s = 2\pi rk(T_r - T_R). \tag{8.7}$$

In conditions of thermal equilibrium, when the supply of energy just balances the dissipation, the maximum permissible specific loading W_s will be

$$W_s = E_s/\pi r^2 = 2k(T_{rx} - T_R)/r, \tag{8.8}$$

where T_{rx} is the maximum permissible temperature of the spot, determined by the melting-point of the target. Since W_s is inversely proportional to r, the current density in the incident electron beam can be increased as the spot size is diminished. When R, effectively the target thickness, is not large compared with r, (8.7) is no longer valid; Oosterkamp (1948) has given values of the small correction factor by which it must be multiplied, for a range of the ratio R/r.

8.2.2. *Thin target.* In the conditions of projection X-ray microscopy the target is so thin as to be comparable with (or less than) the electron range (Fig. 8.2*b*). Energy is delivered to it more or less uniformly throughout a cylindrical volume of radius equal, in first approximation, to the spot size *r* and of length equal to the target thickness *d*. Heat flow is then effectively two dimensional,

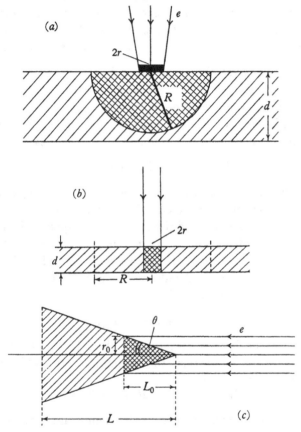

Fig. 8.2. Heat dissipation in X-ray target: (*a*) thick target; (*b*) thin target; (*c*) point anode.

to the target boundary of radius R at temperature T_R. If T_r is the average temperature in the spot, the rate of heat dissipation E_t is given by

$$E_t = 2\pi k d(T_r - T_R)/\log_e(R/r). \qquad (8.9)$$

The temperature at the centre of the spot will be rather higher than T_r by an amount of order $0.1 T_r$.

The rate of heat dissipation thus depends directly on target thickness and changes only slowly with spot size and target radius, both of which occur in the logarithmic term. On simple assumptions about the loss of energy by collision along the electron path, Langner (1957) has shown that the rate of supply in these conditions is also proportional to target thickness d, so long as d is less than the electron range. Hence the temperature rise in the spot is in first approximation independent of target thickness in such very thin targets. The permissible target loading W_t, now depends differently on r:

$$W_t \propto d/r^2 \log_e(R/r). \qquad (8.10)$$

But as d must be reduced roughly in proportion to r, to offset electron diffusion in the target (§ 3.6.3), the beam current density is again approximately proportional to $1/r$.

For comparing the relative merits of thick and thin targets, in connexion particularly with electron scanning systems (cf. § 8.3), we have from (8·7) and (8.9):

$$E_s/E_t = (r/d)\log_e(R/r). \qquad (8.11)$$

The target thickness is normally about equal to the spot diameter in the projection method ($2r = d$), and with $R = 1$ mm., $r = 1\mu$,

$$E_s/E_t = \tfrac{1}{2}\log_e 1000 = 3\cdot 5.$$

A solid target thus has no very great advantage in heat dissipation over a window target.

8.2.3. *Needle target.* Some types of tube make use of a point anode (cf. § 3.4.2). Assuming it to be a cone of uniform semi-angle θ and length L, it can be shown that the rate of heat dissipation in the axial direction is:

$$E_c = \pi k(T_1 - T_2)\tan^2\theta/(1/L_0 - 1/L), \qquad (8.12)$$

if energy is uniformly supplied by the electron beam up to an axial distance L_0 from the tip, at temperature T_1; T_2 is the temperature at distance L from the tip (Fig. 8.2c). With $L \gg L_0$, and since $L_0 \tan\theta = r_0$, the radius of the needle at distance L_0, this becomes

$$E_c = \pi k r_0(T_1 - T_2)\tan\theta. \qquad (8.13)$$

Defining the specific loading now as $W_c = E_c/\pi r_0^2$, since the X-ray spot size is primarily determined by r_0, the maximum value of W_c for thermal equilibrium is again inversely proportional to spot size.

In comparison with a thin target of thickness d, a conical point is at some disadvantage; from (8.9) and (8.13), for the same temperature difference and with $2r_0 = d$,

$$E_c/E_t = \tan\theta \log_e(R/r)/4. \qquad (8.14)$$

The point anode used by Rovinsky, Lutsau & Avdeyenko (1957) had a tip angle θ of $10°$. With $R/r = 10^3$, as before, (8.14) then gives:

$$E_c/E_t = 0.35$$

so that the point anode is inferior to the foil target in heat dissipation by about the same factor as the latter is inferior to the solid target. It should be noted that dissipation in the conical point is independent of its length, so long as only the tip is irradiated with electrons ($L \gg L_0$).

8.2.4. *Heat balance in a thin target.* Since the target loading for thermal equilibrium is approximately proportional to the reciprocal of spot size, for all target geometries considered above, it would appear that the incident current density could be increased continuously as spot size is diminished. The total energy input E, and so the total current I at fixed voltage, would then be proportional to r. In fact, as explained earlier (§ 3.5.2), this is possible only within a restricted range of spot size, owing to the effect of spherical aberration in the focusing system and the limited cathode current density that can be drawn consistent with reasonable length of life. Below a certain spot size, depending on the lens and cathode properties, the current that can be delivered is proportional to $r^{\frac{8}{3}}$, by (3.24).

The optimum conditions of thermal equilibrium are especially important in the projection method of microscopy (§ 3.6), where maximum X-ray output is needed for keeping exposure short but where too high a target temperature will lead to its being punctured. The permissible maximum T_{rx} is usually taken as $0.75T_m$, if T_m is the melting-point of the target metal. The optimum thermal conditions for given spot size r ($= r_2$) can then be found by equating the relation for rate of heat loss E_t (8.9) and that for energy supply ((3.24), multiplied by the beam voltage V_1). In Fig. 8.3 these relations have been plotted against r for typical experimental conditions: $V_1 = 10$ kV, $V_c = 0.25$ V, $C_s = 1$ mm., $R = 1$ mm., $T_R = 100°$ C. and two values of cathode current density, $\rho_c = 1$ A/cm.^{2} and 10 A/cm.^{2} respectively. The heat dissipation curve is drawn for a copper target of thickness $d = 1\mu$, with $T_r = 800°$ C.

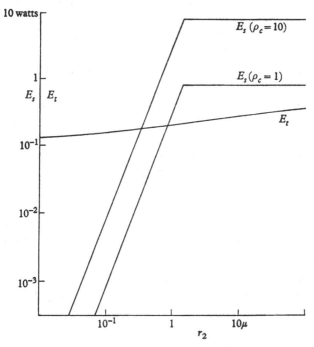

Fig. 8.3. Variation with spot radius r of energy supply E_s and heat dissipation E_t, for copper target, 10 kV electron beam and two values of cathode current density ρ_c.

and $k = 3.9$ W/cm./° C.; the curve for a 1μ tungsten target will be almost identical, with $T_r = 2000°$ C. The maximum permissible current I_x at any given spot size can be at once deduced from the graph; values of I_x for a variety of operating conditions are given by Long (1959). Below a radius of 0.9μ for $\rho = 1$ A/cm.^{2}, or 0.35μ for $\rho = 10$ A/cm.^{2}, the rate at which energy can be supplied from the electron gun and lens system becomes less than the potential rate of heat dissipation in the conditions assumed. The output of X-rays then falls with spot size more rapidly than (8.10) would indicate.

8.3. Electron range in the target

The distance to which electrons penetrate in the target determines the extent of the X-ray source, when the electron spot is very small. With sufficiently good approximation for our present purpose, this spread may be taken as equal to the range of electrons of given incident velocity in the metal of which the target is composed. The electron velocity is normally given in terms of the beam voltage V_1 and experiments have shown that the most probable exit voltage V_x of electrons from a layer of thickness x follows the Thomson–Whiddington relation:

$$V_1^2 - V_x^2 = a_1 x. \qquad (8.15)$$

Defining the range as that thickness x_r for which the mean exit voltage is zero, at given entrance voltage,

$$V_1^2 = a_1 x_r. \qquad (8.16)$$

This must be taken as a very approximate relation, since the value of a_1 varies slightly with incident voltage and target material, and the exponent of V_1 is usually found to be somewhat less than 2.

The experimental data is sparse and to a large extent not comparable, since different conditions of measurement of exit velocity are used. The most extensive investigations have been made by Whiddington (1912, 1914) and Terrill (1924, 1930); the most recent, and probably the most accurate, measurements have been made on aluminium by Lane & Zaffarano (1954), Spear (1955) and Young (1956, 1957). Reviewing all the evidence, it seems reasonable to adopt values of a_1 for aluminium at 2·5, 10 and 50 kV of 5, 10 and 20 × 10^{11} (in V^2 cm.$^{-1}$) respectively. The value for 10 kV is about twice that given in Whiddington's original work and corresponds to a range of 1·0μ.

The variation of a_1 through the periodic table is small and Whiddington suggested it depended on the root of the density $(\rho)^{\frac{1}{2}}$. It seems more probable, on theoretical grounds, that it depends on the density of electrons in the scatterer and hence on the quotient $\rho Z/A$, where Z is its atomic number and A its atomic weight, so that (8.16) may be rewritten in terms of an atomic constant a_2:

$$V_1^2 = a_2 x_r \rho Z/A. \qquad (8.17)$$

On this assumption the value of a_1 for copper would be three times, and that for gold six times, that for aluminium. Such experimental data as exist indicate that the actual values are somewhat smaller, and that the 'constant' a_2 falls by about 0·5 over the whole range of the periodic table (cf. Paul & Steinwedel, 1955).

On the basis of this rather scanty data, and with all the reservations made above, Table 10.4 gives the range factor for electrons in a number of metals at voltages of 2·5, 10 and 50 kV. In comparing these values with other sources, it should be noted that data for a_2 are sometimes quoted as if they were those for a_1, and vice versa, and that some authors (e.g. Terrill, Spear) prefer to work in terms of a factor b defined by

$$V_1^2 = b\rho x_r,$$

so that $b = a_1/\rho = a_2 Z/A$. Whiddington's original results are calculated in terms of velocity, not voltage, so that the conversion factor depends slightly on the voltage employed, owing to the relativistic effect. At 10 kV, the value of his factor a_w is given by

$$a_w = 1\cdot 2 \times 10^{31} a_1.$$

The restriction which electron penetration imposes on target thickness, and hence on intensity of X-ray production when a very small X-ray source is required, are discussed in § 3.6 and § 8.5.

TABLE 8.1. *Value of the range factor a_1 at different incident electron voltages*

kV	C	Al	Cu	Ag	Au
2·5	0·45	0·5	1·6	1·8	$3\cdot0 \times 10^{12}$
10	0·9	1·0	3·1	3·5	$6\cdot0 \times 10^{12}$
50	1·8	2·0	6·2	7·0	$12\cdot0 \times 10^{12}$

8.4. Spectral distribution

The incident electron beam excites both a continuous and a line spectrum, the form of which is shown in Fig. 8.4, plotted against wavelength. The position of the foot of the curve and the total intensity depend on the voltage and on the nature of the target. The line spectrum is characteristic of the element bombarded and is simple compared with optical spectra. The lines occur at positions determined by the energy difference between the outer and inner electron levels in the atom and are labelled correspondingly the K, L, M, etc. series, with subscripts α, β, I, II, etc. for the sub-lines (see, for instance, Clark, 1955). If the wavelength (in

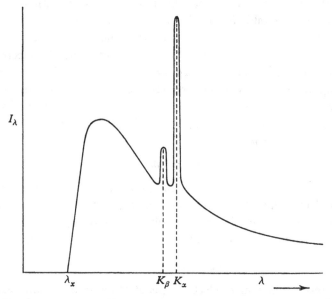

Fig. 8.4. Intensity of X-ray emission (per unit wavelength interval) in function of wavelength, for a thick target.

Ångström units) of the corresponding absorption edge is λ_a, then the minimum voltage (in kilovolts) needed for its excitation V_a, is given by

$$\lambda_a = 12 \cdot 398/V_a. \tag{8.18}$$

The wavelengths of the line spectra of the elements are tabulated in standard works of reference; recent and particularly full data are given by Sandström (1957) and Zingaro (1954 a, b). The dependence of the intensity of a line on V_a and on the beam voltage V_1 is discussed in the next section.

The continuous spectrum is most simply described in terms of frequency ν rather than wavelength (Fig. 8.5). It stretches from

an indeterminate lower limit of frequency to an upper limit ν_x
determined by the electron beam voltage V_1 according to the
quantum relation:
$$E = eV_1 = h\nu_x, \tag{8.19}$$

where E is the energy of the quantum, e the electronic charge and
h Planck's constant (6.625×10^{-27} erg/sec.). If we consider the

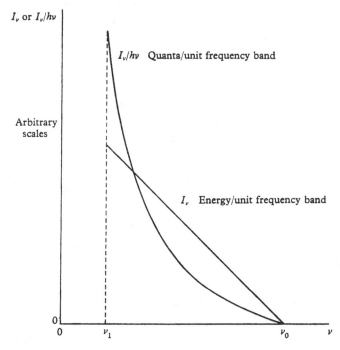

Fig. 8.5. Intensity of X-ray emission (I_ν) and of number of emitted quanta ($I_\nu/h\nu$),
per unit frequency interval, in function of frequency ν, for a thick target.
(Duncumb, 1957a.)

interaction of electrons of given initial voltage V_1 with a very thin
target, quantum mechanical theory suggests and experiment bears
out that the probability of emission (per unit frequency interval)
is constant for all X-ray frequencies below the limit ν_x. The result
would be a uniform distribution of intensity with frequency up to the
limiting value, i.e. a line parallel and close to the abscissa in Fig. 8.5.
An electron has a reduced energy after each act of emission, so that
the total emission from a thick target will be built up step-wise of

contributions starting from successively lower limiting frequencies, resulting in the straight line of Fig. 8.5. Such a distribution, in which intensity at given frequency $(I_\nu d\nu)$ is proportional to $(\nu_x - \nu)$, was found by Dyson (1956, 1959a; Cosslett & Dyson, 1957) to hold also for moderately thin targets ($\sim 1\mu$) of copper, in the forward direction, at voltages in the range used in projection microscopy (6–12 kV). There is some slight variation in intensity with angle, and at about 90° to the incident beam the relation found experimentally by Kuhlenkampff (1922) for a massive target holds good:

$$I_\nu = AZ(\nu_x - \nu) + BZ^2,$$

where A and B are constants. The intensity is then slightly higher than shown in Fig. 8.5, and the lines curve down towards the high frequency limit. This departure from linearity was found with an electron opaque target of gold even in the forward direction. The angular variation is discussed in more detail below.

If the distribution of intensity is plotted against wavelength instead of frequency, the more familiar form of Fig. 8.4 is obtained, which shows a peak near to the short-wave limit owing to the rapid fall in the number of quanta per unit wavelength interval as wavelength increases. The peak wavelength is about 1·5 times the limiting wavelength.

8.5. Efficiency of X-ray production

The efficiency of conversion of electrons to X-ray energy is very low, and in most circumstances is much smaller for the characteristic than for the continuous spectrum. It also depends differently on operating voltage in the two cases. The essential experimental relations only will be given here; details both of the measurements and of the state of theoretical interpretation will be found in recent reviews (Schaafs, 1957; Stephenson, 1957).

8.5.1. *The continuous spectrum.* The efficiency of production η_c of the continuous spectrum, as the ratio of energy in the X-ray beam to that in the incident electron beam, follows the relation

$$\eta_c = kZV_1, \qquad (8.20)$$

where Z is the atomic number of the target element, V_1 the beam voltage and k a numerical factor of order 10^{-9} when V_1 is in volts. As the electron beam has energy $V_1 I$, where I is the target current, the power in the X-ray beam E_c is given by

$$E_c = kZV_1^2 I \qquad (8.21)$$

so long as the target is thick enough to stop all the electrons. This expression holds down to about 2 kV, below which a third-power law seems to be applicable (cf. Stephenson, 1957).

The value of k is given by Compton & Allison (1935) as $1 \cdot 1 \times 10^{-9}$, averaged over all angles and wavelengths in the emitted beam. By numerical evaluation from the quantum theory of Sommerfeld (1931), Kirkpatrick & Wiedmann (1945) found $1 \cdot 3 \times 10^{-9}$ and this value has been confirmed by Dyson (1956, 1959 a) by direct measurements on thin targets of aluminium, copper, gold and tungsten at 6–12 kV. The efficiency of production is then $3 \cdot 8 \times 10^{-4}$ for copper and $9 \cdot 6 \times 10^{-4}$ for tungsten, at 10 kV. For a beam current of 10μA the total power in the X-ray beam will be 38 and 96μW respectively at this voltage.

When the target is thinner than the electron range, (8.20) must be combined with the expression for energy loss (8.15), giving

$$E_c = kZI(V_1 - V_x)V_1 = ka_1 ZIx \left(\frac{V_1}{V_1 + V_x} \right), \qquad (8.22)$$

where x is the target thickness. The energy efficiency is then

$$\eta_{ce} = ka_1 Zx/(V_1 + V_x) \qquad (8.23)$$

being directly proportional to target thickness and roughly inversely proportional to beam voltage. Exposure time in photographic recording is related to the total power in the X-ray beam, and it is somewhat surprising that E_c is in these circumstances almost independent of V_1, (8.22), so long as the target is thinner than the range at the minimum voltage used.

The number of quanta produced per electron, the quantum efficiency η_{cq}, is of greater importance than the energy efficiency wherever the X-rays are recorded by counters rather than by a photographic emulsion. Even in the latter case, evidence is in favour of the effect of the beam depending on the quantum intensity and not on the total energy (Dyson, 1957). It is known from experiment (Compton & Allison, 1935) that the intensity of a characteristic line follows a power law in which the exponent is 1·65 in good approximation (cf. (8.28)). It has been shown by Duncumb (1957 a) that a similar relation holds for the number of quanta n_c emitted in the continuous spectrum from the high-frequency limit set by the beam voltage V_1 (8.19) to the lower limit, corresponding to a voltage V_c, beyond which quanta either do not escape from the target or fail to produce a response in the recording system:

$$n_c = 0\cdot40(eKZI/h^2)\,(V_1-V_c)^{1\cdot65}/V_c^{0\cdot65}, \tag{8.24}$$

where K is a constant, e the electronic charge, h Planck's constant. This expression fits the experimental results to within 5 % over a range of 2–8 in the value of the ratio V_1/V_c, and to within 20 % over the range 1·5–20. The variation in n_c with voltage is shown in Fig. 8.5. The number of electrons falling on the target is I/e, so that the quantum efficiency η_{cq} in the continuous spectrum is

$$\eta_{cq} = 0\cdot40(e^2/h^2)KZ(V_1-V_c)^{1\cdot65}/V_c^{0\cdot65} \tag{8.25}$$

$$= 1\cdot04\times10^{-9}Z(V_1-V_c)^{1\cdot65}/V_c^{0\cdot65}. \tag{8.26}$$

From Fig. 8.6 it is seen that the quantum efficiency is greater than the energy efficiency except at low voltage, close to V_c. The value of the latter is somewhat arbitrary, depending on the experimental arrangement and particularly whether a massive or a transmission target is used. Taking a value of 2 kV (cf. Fig. 8.5), the values of the quantum efficiencies for copper will be 36×10^{-4} and $6\cdot1\times10^{-4}$ quanta per electron at 25 kV and 10 kV respectively; the corresponding values for the energy efficiency are $9\cdot6\times10^{-4}$ and $3\cdot8\times10^{-4}$.

8.5.2. *The characteristic spectrum.* The contribution of the line to the total spectral energy has been assumed small in the above discussion. In practice it is always less than that from the continuous, but the ratio rises as the target voltage is raised above the minimum needed to excite a given line. We shall consider only the energy in the $K\alpha$-line, as that in the $K\beta$-line is some ten times less and that in the L lines smaller still, since they are much more heavily absorbed in the target itself. In first approximation experiment establishes the simple relation (Compton & Allison, 1935):

$$E_k \propto (V_1-V_k)^{1\cdot65}$$

for the energy in the $K\alpha$-line, if V_k is the minimum voltage needed for its excitation and V_1 is the beam voltage. On theoretical grounds (Dyson, 1956) the relation is

$$E_k = AZV_1V_kI[\log_e(V_1/V_k)-(1-V_k/V_1)]Rw_\alpha, \qquad (8.27)$$

where A is a constant, I is the beam current, w_α the fluorescent yield for $K\alpha$ radiation, and R the ratio of the probability for K

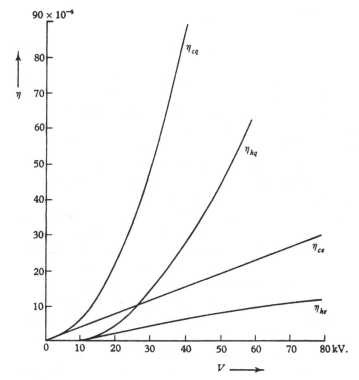

Fig. 8.6. Efficiency of X-ray emission for copper in function of voltage. η_{ce} and η_{ke} = energy efficiencies for the continuous and characteristic radiation, respectively; η_{cq} and η_{kq} = quantum efficiencies of the same.

ionization to that for the production of a quantum of energy greater than V_k in the continuous spectrum. The value of R varies somewhat with atomic number and is equal to 3·5 for copper; it has been shown to be constant over a wide range of voltage (Stoddard, 1934). Comparison with experimental measurements shows that (8.27) can be closely approximated, to the accuracy quoted in

230 X-RAY MICROSCOPY

connexion with the continuous spectrum, by the following expression (Duncumb, 1957a):

$$E_k = 1 \cdot 04 \times 10^{-9} ZIRw_\alpha (V_1 - V_k)^{1 \cdot 65} V_k^{0 \cdot 35}. \tag{8.28}$$

Comparison of (8.28) with (8.21) now gives the ratio of the energy in the $K\alpha$-line to that in the continuous spectrum as a whole. For copper, with the value of $0 \cdot 395$ for w_α (cf. Dyson, 1956), we have

$$E_k/E_c = 1 \cdot 1 V_k^{0 \cdot 35} (V_1 - V_k)^{1 \cdot 65}/V_1^2$$

$$= 1 \cdot 1 (U - 1)^{1 \cdot 65}/U^2 \tag{8.29}$$

on writing $U = V_1/V_k$. Hence the energy in the $K\alpha$-line is less than that from the continuous spectrum as a whole. For copper $V_k = 9$ kV, so that the ratio is $0 \cdot 009$ at 10 kV, $0 \cdot 38$ at 25 kV, and reaches a very flat maximum of $0 \cdot 45$ at about 60 kV, if the validity of (8.28) can be trusted so far. Assuming a $1 \cdot 5$ instead of a $1 \cdot 65$ power law, Guinier (1952) obtained a rather smaller value for the ratio. In so far as the derivation of (8.29) rests on experimental values for this power law, for R and for w_α, it must be regarded as an empirical relation. In practice, the measured values for the ratio E_k/E_c have to be corrected for differential absorption in the path between target and recording apparatus, and for variation of recording efficiency with wavelength. Possibly because of differences in the assumptions made in these corrections, and in the experimental conditions, the values reported for the ratio by different authors differ widely. That of Parrish & Kohler (1956) for copper, $0 \cdot 35$ at 30 kV, agrees well with (8.29), but Bendit (1958) finds a value of $2 \cdot 2$ and Arndt & Riley (1952) one of $5 \cdot 4$, which would correspond to an efficiency of $K\alpha$ production nearly an order of magnitude greater than that given by (8.27). The difference between the energy efficiency and the quantum efficiency is too small to affect the comparison appreciably (cf. (8.29) and (8.31)).

On the whole, it must be concluded that even at the optimum operating voltage the output of K radiation from a target will not exceed the integrated intensity of the continuous spectrum. The variation of the ratio with voltage shows that little is gained by operating at more than about four times the excitation voltage V_k.

Since the cost and insulation troubles of the apparatus rise rapidly with voltage, it is usual to operate an X-ray tube with copper target at 30–35 kV for the $K\alpha$-line. The optimum value of voltage ratio U is almost independent of the nature of the target element, but the optimum operating voltage $(U \times V_k)$ will rise with atomic number, since $V_k \propto (Z-1)^2$ according to Moseley's rule.

The ratio of the energy efficiency for the $K\alpha$-line η_{ke} to that for the continuous spectrum η_{ce} is the same as the ratio of the actual energies, so that the value of η_{ke} can be found by substituting (8.20) into (8.29). The calculated value is $3 \cdot 3 \times 10^{-4}$ at 25 kV and $0 \cdot 080 \times 10^{-4}$ at 10 kV for copper (cf. Fig. 8.6). The quantum efficiency η_{kq} for $K\alpha$ production is a more important factor; with n_k and n_e for the numbers of $K\alpha$-quanta and electrons respectively, it is given by

$$\eta_{kq} = n_k/n_e = \eta_{ke}V_1/V_k$$
$$= 1 \cdot 04 \times 10^{-9}ZRw_\alpha V_k(U-1)^{1 \cdot 65}. \qquad (8.30)$$

For copper it has the values $9 \cdot 2 \times 10^{-4}$ and $0 \cdot 09 \times 10^{-4}$ quanta per electron at 25 kV and 10 kV respectively, and these agree well with the experimental measurements made by Worthington & Tomlin (1956) and Dyson (1956, 1959b).

The number of $K\alpha$-quanta is thus smaller than the total in the continuous spectrum, at given voltage. The ratio for copper, which is also that of the quantum efficiencies for the two processes, is

$$n_k/n_c = \eta_{kq}/\eta_{cq} = 1 \cdot 35 \left\{\frac{V_1 - V_k}{V_1 - V_c}\right\}^{1 \cdot 65} \left\{\frac{V_c}{V_k}\right\}^{0 \cdot 65}. \qquad (8.31)$$

It has the value $0 \cdot 014$ at 10 kV, $0 \cdot 26$ at 25 kV, and $0 \cdot 40$ at 45 kV. Unlike the relative energy efficiency it does not reach a maximum (within the present approximation), but tends to a value of $0 \cdot 5$ at very high voltage. The ratio of the number of K quanta to that in a narrow band of the continuous spectrum at the same wavelength is discussed in § 7.1.

It follows from this discussion that in all circumstances the efficiency, whether measured in terms of energy or quanta, remains of order 10^{-4} to 10^{-2}, being higher for elements of high Z and at high voltage. The quantum efficiency can be increased appreciably if experimental conditions are arranged to record quanta of energy less than the cut-off limit of 2 kV assumed above. It must be noted, however, that even if all quanta can be utilized to a voltage of 0·2 kV, beyond the carbon $K\alpha$-line, the increase in the number of quanta will only be four times over that given by (8.24). In addition, the output at low voltage appears to fall off more rapidly than assumed above (cf. Stephenson, 1957).

As there are approximately 6×10^{18} electrons per coulomb, a quantum efficiency of 10^{-3} quanta per electron represents an emission of 6×10^{15} quanta per sec. for a beam current of 1 A, or 6×10^9 per sec. for $1\,\mu\mathrm{A}$. These are emitted at all angles from the target, almost isotropically in most conditions of practical interest (see below), and they cannot readily be focused into a directed beam as is possible with electrons. For a system of known aperture of collection the important quantity is thus the emission per unit solid angle. A total emission of 6×10^9 quanta corresponds to approx. 5×10^8 quanta per steradian, of all energies. Usually only a narrow energy band, in a narrow solid angle, will be collected in quantitative microanalysis. Long (1959) has evaluated the output from targets of aluminium, copper and gold, at 10, 20 and 30 kV, in terms of the number of quanta per sec. per $1°$ aperture per $\mu\mathrm{A}$ per electron volt bandwidth.

8.6. Angular distribution

The variation of X-ray intensity with angle, measured from the forward direction of the incident beam, has been investigated for thin targets by Kuhlenkampff and his school (cf. Stephenson, 1957), for moderately thin targets by Dyson (1956, 1959a; Cosslett & Dyson, 1957) and for massive targets by Bouwers & Diepenhorst (1928), and Oosterkamp & Proper (1952). A minimum is found at 0°, with a rise towards larger angle and a subsequent fall as self-absorption in the target becomes appreciable. The relative magnitude of these effects depends on the electron voltage and the target material, and are different for different wavelengths in the emitted beam. Classical theory would predict zero radiation in the continuous spectrum at 0° and a maximum at 90°, after correction for absorption. The quantum mechanical theory, proposed by Sommerfeld (1931) and worked out in detail by Kirkpatrick & Wiedmann (1945) for thin targets, gives a distribution much more closely in accord with experiment (Amrehn & Kuhlenkampff, 1955). The characteristic spectrum is isotropic in distribution.

Typical polar distributions found by Dyson for the continuous spectrum from electron opaque targets of gold and aluminium are shown in Figs. 8.7 and 8.8, for two voltages and for a number of frequencies down to about half the limiting frequency. The upper

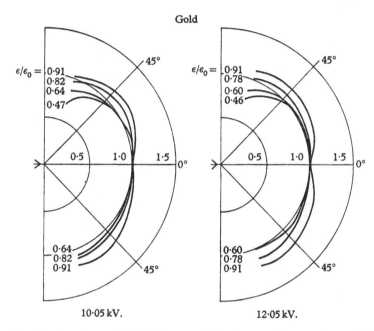

Fig. 8.7. Angular variation in X-ray emission from a gold target ($0.21\,\mu$ thick) at two voltages and for a number of emitted frequencies (expressed as fractions of the maximum frequency, ν_x: $\epsilon/\epsilon_0 = \nu/\nu_x$). Upper quadrant: observed intensity; lower quadrant: corrected for absorption in the target. (Cosslett & Dyson, 1957.)

quadrants show the experimental values, which in the lower quadrants have been corrected for absorption in the target. Measurements made also with beryllium and tungsten confirmed that there is a systematic increase in anisotropy with decreasing atomic number. For gold and tungsten, X-rays are emitted almost isotropically. For aluminium the intensity at 60° is about 60% greater than in the forward direction; for beryllium the increase is about 100%. With still thinner targets the anisotropy is still greater and persists even for the softer components, which are virtually isotropic in electron opaque conditions.

The angular distribution from massive targets is very similar to that shown for moderately thin targets, but is, of course, only available on the front of the target. It is almost isotropic, except at very high angles where absorption becomes great. Since X-ray illumination is a volume and not a surface effect, it does not obey

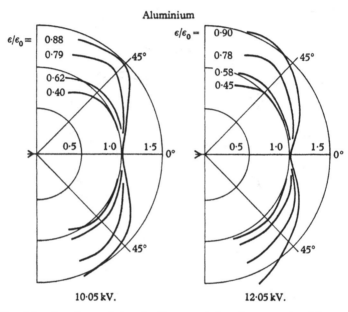

Fig. 8.8. Angular variation in X-ray emission from an aluminium target ($2 \cdot 9\mu$ thick) at two voltages and for a number of emitted frequencies (expressed as fractions of the maximum frequency ν_x: $\epsilon/\epsilon_0 = \nu/\nu_x$). Upper quadrant: observed intensity; lower quadrant: corrected for absorption in the target. (Cosslett & Dyson, 1957.)

Lambert's law ($I \propto \cos\theta$) as does that from a light source. Hence an increase in effective brightness (intensity/projected area) can be obtained by viewing the surface of an extended emitter at an angle (Fig. 8.9). For a given cross-section of the issuing beam, the brightness will then be proportional to $1/\cos\theta$ up to the limit determined by absorption. For this reason a focal line rather than a focal spot is preferred in tubes with massive targets, the accepted beam issuing at a mean angle of about $10°$ to the surface. This arrangement also provides slightly better heat dissipation in the target as compared with a round or square spot (cf. de Graaf & Oosterkamp, 1938). With transmission targets, however, as used

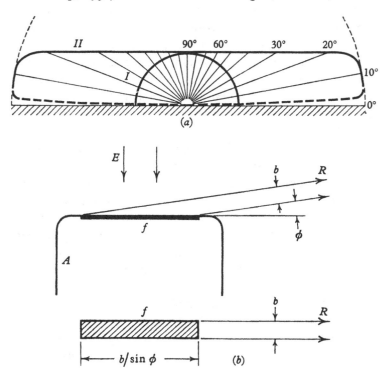

Fig. 8.9. (a) Angular variation of brightness of emitter (intensity/projected area) for surface radiation (I) and volume radiation (II). (b) Foreshortening of line focus (width b) by observation at angle φ to surface. If the length of the focus is b/sin φ, the observed focus will be a square of side b. (de Graaf & Oosterkamp, 1938.)

in projection microscopy, little advantage is gained by siting the specimen off the axis of the incident electron beam so long as the target is electron opaque. If the target thickness is equal to the electron range there is a case for using a line focus with foreshortening, especially for microdiffraction (cf. § 14.1). In future high-resolution work, where a very thin target may have to be used to minimize electron diffusion, a considerable advantage would accrue from placing the specimen at 60° to the forward direction.

8.7. Filtration and other means of monochromatization

It is often necessary, especially for absorption spectrometry, to select a particular wavelength (in practice, a narrow band of wavelengths) from the spectrum issuing from a target. The simplest method is to use a thin foil as a filter; for a higher degree of monochromatization correspondingly more complicated arrangements of polished or crystal reflectors must be used. It is important to obtain as high an intensity as may be compatible with the desired purity of the selected radiation.

8.7.1. *Filters.* A thin foil will appreciably narrow the wave-band in the transmitted beam, if it is composed of an element having an absorption edge suitably situated with respect to the peak of the spectrum issuing from a given target. Much of the radiation of wavelength shorter than that of the edge will be eliminated (Fig. 8.10); that in a narrow band beyond the edge will be almost completely transmitted, as the absorption coefficient falls to a very small value (cf. Table A. 1, in the Appendix). When the target voltage is high enough for a characteristic line to be excited with appreciable intensity, the material of the filter foil will be chosen with respect to the line wavelength. In many cases an element is available with an absorption edge which will suppress the $K\beta$- but pass the $K\alpha$-line. Tables showing the element and thickness best suited to a number of the most commonly used target materials are given by Glocker (1958) and by Peiser, Rooksby & Wilson (1955). With a copper target, for instance, a nickel foil $8\cdot5\mu$ thick will reduce shorter wavelengths to 11 % or less, whilst reducing the $K\alpha$-line only to 66 % of its initial intensity. By combining the selective recording properties of a proportional counter with such a filter, Arndt & Riley (1952) were able to obtain a beam in which 99 % of the transmitted energy was in the copper $K\alpha$-line.

In microspectrometry a very narrow band of wavelengths must on occasion be selected from the continuous spectrum, when no line is conveniently placed. Ross has developed a method of doing so by successive use of foils of materials differing by one in atomic number, the 'balanced filter' method (cf. Kirkpatrick, 1939, 1944). Küstner (1931, 1932) devised a different method, using a single filter in two different positions successively, to eliminate the background of scattered continuous radiation from selected fluorescent line spectra.

8.7.2. *Self-filtration.* As pointed out in § 2.2, the characteristic emission lines of an element always fall at a slightly longer wavelength than the corresponding absorption edge. A target is thus

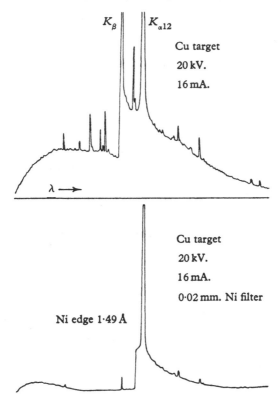

Fig. 8.10. Monochromatization of radiation from copper target by means of a nickel filter 0·02 mm. thick. (Wallgren, 1957.)

highly transparent to its own line radiation, additionally so because some of the absorbed energy will be re-emitted as fluorescent radiation at the line wavelength. By using a copper target some twenty times thicker than the electron range, Bessen (1957 a, 1959) has obtained a remarkably high degree of monochromatization (Fig. 8.11).

8.7.3. *Total reflexion.* Total external reflexion of X-rays at a solid surface is highly wavelength dependent (§ 4.1). Ehrenberg & Spear (1951 *b*) therefore suggested its use to select a soft wave-

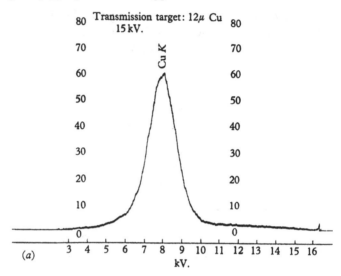

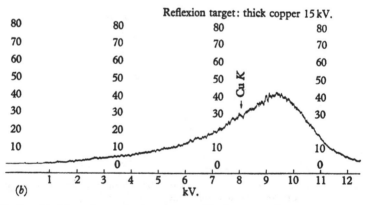

Fig. 8.11 (*a*). Monochromatization of copper *K* radiation (1·5 Å) by 'self-filtration' in a transmission target of copper 12 *µ* thick, compared with (*b*) the broad band obtained from a solid target. (Bessen, 1957 *a*.)

length from the spectrum issuing from a tube operated at high voltage. In this way high intensity is obtained in the desired band, since the initial intensity depends on V_1^2 (8.21) or on $(V_1 - V_x)^{1\cdot65}$ in the case of a line (8.27), and the efficiency of reflexion is high for

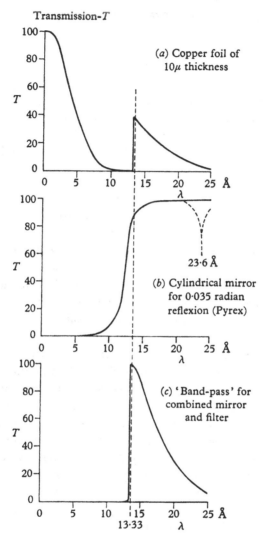

Fig. 8.12. Monochromatization of copper L radiation (13.3 Å) by a combination of a copper filter (10μ thick) with total reflexion from a glass mirror. (Henke, 1957.)

a particular wavelength and has a sharp angular cut-off (cf. Figs. 4.2 and 4.3). The hard component penetrates the reflector and is lost. Ehrenberg & White (1957) were able to obtain by this means contact microradiographs of biological specimens showing high contrast, because of the soft radiation, and yet at reasonably short exposure in spite of the low gathering power of a plane mirror.

Henke (1957) has used a combination of filtration and total reflexion to obtain a highly monochromatic beam of X-rays from a copper target. The tube was operated at voltages up to 10 kV, compared with the minimum of 950 V needed to excite the $L\alpha$-line, and an intense beam with peak at the wavelength of the latter was obtained (Fig. 8.12). In order to collect as large a cone of radiation as possible, he used reflexion from the interior of a capillary tube instead of a plane surface.

8.7.4. *Crystal monochromators.* The method usually used for monochromatization is that of diffraction from a crystal lattice, which is highly selective according to the Bragg relation (§ 5.1). A plane surface will suffice if the initial beam is of high intensity, but in the interests of exposure-time it is preferable to use a curved crystal. For greater efficiency of focusing, doubly curved crystals are now often used, although even then the fraction of the total X-ray beam actually collected is still small (cf. Furnas, 1957). Most of the types of curved crystal discussed in chapter 5, as imaging systems, can be utilized as monochromators; it is only necessary to adjust the curvature, and the relative position of focal spot, crystal and camera, so that an image of the spot (or an aperture in front of it) is formed at the position of the photographic emulsion (or counter). Accounts of the experimental details of the most frequently used systems will be found in the standard texts (Guinier, 1956; Sandström, 1957; Peiser, Rooksby & Wilson, 1955). In the soft wavelength region ruled gratings have to be used, because crystals with large enough spacings to produce diffraction at reasonably small angles are difficult to obtain (Skinner, 1938; Tomboulian, 1957).

Note

Additional material relevant to this chapter will be found on p. 369.

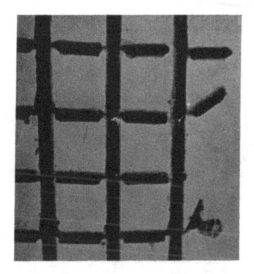

(A) Projection microradiograph of 1500 mesh/in. silver grid, showing Fresnel fringe. Initial magnification, ×150; reproduced at ×1100. Taken at 7 kV, 5 min. exposure, with 0·1 μ gold target. (Nixon, 1955 b.)

(B) Micrograph of silver grid (1500 mesh/in.) taken with the two-mirror X-ray microscope. Initial magnification ×7; reproduced at ×230; wavelength 8·3 Å; exposure 20 min. (McGee, 1957.)

PLATE II

Compound mirror block with cover removed. The specimen is placed at the top of the block and is adjusted with the upper two micrometers. Each of the four remaining micrometers focuses one of the mirrors. (Kirkpatrick & Pattee, 1953.)

PLATE III

Reflexion X-ray microscope of Kirkpatrick and Pattee, incorporating two pairs of mirrors. (Kirkpatrick & Pattee, 1953.)

PLATE IV

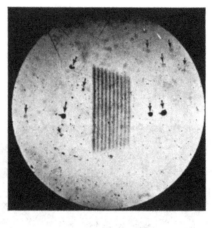

(a)

(b)

X-ray images from the curved crystal microanalyser: (a) Test sample of alternate iron and copper laminations, 100μ wide, taken in the optimum position of Fig. 5.4 with Cu $K\alpha$ radiation; reproduced at $\times 6\cdot 70$. (b) Sample of ore imaged (left) with Cu $K\alpha$ radiation (at $\times 4$), revealing distribution of copper over the surface, and (right) with visible light for comparing resolving power. (von Hámos, 1953.)

PLATE V

(A) Scanning microphotometer and data analyser for contact microradiography.
(Hallén & Hydén, 1957.)

(B) Bone microtome using a 125 μ carborundum wheel rotating at 7000 rev./min.
to cut sections down to 25 μ thick. (Clark & Iball, 1957.)

PLATE VI

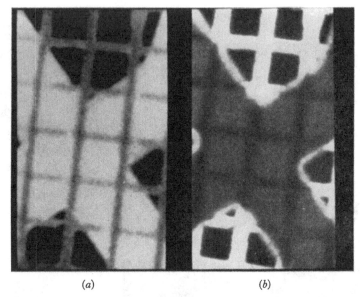

(a)   (b)

(A) Images of copper grid (200 mesh/in.) and silver grid (800 mesh/in.) formed by scanning microanalyser at selected X-ray wavelengths: (a) with copper K radiation, and (b) with silver L radiation. Photographed at ×400; reproduced at ×400. (Duncumb & Cosslett, 1957.)

(B) Scanning microanalyser at Tube Investments Research Laboratories, Hinxton, Cambridge.

PLATE VII

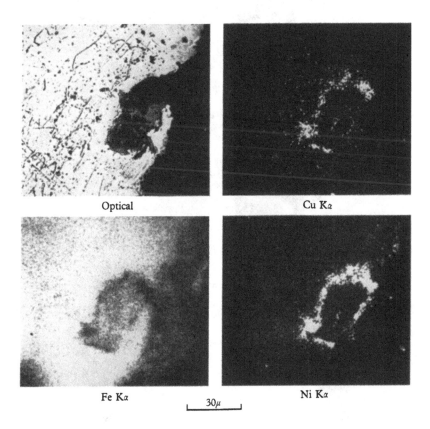

Surface of a mild steel, examined by optical and X-ray-scanning microscopy.
The X-ray micrographs, taken with characteristic line radiations, show the
distribution of copper, iron and nickel respectively. Reproduced at ×500.
(Melford & Duncumb, 1958.)

PLATE VIII

(A) Philips 5 kV tube for contact microradiography.

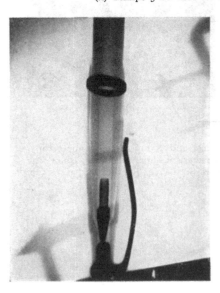
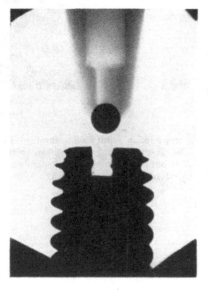

(a)   (b)

(B) Radiographs of a ball-point pen at 50 kV, photographed from the fluorescent screen at 10 cm. distance: (a) Contact image; (b) with initial X-ray magnification of ×8. (Nixon, 1956b.)

PLATE IX

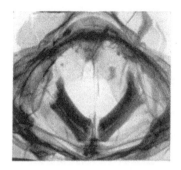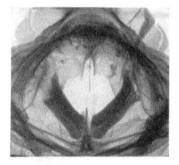

(A) Stereographic X-ray micrographs obtained by tilting the specimen, an air-dried cicada abdomen. (Nixon, 1956a.)

(B) General Electric projection X-ray microscope with electrostatic lenses. (Courtesy of General Electric Company, Milwaukee.)

PLATE X

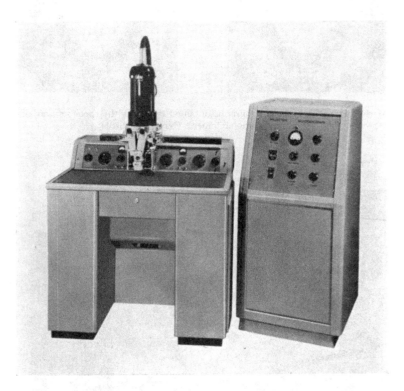

Philips projection X-ray microscope, with magnetic lenses, a modification of the EM 75 electron microscope. (Bessen, 1957a.)

PLATE XI

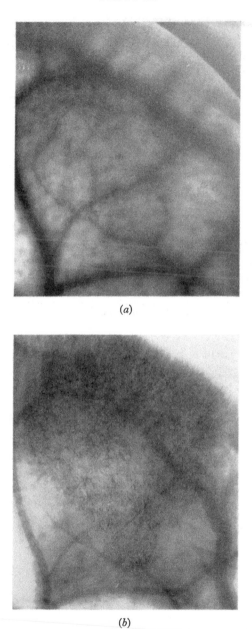

(a)

(b)

Living rabbit ear after injection of contrast medium into the blood stream (a) just before freezing, (b) 50 min. after freezing (projection microradiograph × 5). (Saunders, unpublished, 1957.)

PLATE XII

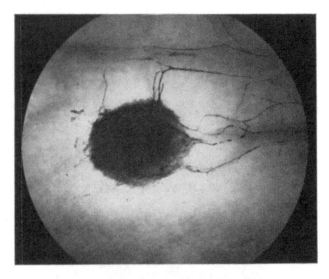

(A) Rabbit ear showing lymphatic drainage. The dark area represents the region injected, and the drainage channels lead away from this region; note the non-return valves in each tube (projection microradiograph ×5). (Saunders, unpublished, 1957.)

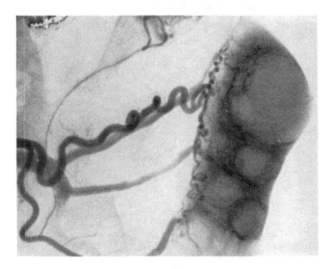

(B) Living rabbit ovary showing the blood vessel distribution (contact microradiograph). (Engström, 1956.)

PLATE XIII

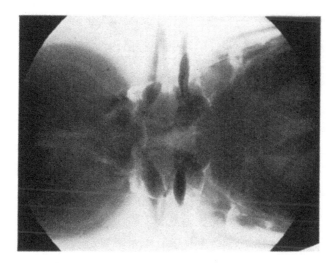

(A) Undried beetle, showing contrast between the air-filled trachael network and the tissues with high water content (projection microradiograph × 20; copper target at 20 kV). (Smith, 1957.)

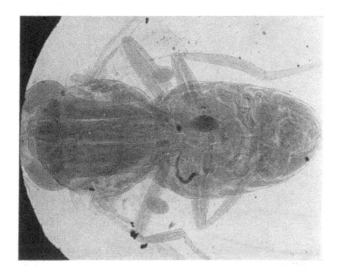

(B) Freeze-dried fruit fly (*Drosophila melanogaster*) with contrast determined by the different thicknesses and densities of material (projection microradiograph × 47; tungsten target at 10 kV). (Cosslett & Nixon, 1952*b*.)

PLATE XIV

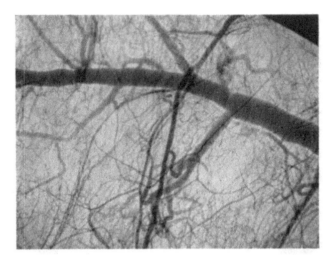

(A) Shunt between an artery and a vein in rabbit ear (projection microradiograph ×20). (Saunders, unpublished, 1957.)

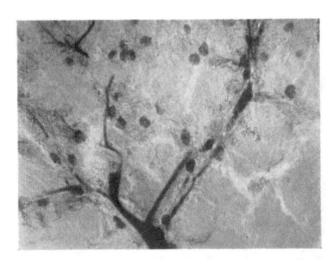

(B) Thin section of rat kidney after micropaque injection and freeze drying. Small capillaries and glomeruli appear black (projection microradiograph ×30; 15 kV; 3 min. exposure). (Saunders, unpublished, 1957.)

PLATE XV

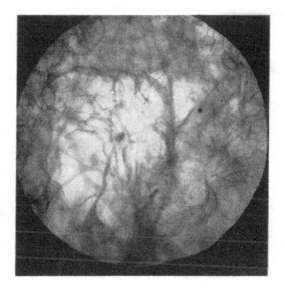

(A) Human lung, air dried, showing air sacs and septa (projection microradio-graph × 10, taken at 15 kV with 5 min. exposure). (Saunders, unpublished, 1957.)

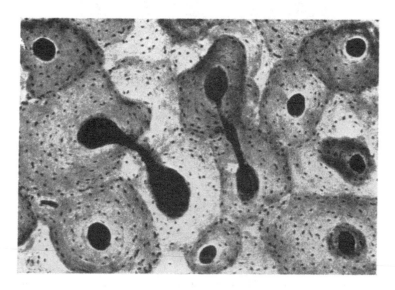

(B) Thin section of bone showing distribution of mineral salts (contact microradiograph ×95). (Engfeldt, 1954.)

PLATE XVI

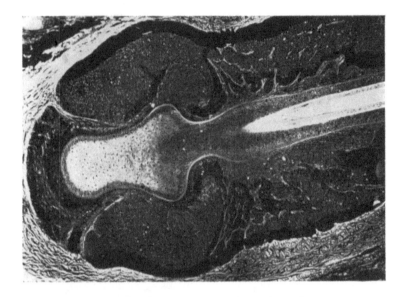

(A) Section through hair follicle of the rat, showing mass distribution (contact microradiograph × 115). (Engström, Lundberg & Bergendahl, 1957.)

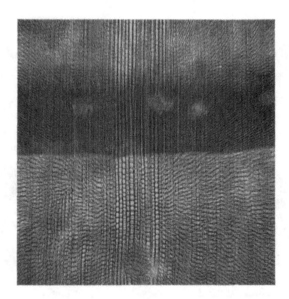

(B) Pine wood growth rings. The dark band shows the denser summer wood compared to the lighter band formed when the cells are growing faster during the spring (projection microradiograph × 30). (Jackson, unpublished, 1956.)

PLATE XVII

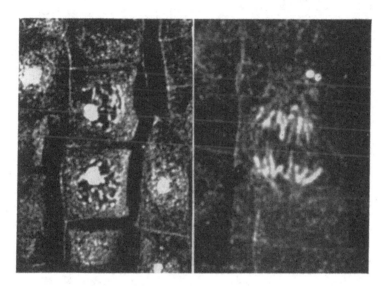

Onion root tip cells, showing chromosomes (contact microradiograph 1 kV, on Kodak MR plate, reproduced at ×450 (left) and ×670 (right)). (Engström & Greulich, unpublished.)

PLATE XVIII

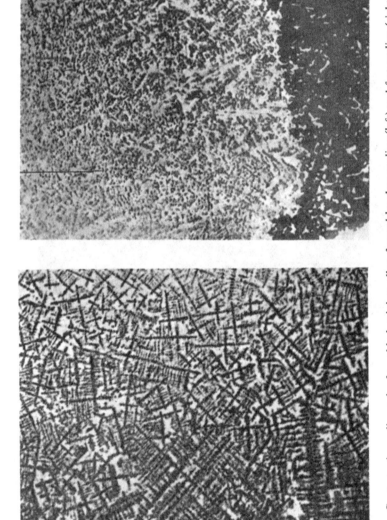

Contact microradiograph of a gold-aluminium alloy, formed by slow cooling (left) and fast cooling (right). Magnifications left × 5, right × 3·5. (Heycock & Neville, 1900.)

PLATE XIX

Manganese sulphide inclusions in cast iron. With cobalt radiation (left) they appear white; with copper radiation (right) they are not seen. Graphite flakes appear dark in both micrographs (contact microradiograph ×120). (Andrews & Johnson, 1957.)

PLATE XX

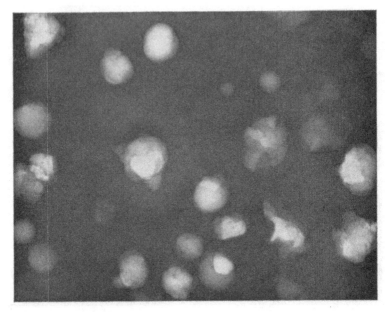

(A) Nodular cast iron (3 % carbon). The light areas are graphite nodules about 25μ in diameter (projection microradiograph, × 280, 20 kV, 5 min. exposure). (Nixon & Cosslett, 1955.)

(B) Iron ore sinter showing dendrites of iron oxide growing in a slag pool (contact microradiograph × 250). (Andrews & Johnson, 1957.)

PLATE XXI

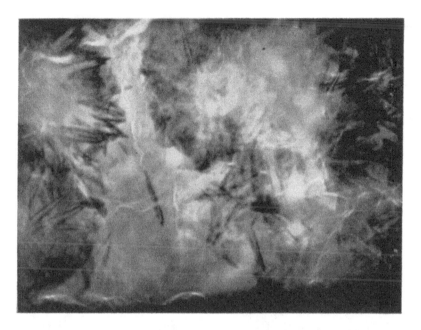

(A) Thucolite ore containing pitchblende and hisingerite. The pitchblende appears dark, the hisingerite grey and the white lines are cracks in the 25μ section (projection microradiograph $\times 500$; 20 kV; copper target; 4 min. exposure). (Jackson, 1957c.)

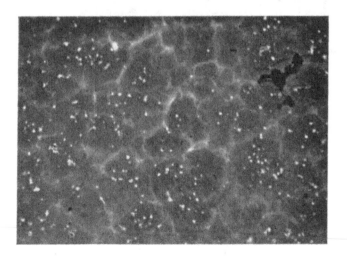

(B) Cast magnesium alloy, showing precipitates and grain boundary segregation; cobalt radiation (contact microradiograph $\times 100$). (Sharpe, 1957a.)

PLATE XXII

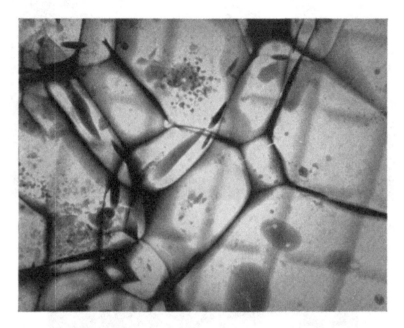

Aluminium, 5 % tin alloy, ½ mm. thick. The tin appears dark and has mostly been drawn to the grain boundaries, in which a few small cracks (white) are visible (projection microradiograph × 400; 20 kV; 5 min. exposure). (Nixon, 1956*a*.)

PLATE XXIII

(A) Porous bronze (95 % copper, 5 % tin). The tin appears in dendritic form and small white circles show the position of voids within the darker tin (projection microradiograph × 1000; 20 kV; 10 min. exposure). (Nixon, 1956*a*.)

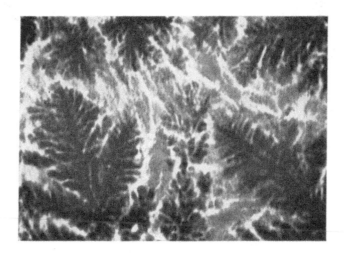

(B) Diffusion zone at an aluminium-brass interface (contact microradiograph × 140; manganese radiation). (Sharpe, 1957*a*.)

PLATE XXIV

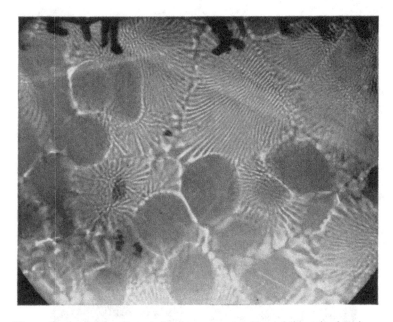

Zinc, 5 % aluminium alloy; dark areas represent zinc and the striped regions a eutectoid mixture of the two (projection microradiograph × 320; 20 kV; 4 min. exposure). (Nixon & Cosslett, 1955.)

PLATE XXV

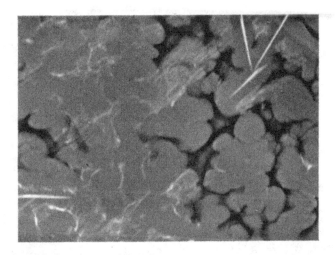

(A) Cast aluminium alloy showing Al_3Ti platelets, intermetallic phase and segregation (contact microradiograph, × 140; manganese radiation). (Sharpe, 1957a.)

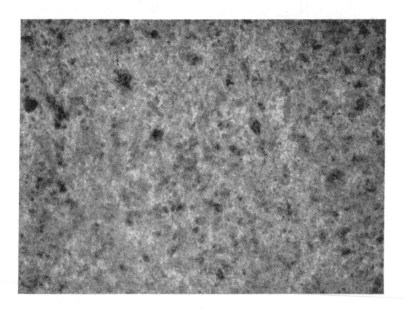

(B) Paper containing $CaCO_3$ and TiO_2. Exposure-time 2 min. at 3 cm. film distance and × 18 X-ray magnification; reproduced at × 95 (projection microradiograph). (Isings, Ong, Le Poole & van Nederveen, unpublished, 1957.)

PLATE XXVI

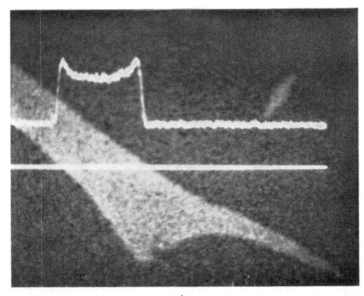

(a)

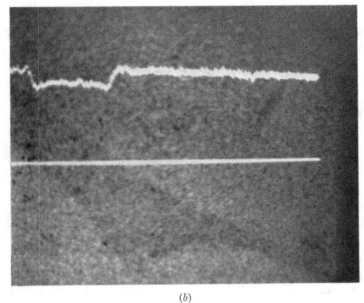

(b)

Variation in concentration (upper line) of (a) nickel, and (b) iron in a meteorite, along a path (lower line) crossing a taenite inclusion (initial magnification, ×250, reproduced at ×460). (Agrell & Long, 1960a.)

PLATE XXVII

(A) Foraminiferon, showing central chamber, radiating septa and marginal cord (contact microradiograph × 21). (Hooper, 1958.)

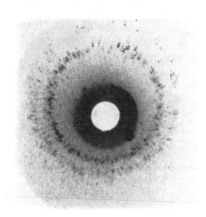

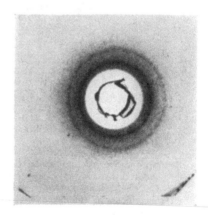

(B) Transmission spotty ring pattern from a copper foil 3μ thick. (Nixon, 1957.)

(C) Back-reflexion spotty ring pattern from the same specimen as in B. (Nixon, 1957.)

PLATE XXVIII

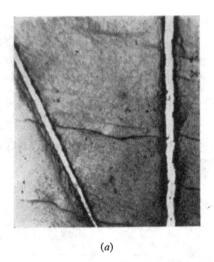

(a)

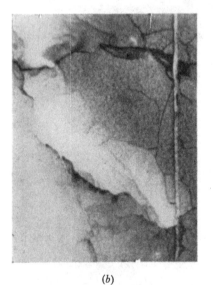

(b)

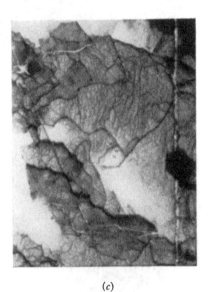

(c)

X-ray diffraction micrograph of the same area of an aluminium crystal after slow deformation for (a) 3 days at 300° C., 1·45 % elongation; (b) 8 days at 300° C., 3·8 % elongation; (c) 13 days at 300° C., 7·38 % elongation. All at × 16 magnification. (Honeycombe, 1951.)

PLATE XXIX

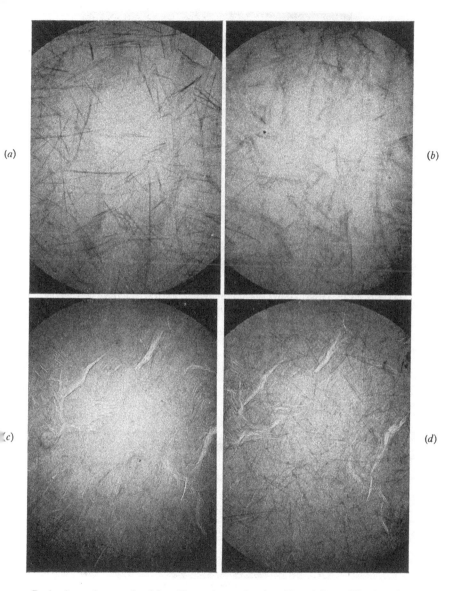

Projection micrograph of beryllium sheet, showing Kossel lines. The length and sharpness of the lines change when the metal is mechanically deformed and then recrystallized. (a) Initial condition; (b) after 2 % compression; (c) after 13 % compression; (d) after recrystallization. Original magnification, × 40 at front face and × 11 at rear face of specimen; enlarged 2·3 times in reproduction. (Dr J. Sawkill, Tube Investments Research Laboratory, Hinxton, Cambridge.)

PLATE XXX

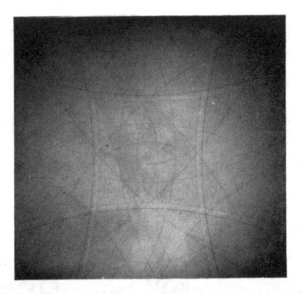

(A) Kossel lines from a thin flake of lithium fluoride with the 100 plane parallel to the plate; obtained with an X-ray source 1–2μ in diameter. (Dr J. V. P. Long and Dr S. B. Newman, unpublished.)

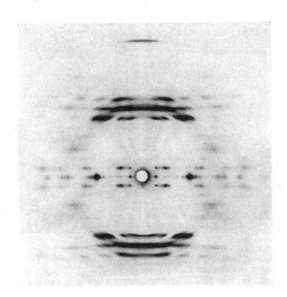

(B) Diffraction pattern from a fibre of bile acid and amino acid obtained with the Hilger micro-focus unit. (Dr A. Rich and Dr H. Davies, Nat. Inst. Health, U.S.A.)

PLATE XXXI

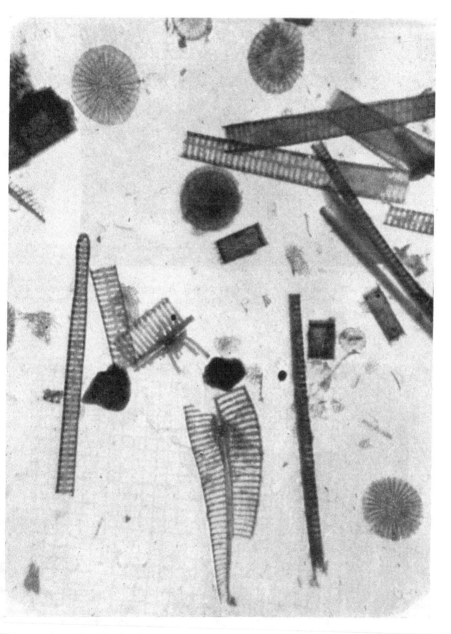

Diatoms imaged with the X-ray conversion microscope; exposure-time 3 min. (initial magnification, ×400; reproduced at ×1400). (Huang, 1957.)

PLATE XXXII

(A) X-ray projection micrograph of 1500 mesh/in. silver grid obtained by focusing with the method of Fig. 15.5. Target-film distance 1 cm., exposure-time 20 min. (Ong & Le Poole, 1958.)

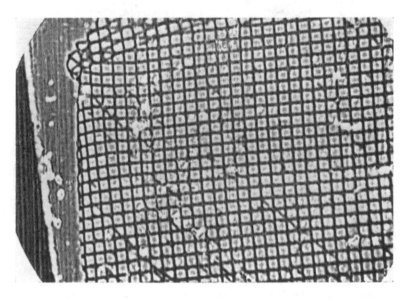

(B) Relief X-ray micrograph of 1500 mesh/in. silver grid, formed on a Saran plastic film; oblique illumination, exposure time 8 hr. (Pattee & Warnes, unpublished, 1956.)

CHAPTER 9

SPECIMEN PREPARATION TECHNIQUES

9.1. Biological and medical specimens

A wide range of specimens has already been studied with the
X-ray microscope, and we shall describe here only the main
preparative methods and the precautions to be observed in
carrying them out. A more detailed account of particular applications will be found in chapters 12 and 13. Many of the specimen
techniques of light and electron microscopy can be carried over
almost unchanged into X-ray work.

9.1.1. *Freeze-drying.* The water content must be removed from
most biological material in order to provide adequate contrast
between the tissue structures. Living specimens will be discussed
later (§ 9.1.6). Drying down under normal temperature and pressure causes considerable distortion and the resulting specimen,
although giving an excellent X-ray micrograph, unfortunately
bears little relationship to the original object. On the other hand,
freeze-drying removes the water with very little distortion and is
one of the best specimen preparation methods for X-ray microscopy. Errors introduced by other procedures have been discussed
by Nürnberger & Engström (1956). The freeze-drying method was
originally developed by Gersh (1932) and later adopted for electron
microscopy as an alternative to the more normal methods of fixation. Freeze-drying not only fixes the specimen, but also renders
it more transparent to X-ray radiation.

In practice the specimen is placed in liquid air, instantly frozen,
and then placed in a vacuum and held at about − 30° C. for several
hours while the water is removed by sublimation. The use of iso-
pentane at liquid air temperature avoids excessive boiling and
incomplete wetting, which cause the freezing of large specimens
in liquid air to be too slow. The length of time required varies
with the size of the specimen (approximately as the cube of the
diameter) being a few minutes for virus particles (for electron

C & N

microscopy), but several hours for specimens about 1 mm. in size and several days for objects of several mm. cube. These larger pieces are difficult to dry completely and the 1 mm. size is a better upper limit for normal use. The apparatus consists of a rotary mechanical oil-pump and a diffusion pump, evacuating a vessel that can be placed in a freezer and left running unattended day and night (Gersh, 1948; Smith, 1958). Once dried the specimen is kept in a desiccator and should not deteriorate over many months, depending on the completeness of the original drying. Rapid freezing, as opposed to freezing with ice or refrigeration, gives very much smaller ice-crystals and so less rupturing of the specimen. These effects are only just visible with the electron microscope and so far have not been seen with the X-ray microscope.

Another method of drying without distortion is the 'critical point' method introduced by Anderson (1953, 1956). It takes only an hour for very small specimens, but the equipment is more complicated than for freeze-drying; it is hardly to be recommended for X-ray microscopy at present.

The larger frozen-dried specimens can be supported across the holder, in the same way as undried specimens. Smaller objects can be frozen directly on the grid (plus film) that is used in the microscope and in this way there is no need to transfer the specimen after treatment.

9.1.2. *Embedding and sectioning.* Biological materials, such as large bones or whole organs, are too thick for X-ray microscopy and must be cut into smaller sections. Usable specimens for qualitative work can sometimes be produced by merely cutting with a razor blade through soft dried tissue, giving sections of 0·5 mm. or less. Here, as in all sectioning, it is most important to have the two sides of the section parallel so that the density across the negative is uniform. For thinner samples a normal hand microtome will cut down to several microns and some instruments, such as the Cambridge Rocker type, will cut down to 1 μ. It is not necessary to use the type of ultra-microtome developed for electron microscopy. In the X-ray microscope the penetration is so much greater that 1 kV radiation yields less contrast than would a 10 kV electron beam. The optimum thickness is related to the

X-ray wavelength (or kV) in the same way as in X-ray diffraction, where the sample must be thick enough to diffract some of the beam and yet not thick enough to absorb the beam appreciably after diffraction. Engström & Greulich (1956) have shown that an optimum thickness t_x exists for X-ray contrast in an openwork specimen (cf. § 2.3), in which all details have much the same absorption coefficient μ:

$$t_x = 1/\mu. \tag{9.1}$$

The absorption coefficient is related to the wavelength and in a practical case they find that for a 2μ section of biological material about 1 kV should be used on the X-ray tube.

The material for sectioning must be embedded in a solid medium, which is usually removed before the X-ray exposure is made. The classic materials, such as paraffin wax, have been used as well as the newer plastics introduced for electron microscopy. Examples of both types are given by Mosley, Scott & Wyckoff (1956). A 10μ section of frozen-dried mouse kidney, paraffin embedded, was cut on a conventional rotary microtome, deparaffinized and radiographed with a projection X-ray microscope. A 5μ section of decalcified human dentin, methacrylate embedded, was cut on a rotary microtome with a glass knife, the plastic removed and a thin formvar film used for support during the exposure.

Sections of harder material cannot be cut with a normal microtome and other methods have been developed. In one application for microradiography and autoradiography, Jowsey (1955) uses a milling machine with a 0·5 mm. cutter, at 1000 r.p.m. and a feed rate of 25μ/rev. The sections are cut to 100μ and can be ground down on glass to 50μ. The undecalcified bone samples were embedded in perspex, which was polymerized by heating. The large size of the cutter means that an equivalent amount of material is lost at each cut and it is difficult to obtain proper serial sections.

A thinner blade has been used by Clark & Iball (1955, 1957) for bone sections for X-ray diffraction studies, and this method should be the best for X-ray microscopy as well. They use a 125μ resin-bonded carborundum wheel, which is available commercially. The wheel is flexible when not rotating but becomes rigid at the operating speed of 7000 r.p.m. A photograph of this unit is shown in Pl. V B. The sections can be cut as thin as 25μ directly, so that no hand grinding is needed afterwards; with the thin wheel serial sectioning is quite feasible. The embedding medium in this case was a cold setting polyester, 'Marco resin', that will polymerize at 20° C. in a few hours.

Hand-grinding, without sectioning, is practicable with some specimens. The technique is similar to the methods used for metallurgical samples, as described in § 9.2.1. Improved procedures, allowing continuous control of thickness, have been developed by Hallén & Röckert (1958, 1960) and by Molenaar (1960).

9.1.3. *Injection and staining.* X-ray microscopy can be used to detect, in biological material, all the naturally occurring elements heavier than oxygen. Hence it is also possible to detect artificially introduced heavier elements, which are confined to one area or type of specimen structure. Many stains have been developed for light microscopy, with the emphasis on colour. With X-rays the need is for high absorption, and fortunately some suitable stains are found among the salts of heavy metals. Some of the early work on contact microradiography involved staining with gold salts, and high resolution was obtained (Lamarque, Turchini & Castel, 1937; Lamarque, 1938). More recently stain historadiography has been practised by Bohatirchuk (1957) by combining optical observation of a stained section with the microradiograph of the same area. The need for contrast in large-scale medical radiology has been met by the production of several excellent radiopaque media for injection. They are also of great value for living specimens and will be discussed here, although living organisms form the subject of a later section.

Osmium tetroxide has been used as a stain for contact microradiography by Recourt (1957b), who investigated its rate of penetration in a series of different external conditions, such as variations of temperature and concentration. Silver is selectively absorbed by nerve structures and should be a useful X-ray stain for such specimens. Lead acetate has been used with botanical subjects by Barclay & Leatherdale (1948): plants were left standing in a saturated solution for three days, to allow absorption into the vascular system without intracellular diffusion. This method is perhaps as much injection as staining but the uptake occurs over a longer time than the usual almost instantaneous injections. Lead naphthenate has been used for injection of the trachae of small insects, *parthenothrips dracaenae* (Nixon, 1952) following a vacuum injection technique of Wigglesworth (1950) developed for the light microscope. The insect is exposed to a vacuum of a few mm. Hg and then dipped into lead naphthenate while still under vacuum. The liquid runs into the very small trachae and other openings of the body that have been evacuated. The insect is then placed in an atmosphere of hydrogen sulphide, which produces a dense black precipitate, lead sulphide, easily seen in the light microscope. For X-ray microscopy it was unnecessary to reduce the naphthenate to the sulphide, since adequate X-ray contrast is given by the lead, no matter what its state of combination.

Direct injection into the blood vessels of living and dead tissue has been tried by many workers. A suction-injection method used by Saunders (1957a) has some similarity to the vacuum technique above. The method was first used by Kramer (1951) for injecting Indian ink into the tooth pulp vessels, after which the specimen was decalcified and cleared. For microradiography, the optically opaque Indian ink is replaced by X-ray opaque barium solutions. As the dental pulp vessels are extremely small, pressure injection could not be used to force in the contrast medium. Instead, a small hole was drilled in the top of the freshly extracted tooth, which was then placed in the contrast liquid, after the tip at the bottom had been cut off. Suction through the hole draws the medium into the minute vessels and then the overlying dentin and enamel are ground away, stopping short of the pulp cavity, before making the X-ray exposure.

Blood vessels in tissue and muscle have been directly injected with such contrast media as Thorotrast, Micropaque, Umbradil and other proprietary compounds used in radiology. With the displacement method the injection pressure is well above physiological values and may lead to changes in the vascular pattern or size of vessels. Injection into the living blood stream at normal pressures is less disturbing to the true nature of the specimen but the circulating blood supply dilutes the contrast and reduces the high resolution of the previous method. A number of specific techniques are given by Barclay (1951), who called this subject 'micro-arteriography', and by Bellman (1953) under the heading 'micro-angiography'.

Particle size of the contrast medium is important, since the smallest capillaries are about $1\,\mu$ in size and must be filled. Conversely, if the size of the particles is too small there is a tendency to diffuse through the walls of the vessels, giving unwanted contrast outside of the network. Some silver solutions will coagulate and block the blood stream and other injectants cause shock in the experimental animal, changing the vascular region from the normal condition.

9.1.4. *Mounting the specimen for contact microradiography.* The degree of care needed in mounting the specimen close to the photographic emulsion will depend on the resolution sought and on the intensity of the available X-ray beam. In all cases the mounting procedure, and any previous preparation of the emulsion, must be carried out in the darkroom; orange-red light (Wratten no. 1 safelight) can be used with M.R. plates. For many purposes an exposure-time of tens of minutes can be tolerated, so that it is enough to cover the emulsion with a light-tight layer of black paper, press the specimen against it and expose at a sufficiently great distance from the source to bring the geometrical blurring below the desired limit, according to (2.17). For high-resolution work, however, the specimen must be brought into more intimate contact with the emulsion (cf. § 2.4). With metallurgical or geological sections this usually offers less difficulty than with biological sections. In the method originally used by Lamarque (1938) the section, which may be as thin as a few microns, was transferred to a glass slide from the surface of the solvent used to remove the embedding material, and then to the emulsion by pressing the latter against the slide. After exposure the section was washed off in water and the emulsion developed. Barclay (1951) preferred to interpose a plastic film about $25\,\mu$ thick between section and emulsion, and found that 'Styrafoil S' (BX Plastics Ltd) had the best properties, being completely non-hygroscopic. The section was floated on to the surface of the emulsion, which had been previously soaked in water for 10 min. and covered with the plastic foil, and spread with a camel-hair brush. The highly electrostatic properties of the foil could be minimized by breathing gently on it before handling.

The standard technique employed in Engström's laboratory (Engström & Lindström, 1950; Lindström, 1955) is to mount the specimen on a plastic film spread across the aperture in the metal disk that serves to define the field and also, in quantitative work, carries the reference system for absorption calibration (Fig. 6.3). The specimen section is transferred to the previously prepared film in a drop of distilled water. The free side of the section may then be pressed directly against the emulsion. The method preferred by Mosley, Scott & Wyckoff (1957) is to transfer the section, after flattening on warm water, to the surface of absolute alcohol. From this it is picked up on the emulsion side of the photographic plate, which does not swell in alcohol as it does in water, and then firmly attached to the emulsion by slight warming over a hot plate. The emulsion is coated initially with a very thin layer of formvar, as used in preparing films for electron microscopy. Carr (1957), on the other hand, states that an intervening film can be dispensed with when frozen-dried preparations are used.

For high-resolution work, the emulsion may be coated with a very thin layer of nitrocellulose (about 1000 Å thick) by dipping the plate in a 1 % solution of celloidin in ether-ethanol for 20 sec. and then drying vertically for at least ½ hr. (Greulich & Engström, 1956). This layer both protects the emulsion from mechanical damage and aids the removal of the section after exposure. The following revised procedure is due to Greulich, Bergendahl & Ekblad (1957). It allows eventual recovery of the tissue for ordinary histological examination, thereby increasing the scope of the microradiographic studies. The solutions and solvents employed should be *absolutely* dust-free. All but the first two steps are carried out under the illumination of an appropriate darkroom safelight.

Step 1. An ordinary paraffin section is freed of wrinkles by floating it on warm distilled water in the routine manner. With the aid of a fine sable brush, the section is then transferred on a clean glass slide to cold distilled water.

Step 2. The section is then floated upon successive changes of

50% and 70% ethanol, and finally upon absolute ethanol. Some care must be exercised to keep the section at the surface of the absolute ethanol for subsequent mounting on the emulsion.

Step 3. Fine-grained photographic plates (Kodak Ltd, Maximum Resolution Plate, or Eastman Kodak Spectroscopic Plate, no. 649) are cut to the size required by the X-ray unit, usually 1–2 cm. square. These pieces are coated with a thin layer (0·1 μ) of nitrocellulose by twice dipping them for one minute in a 0·5% celloidin solution in ether-ethanol. One minute is allowed to elapse between immersions. The pieces are dried vertically overnight before use.

Step 4. The section is mounted on to the emulsion from absolute ethanol. Ethanol is used as the mounting medium rather than water in order to prevent hydration of the emulsion, which would later result in the tissue sample adhering permanently. Moreover, the ethanol dries rapidly so that the subsequent steps can be carried on almost immediately.

Step 5. Within 5–10 min. after the section is mounted, it should be deparaffinized in benzene for 3–5 min. In this step, the emulsion-specimen complex is inverted in the solvent, so that dissolving paraffin will run outwards from the emulsion, thus minimizing the amount of residual paraffin absorbed by the emulsion.

Step 6. The sample is microradiographed.

Step 7. In order to remove the tissue section for later histological study, the exposed sample-emulsion complex is covered with cellulose tape and placed in acetone for 3–5 min. Acetone dissolves the celloidin layer between the emulsion and the specimen, but it does not affect the adhesive qualities of the cellulose tape. Therefore, when the tape is removed, the section adheres to it. The tape and section may then be treated by ordinary histological methods. It should be noted in any case that the celloidin membrane must be removed from the emulsion, for it prevents the penetration of the X-ray developer. At no time should the emulsion be allowed to dry during the procedure.

Step 8. The emulsion is partially hydrated by immersion in 10-sec. changes of 95%, 70% and 50% ethanol and is then developed in Kodak D-19b, full-strength, for 5 min. at 18° C., and is finally fixed and washed according to routine photographic procedure. When the microradiogram is dry, it may be mounted under a cover-glass with Canada balsam or clarite.

No harmful effects of the above technique on the response of the photographic emulsion have been observed.

9.1.5. *Mounting the specimen for projection or reflexion X-ray microscopy.* Many of the biological specimens need no preparation and can be exposed to the X-ray beam without special support, such as parts of plant leaves, stems, flowers, feathers, insect wings, natural fibres of plant or animal origin and other small, rigid and relatively dry objects. These can be mounted on the specimen holder so that the area of interest is over a central hole, to avoid X-ray absorption in a supporting film. Mounting of specimens is best done under a stereoscopic binocular microscope with a triple objective holder giving three magnifications up to × 100 and with wide field eyepieces. Another aid to mounting is a micro-manipulator or micro-dissector of the type described by Barer & Saunders-Singer (1951) and Harding (1957); both are available commercially. These instruments are based on the pantograph principle and give a reduction of a few times for any three-dimensional movement of either hand. In the X-ray microscope it is not always possible to search the field of view for the best place to photograph, as in light and electron microscopy, and so the specimen must be initially mounted in the correct position with the aid of the stereoscopic binocular microscope.

Other smaller specimens such as pollen grains, small seeds, fragments of bone, diatoms, etc., may still need no preparation, but must be mounted on a supporting film. Thin plastics such as Mylar are available in large sheets, 6μ thick, and can be used for specimens of high contrast. With thinner objects at lower kilovolts a thinner support must also be used, such as a plastic film cast on water and picked up onto a fine mesh grid. These films are a few hundred Ångströms thick and give no noticeable absorption except at very long wavelengths. The still thinner carbon films, as used in electron microscopy (Bradley, 1954), may be used for ultra-soft X-ray microscopy.

9.1.6. *Living organisms.* The injection of living specimens, already discussed above, is one of the most important medical applications of the X-ray microscope (chapter 12). An essential part of the technique is to reduce any gross movements during the exposure by anaesthetization of the animal, which follows normal medical practice. Pulse movements will nevertheless still occur and the shortest possible exposure is needed. Ciné methods, especially when coupled with an X-ray image intensifier, may help to avoid loss of resolution from this residual movement. The usual subject for blood-vessel studies is the rabbit ear, but Engström (1956) has managed to prepare a rabbit ovary in such a way that it could be partially removed from the body for radiography at intervals, to study the growth of the blood supply. This was achieved with contact microradiography, but it seems that the projection method would be better in practice since the resolution does not then depend on close contact between the living, possibly wet, material and the photographic plate.

Other living specimens can be used such as insects, in which excellent contrast is provided by the air sacs in the body, without the need for contrast media (Nixon, 1952). In general, the resolution is not as good as with a killed and dehydrated specimen for a variety of reasons; a fresh frozen-dried specimen is preferable unless some dynamic change is to be observed in the living organism.

9.2. Metallurgical and industrial specimens

Many of the specimens under this general heading can be used for X-ray microscopy with no preparation other than dispersion on the supporting film. Dusts, fibres, cloth, paper, inclusions in plastics and similar articles will yield information when placed in the X-ray beam. For metal specimens it is necessary to prepare thin sections similar to those for biological samples described above. In general it is not possible to cut a metal section thin enough for X-ray microscopy and so the samples are ground down with abrasive from thicker pieces. Very recently there has been some success in cutting metals with a diamond knife for electron microscopy and this method may be possible in the future for X-ray microscopy as well. Until then thin samples must be prepared in more laborious ways.

9.2.1. *Sectioning by grinding.* The two-step process of Betteridge & Sharpe (1948 a), illustrated in simplified form in Fig. 9.1, is a convenient method to use. They prepare a lathe-turned recess in a brass block o·020 in. (o·5 mm.) deep and wax in the specimen, previously cut to a slightly greater thickness. The block is polished on a metallurgical wheel until the sample is rubbed down level with the block and the block itself is slightly polished. The wax ensures that the temperature of the specimen has not been raised enough to change the structure, as any large rise in temperature will melt the wax and the specimen will leave the block. The second step is to remove the sample and re-wax it into a similar block with

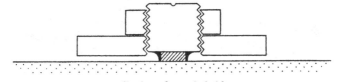

Section of sample holder

Fig. 9.1. Sample holder for grinding metallurgical specimens with parallel sides and 25–50 μ thick. (Sharpe, 1957a.)

a recess of only o·002 in. (5 μ) or less in depth and polish the other side. Then the sample is removed and is ready to be used, without the etching usual in optical metallography. It is most important to keep the two sides of the sample parallel, so that the radiograph has a uniform density distribution across the field, except for true variations due to different elements or internal inclusions. The polishing does not need to be of the same high quality as in metallography since it is the internal structure that is being studied and not just the surface. Any large scratches will be straight lines on this micro-scale and will appear more transparent to the X-rays, and therefore light in the positive print. These two effects make it possible to recognize scratches without confusion with the detail of the specimen.

Sometimes a thin specimen is obtained by cutting a wedge and looking at the thin edge. This method produces the undesirable gradation of intensity eliminated above; the final result is not as easily interpreted. Compared with biological specimens greater thinness of metallurgical specimens is necessary because of the higher atomic numbers of the elements. Aluminium can be used at 0·5 mm. but steel should be reduced to 50μ. For this reason stereographic methods are more successful with the lighter metals, where there is more specimen depth to be seen. In contact micro-radiography the metal foil or section must be brought into close contact with the emulsion if high resolution is sought, but this is usually done in the dry state. A specimen holder, as shown in Fig. 10.1, is useful; a metal tube serves both to limit the field and to press the specimen against the plate. The inner end of the tube is covered with black paper to facilitate handling of the holder in

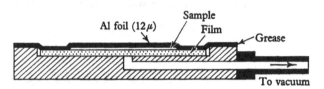

Fig. 9.2. Sherwood vacuum camera for microradiography with thin specimens. When the camera is evacuated the thin aluminium foil presses the sample into close contact with the fine-grained film. (Engström, 1956.)

the light. Owing to preparative difficulties metallurgical samples are in general thicker than biological sections, being often 0·1–0·2 mm. or more thick. When very high resolution is sought with thin specimens, the vacuum cassette devised by Sherwood (1947) is useful (Fig. 9.2). The photographic plate or film rests in a recess in a metal plate, through the back of which a connexion runs to a vacuum pump. The sample is placed on the emulsion and covered with an aluminium foil about 12μ thick, which extends across the lip of the recess and is sealed to the metal rim with vacuum grease. On evacuation with a simple pump, the foil presses the sample firmly against the emulsion.

9.2.2. *Shadow-casting and replica methods.* These two standard methods of electron microscopy have been used with success for X-ray microscopy by Le Poole & Ong (1956, 1957). A heavy layer of gold, deposited at a low angle, was used to increase contrast from diatoms and spermatozoa, and could be used also for dusts and other more transparent specimens. The very large depth of field of the projection X-ray microscope has been exploited with replica methods for studying fracture surfaces; with a deep fracture the full depth of failure cannot be focused with a light microscope. A carbon replica, obtained from an initial plastic replica, is found to follow faithfully all the features of the fracture. This replica is shadowed to enhance contrast and then viewed with the X-ray microscope, preferably stereographically.

Other special specimen preparation techniques may be borrowed from electron microscopy; for instance, for high and low temperature studies, for stress and strain experiments and for fatigue and brittle fracture, so that the specimen is tested while under observation in the X-ray microscope.

Note

Additional material relevant to this chapter will be found on p. 370.

CHAPTER 10

TECHNIQUES OF CONTACT
MICRORADIOGRAPHY

10.1. Apparatus for contact microradiography

10.1.1. *General considerations.* The basic principles of the contact method of X-ray microscopy have been outlined in chapter 2; the present chapter will deal with its practical application. The essential requirements consist simply of a photographic plate of ultra-fine grain and an X-ray tube, preferably with a small focal spot. Normal X-ray diffraction tubes with copper targets, if operated at 10–20 kV instead of 45 kV, will give soft enough radiation for use with metallurgical and the thicker biological specimens at a resolution of a few microns. Similarly a grain inspection unit or low voltage skin therapy tube, especially those with beryllium windows, can be used at 10 kV or higher. The effective X-ray source is about 1 mm. in width in this type of tube and so the distance from the tube to the photographic plate must be large to reduce geometrical blurring. The relationship given earlier (2.18) shows the minimum distance of the specimen from the X-ray source: for a resolution of 1μ and a specimen thickness of 0·5 mm. the photographic plate must be 50 cm. from the tube. Long exposures are then needed owing to the low sensitivity of the emulsion and the absorption in air of the softer radiation, as well as to the great distance from the tube. Despite these disadvantages most of the 300 papers in the Kodak Bibliography of Microradiography (1955, 1957) describe results obtained with converted X-ray tubes originally designed for other uses.

For more refined work, especially for quantitative and semi-quantitative investigations, account must be taken of the points made in chapter 2. The effective wavelength employed, and therefore the voltage and target of the X-ray tube, must be selected to suit the amount and nature of the elements to be detected, the specimen must be as close as possible to the emulsion, and the size and distance of the focal spot must be controlled. As high-resolution photographic emulsions have a very slow response, the highest attainable X-ray intensity at the plate is desirable: an intense source close to it. A point focus tube and very short camera will, therefore, offer the best working conditions (§ 2.4), although a limit is ultimately set by other factors, such as the size of the uniformly illuminated field. Apparatus rationally designed for contact microradiography on these lines is only now being developed.

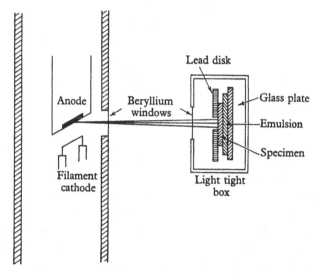

Fig. 10.1. Geometrical arrangement for contact microradiography. The line focus in the X-ray tube is foreshortened to a square in the direction of the beryllium window. The emerging X-rays pass through another window of low absorption before reaching the limiting lead diaphragm, specimen and emulsion.

The arrangement of the apparatus when a conventional type of tube is used is shown schematically in Fig. 10.1. The X-ray tube has a line focus, foreshortened to square form in the direction of 'view', in the interests of high intensity (§ 8.6). The beam emerges through a thin beryllium window, of low absorption, and falls on the 'sandwich' comprising the photographic emulsion (usually on a glass plate), the specimen, a limiting diaphragm of lead and a light-tight cover. Commercial X-ray tubes are almost all sealed off and thus have a given target and fixed X-ray spectrum. Several experimental tubes have been made with interchangeable targets, or with a rotatable target having different elements on its facets.

If the selected wavelength is longer than 2–3 Å, it is necessary to evacuate the path between tube and specimen or to fill it with an inert gas such as helium at low pressure. When good monochromatization is required a foil of a suitable element may be placed in the beam as filter (cf. § 8.7.1); if fluorescent radiation is preferred, the foil will be replaced by a piece of the selected secondary emitter. If a window is dispensed with, tube and camera being evacuated together, a foil of some sort must still be placed between target and camera to prevent fogging of the emulsion by light from the filament. In the ultra-fine focus projection tube, where the target is a thin foil normal to the beam (§ 3.4.3), this requirement is automatically fulfilled. X-rays transmitted through the target form the image in this case—filament, target and specimen being all in line.

The first tube and camera specially designed for microradiography appears to have been made by Goby (1913 a, b). Dauvillier (1927) was the first to put into practice the requirements for using soft radiation. He constructed a tube with an aluminium window, operated at 3–7 kV, and evacuated together with the camera. Many tubes and cameras for special purposes have been devised since then and we shall describe only those designed primarily for microradiography. It is convenient to subdivide according to wavelength range (or tube voltage), commencing with the more penetrating radiations.

10.1.2. *Apparatus for hard radiation* (< 2 Å). The investigation of highly absorbing materials, such as most metallurgical specimens, geological sections and certain biological materials such as bone, requires radiation of wavelength shorter than about 2 Å and a voltage greater than 10 kV for its efficient excitation. For these applications a commercial X-ray tube is usually suitable, particularly those designed for crystal analysis by diffraction (Philips, Machlett) which have a smaller focal spot (∼ 1 mm.) than those employed for macroscopic radiology. The use of such standard equipment has been described by Clark (1955), Sharpe (1957 a, b) and others in metallurgy, and by Bohatirchuk (1942), Barclay (1951) and Engström & Wegstedt (1951) in medical microradiography. For many purposes copper radiation (1·54 Å) is sufficiently penetrating, but targets of chromium (2·28 Å), tungsten (0·21 Å) and molybdenum (0·71 Å) are also frequently used. A voltage several times higher than the minimum needed to excite a particular line must be employed, in the interests of intensity (cf. § 8.5.2), so that 40–60 kV is usual. High-voltage elements and leads are normally totally enclosed, so that the tubes are shockproof; a water supply is needed for cooling the target. Because of these requirements the tubes are somewhat bulky and heavy and must be permanently mounted on a stand, usually with the window vertically above the specimen and plate. Many tubes have two or three windows.

The disadvantage of such commercially available apparatus is that it is permanently sealed off, so that a separate tube is needed for each wavelength. A few models are continuously pumped and have interchangeable targets, such as the Metropolitan-Vickers Raymax and the Hilger and Watts Microdiffraction unit (see § 14.1) which incorporates the tube designed by Ehrenberg & Spear (1951 a). Even so, they are not ideally suited to microradiography, where small tubes with special cameras and rapidly interchangeable targets are needed. Engström (1946) has described a tube for voltages up to 50 kV in which the water-cooled anode ends in a hexagonal block to which targets of different elements are fixed. These can be rotated into position in turn, so that successive images of a specimen can be obtained with different wavelengths without breaking vacuum. Ely (1957) has described a modified form of the Ehrenberg–Spear tube (for 15 kV) in which both target and window can be changed during operation; provision is also made for continuous motion of the target, to minimize overheating, and of the window, to eliminate artefacts in the image due to window structure (see also Ely (1960)).

Any of the tubes described can be adapted to give fluorescent radiation. The secondary target of the selected element is placed close to the exit window of the tube, and the radiation again utilized at a shallow angle of observation. Fig. 10.2 shows the arrangement schematically (Splettstosser & Seemann, 1952). Again it is valuable to have a tube with demountable anode, since the primary target material should be chosen to suit the secondary radiation desired. The efficiency of fluorescent excitation is greatest if the wavelength of the incident radiation is 0·2–0·6 Å shorter than that of the absorption edge for the secondary emitter (Hevesy, 1932). Hence the primary target should be an element rather higher in the periodic table than that of the secondary target, for instance, the K-radiation from copper ($Z = 29$) can be used for exciting the fluorescent K-radiation of iron ($Z = 26$). This conclusion is borne out by the measurements of Rogers (1952), although he also finds that several elements give almost as high a fluorescent output when excited by radiation from a metal of very high Z such as tungsten.

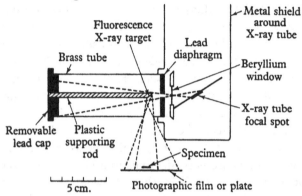

Fig. 10.2. Microradiography with fluorescent X-rays.
(Splettstosser & Seemann, 1952.)

Conversely, the excitation of fluorescence radiation in a metallurgical specimen will produce a general background fog in the photographic emulsion and reduce the contrast otherwise present, and more so in the contact than in the projection method. For this reason, the use of a copper target with a specimen composed largely of iron is not in fact to be recommended. The fluorescent radiation may be eliminated by using a chromium target ($Z = 24$), which has its $K\alpha$ emission at 2·29 Å, well below the energy of the iron absorption edge at 1·74 Å. The difference between fluorescence of the matrix, giving poor contrast, and that of an inclusion, resulting in excellent contrast, is discussed by Sharpe (1957 a).

10.1.3. *Apparatus for soft radiation* (2–10 Å). The production of X-rays of wavelength longer than about 2 Å involves different considerations from that of hard radiation. The self-absorption of the emitted X-rays in the target becomes increasingly greater with wavelength. Taking off the beam at a shallow angle, as in high-voltage tubes, now leads to a diminution instead of an increase in the effective intensity. Since the self-absorption of soft X-rays increases more rapidly with the atomic number of the target than does the efficiency of X-ray production, it is better to use a target of aluminium or copper rather than tungsten or platinum. In the third place, the ratio of line to continuous radiation of the same wavelength is very high (§ 7.1.1), so that it is preferable to use a target with a characteristic emission line in the region of the wavelength needed for a particular investigation.

These considerations have been taken into account by Ehren-berg & White (1957) in modifying the microdiffraction tube of Ehrenberg & Spear (1951 a) for microradiography with soft radiation (§ 14.1.1). In order to get a high intensity, they operate the tube at a much higher voltage than the minimum needed to excite the required line and then select this from the background radiation by total reflexion from an optical flat. As the critical grazing angle depends on the wavelength of the X-rays, the harder radiation is almost completely suppressed (§ 8.7.3). The radiation is taken off at an angle of 45° from the target and an aluminium foil serves as window to the tube, which is continuously pumped. The aluminium $K\alpha$-line at 8·3 Å has been mostly used, and the intensity is such that the exposure-time is reduced to about 10 min. A particular advantage for contact microradiography is that the focal spot is as small as 40μ in diameter, so that relatively thick specimens can be used or thin specimens well separated from the emulsion. When it is 10 cm. from the focal spot, a resolution of 1μ may be obtained from a specimen 2 mm. thick.

A number of other experimental tubes for soft X-rays have been described, particularly for microanalytical purposes (Engström & Lindström, 1950; Brattgård & Hydén, 1952; Clemmons & Aprison, 1953; Engström, 1955 b). Subsequently a sealed-off tube, suitable for microradiography in general, was designed by Combée and is now produced by Philips of Eindhoven (Combée & Engström, 1954; Combée, 1955 and van den Broek, 1957). It is very compact, measuring less than 10 cm. in length and only 3·6 cm. in diameter (Fig. 10.3). The line filament F is 4 mm. from the anode A, which is a tungsten strip set in a copper block. To utilize as much as possible of the excited intensity, the window W is sited 11 mm. from the centre of the focal spot and is of beryllium only $50\,\mu$ thick, this being the thinnest window that can at present be made permanently vacuum-tight. It transmits 80% of silver K-radiation and as much as 20% of aluminium $K\alpha$ (cf. Engström, 1956, Fig. 33). It can thus be used down to a voltage of 2 kV without modification. The H.T. supply is in fact variable between 1 and 5 kV. The focal spot is 0·3 mm. in width and the maximum dissipation is 10 W. Water-cooling is

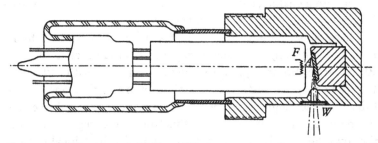

Fig. 10.3. Section through Philips 5 kV contact microradiographic tube, with tungsten anode A set into a solid copper block, air-cooled; $50\,\mu$ beryllium window W. (van den Broek, 1957.)

dispensed with in the interests of portability, but provision is made for air-cooling with a fan or blower. The object-focus distance is normally 1·5 cm. and the camera can be evacuated with a water-jet pump. The exposure-time is then of order 5 min. at a tube voltage of 3 kV. A longer camera is available when a larger field of view is required than that normally covered (3 mm.). The whole unit is extremely compact and light, in spite of the inclusion of all the electrical equipment (Pl. VIII A).

The commercially available model has been modified by Recourt (1957 a) for use at still longer wavelengths (Combée & Recourt, 1957). The beryllium window was removed and the camera and tube continuously evacuated during exposure, the specimen being covered with a thin layer of aluminium by evaporation, to exclude light. Micrographs were made of a number of biological sections at a tube voltage of 1 kV and lower.

10.1.4. *Apparatus for ultra-soft radiation* (> 10 Å). The production of X-rays of wavelength longer than about 10 Å encounters ever greater difficulties, owing to the low efficiency of excitation and the high self-absorption in the target and camera. The use of an evacuated camera and a tube window of an element of low atomic number and extreme thinness (or none at all), becomes essential. Dauvillier (1927) used an aluminium window and voltages down to 3 kV for a wavelength of 8·3 Å. The apparatus of Lamarque & Turchini (1936) could be used for voltages down to 500 V and had a lithium window. The longest wavelength employed was 12 Å, having a minimum excitation voltage of 1 kV. The recent emphasis on microanalysis of biological material has resulted in the design of improved tubes for operation at this and even longer wavelengths. The tube originally devised by Engström (1946) for medium soft to hard radiation, and later redesigned by Engström & Lindström (1950) for soft radiation (at 3 kV), has been further improved by Lindström (1955). The rotatable anode now carries twelve alternative targets, each 1 × 8 × 50 mm.; electrostatic focusing of the electron beam is effected with a negatively biased electrode. The focal spot is 1 mm. by 6·5 mm. Although it can be operated at voltages up to 10 kV the tube is primarily designed for use at 0·5–2 kV (25–26 Å limiting wavelength). At 0·6 kV the maximum current is 10 mA and at 1·5 kV it is 45 mA. For producing a continuous spectrum of soft radiation, platinum or tungsten targets are used. Lindström reproduces an instructive series of micrographs showing how image contrast in biological specimens decreases rapidly above a working voltage of about 1·5 kV. The specimen holder can be evacuated and its distance from the focal spot is variable; it was usually 18·5 cm., at which distance the exposure-time on Maximum Resolution Plates was in the range 6–10 min. A further development of this type of tube has recently been described by Engström, Greulich, Henke & Lundberg (1957).

The tube described by Fitzgerald (1956, 1957), differs from that of Lindström (1955) primarily in being simpler in mechanical design and in incorporating a number of automatic switches and warning devices in the vacuum and electrical systems. A particular advantage of the specimen holder is the large field of view, 2·0 cm. diameter.

A very simple 'midget' X-ray tube for the ultra-soft region has recently been described by Engström & Lundberg (1957). Its

small size is apparent from Fig. 10.4, being 4 cm. in diameter and
less than 10 cm. long. The filament and anode are mounted on
ordinary sparking-plugs, *A* and *B* respectively. The voltage is
variable between 500 and 1500 V; the beam current is about 1 mA
and the focal spot about 0·1 mm. in diameter. The specimen-film
sandwich is mounted at *D*, and an aluminium foil *C*, 1000 Å thick,
protects it from light from the filament. The field of view at the

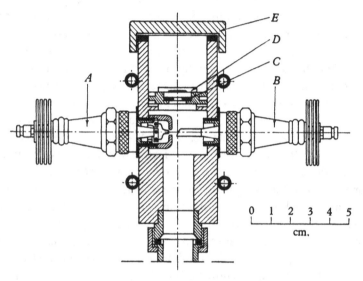

Fig. 10.4. Midget X-ray tube using two sparking plugs (*A*, *B*) as insulators for
cathode and anode. An aluminium foil (1000 Å thick), placed at *C*, acts as light
filter. The specimen and plate are at *D* and the whole camera, sealed by the
end-plate *E*, is at high vacuum. (Engström & Lundberg, 1957.)

specimen is 12 mm. The tube is mounted directly on top of a
diffusion pump, by-pass tubes permitting the anode space and
camera to be evacuated simultaneously to prevent rupture of the
foil *C*. It is pumped down and ready for operation in 30 sec. With
1200 V on the tube, corresponding to a limiting wavelength of
10 Å, the exposure-time was 20 min. at a beam current of 1 mA.
Although primarily designed for dry weight and water content
determinations in biological specimens (§ 6.2), the midget tube is
obviously of general utility in microradiography at ultra-soft
wavelengths.

A more complicated apparatus, designed to give high X-ray intensity and to enable several specimens to be exposed at once, has been described by Henke (1957). It incorporates the method of focusing by total reflexion first used by him in a microdiffraction tube (cf. § 14.1).

10.2. Recording the microradiograph

10.2.1. *Photographic considerations.* The amount of exposure and the subsequent processing of the primary negative will vary according to the experimental requirements. When a living organism is the subject, as in microangiography, a short exposure is essential and a high contrast developer must be used to bring up image detail; grain size is sacrificed to speed. For high-resolution work, however, it is preferable to over-expose the negative and use a slow, fine-grain developer. It then becomes important, especially for quantitative measurements, to know the form of the response curve of the emulsion under different processing conditions. These considerations have been discussed in detail in connexion with the techniques of quantitative microanalysis (§ 6.4). For qualitative microradiography the type of photographic material and the developer will be chosen with an eye to the size and nature of the detail to be resolved. In general, good contrast will be more desirable than linearity of response. High contrast developers produce an increase in the grain size of the photographic emulsion leading to poorer resolution, as noted above in connexion with microangiography. It is preferable to increase contrast by the use of softer X-rays, if possible, so that a fine-grain developer may still be used (see also § 2.5).

10.2.2. *Optical enlargement.* As the primary negative may be capable of recording details on the limit of optical resolving-power, the best possible microscope must be used if full benefit is to be obtained from the contact method. If a cover-slip is placed over the microradiograph, an immersion objective can be used. Lindström (1955) employs apochromatic objectives, either of 4-mm. focal length and 0·95 numerical aperture, or a 2-mm. immersion type of aperture 1·32, with periplanatic oculars (×6 and ×8). The need for the utmost care in focusing follows from what has been said about the minute depth of focus of such a high-power microscope (cf. Henke, 1959). The photographic material used for recording the enlargement requires to be of only normal speed and resolution, such as a lantern type of plate. Barclay (1951) used Ilford Ortholine, Engström & Lindström (1951) Gevaert Replica, and Brattgård & Hydén (1952) Kodak 250 plates. When densitometric measurements are to be made (§ 6.2.2), the whole of the photographic procedures have to be carefully standardized.

10.3. Stereoscopic microradiography

Stereoscopic techniques were first used in microradiography by Goby (1925). Since the penetration of X-rays allows a view of the interior of the specimen, the fullest information can be derived only by making visible the third dimension in the image. In spite of some limitation in depth perception, due to geometrical blurring in thick specimens, stereographic microradiography has been found of value by many workers. If only moderate resolution is acceptable, the section can be thick enough to make visualization of the third dimension possible. With a resolution of $1\,\mu$ the subsequent magnification must be × 100 or more for proper appreciation of image detail, so that the image of a specimen $100\,\mu$ thick, viewed stereoscopically at 25 cm., would appear to be 1 cm. thick. Similarly, for thinner sections at higher resolution the stereoscopic appearance would be about the same, owing to the higher magnification required.

Since the incident X-ray beam is almost parallel, the specimen and plate must be tilted between the two exposures which have to be made. By mounting the specimen on a thin layer of plastic, its transfer to the second emulsion can be facilitated. The procedure is similar to that for electron stereomicroscopy, where the angular aperture of the beam is of the order of 10^{-2} radian and the specimen must also be tilted between exposures. The analysis of electron stereomicrographs has been discussed in detail by Helmcke (1954) in connexion with a quantitative reconstruction of various diatoms. The use of parallel illumination produces views that lack perspective, unlike the point projection system where the stereo-pair is obtained by a translation of the specimen across a divergent cone of rays (see § 11.10).

The exact amount of tilt required will depend on the final magnification and on the tube distance (Engström, 1951 a) as shown in Fig. 10.5. Departure from the optimum tilt-angle will still produce a stereo-pair but the image will be distorted in some way with respect to the original. These distortions are described by Bellman (1953), who used stereo-microradiography in a study of blood vessels filled with contrast medium. To test the quantitative use of the method, he measured the thickness of sheets of mica

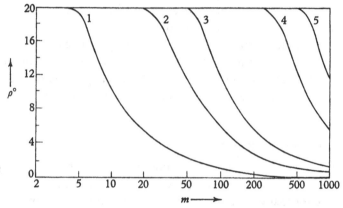

Fig. 10.5. Tilting angle for stereoscopic microradiography, as a function of magnification. The curves 1–5 represent various sample thicknesses for values of a/d of 100, 500, 1000, 5000 and 10,000, where a is the distance between target and film, and d is the thickness of the sample. (Engström, 1956.)

and found the error to be a few percent only. For the best agreement the tilting-angle had to be increased to 30° between exposures so that the plates would show a greater difference for measurement. The negatives were viewed with a special double microscope, so that the stereo effect could be obtained without enlarging and printing and subsequent mounting in a normal viewer. The microscope stages were linked together so that, after the two negatives had been correctly oriented to give a stereoscopic image, the pair could be moved as one unit to search the field of view. Sherwood (1937 a) also used microscopic fusion of the two images but without the locked stages.

The stereographic display of the internal details of a biological or metallurgical specimen adds enormously to the value of the

normal flat print. A good example is the observation by Williams
& Smith (1952) of the grain structure of an aluminium-5 % tin
alloy (cf. p. 326). The sample was 1 mm. thick and was reproduced
at × 125 magnification; the tin (dark) is deposited along the grain
boundaries of the aluminium, giving the appearance of a soap
bubble froth. The authors microradiographed a stained bubble
mixture to show that the angles between the grain boundaries were
similar to those in the solid metal.

The stereo-microradiographic camera is essentially a rigid
device that will hold the specimen and plate firmly at some variable,
but known, angle with respect to the X-ray beam. A typical
laboratory model is that of Clark & Eyler (1943) and a commercial
type is sold by Siemens (Erlangen). General information on
stereo-techniques with light, most of which is applicable to X-ray
work, is given by McKay (1948) and Dudley (1951).

10.4. Microfluoroscopy

The contact microradiographic image may be viewed directly
on a fluorescent screen, as in medical fluoroscopy. The method is
attractive for many purposes, and particularly for routine work in
industry, because of its speed and simplicity. An example of its application is shown in Pl. VIII B (Nixon, 1956 b): direct photography
of the image of a ball-point pen, with or without a small initial
X-ray magnification. Optical magnification of the fluorescent image
has the advantage of conserving the brightness, and allows shorter
exposures than in projection at the same magnification (§ 3.9).

The resolving-power is now primarily determined by the grain
size in the screen, rather than by geometrical blurring or the limitations of the optical system. With a screen of calcium tungstate,
Engström (1947 a) obtained satisfactory viewing conditions at
optical magnifications up to × 50, using an X-ray tube at 5 kV.
For higher voltages he recommended a screen of uranium glass, as
being effectively grainless and so allowing greater magnification.
By observing the local changes in contrast between images formed
at two slightly different wavelengths, on either side of an absorption edge of one of the constituents, Engström was able to carry
out a qualitative analysis of its composition.

Recently Pattee (1958) has prepared a fine-grained screen, the image on which can be usefully enlarged by 1000 times or more, thus making microfluoroscopy practicable at high resolution. Quantitative intensity measurements were made directly on the image, as enlarged through the optical microscope. The fluorescent screen is made by vacuum evaporation of zinc, silicon monoxide and manganese on to quartz cover-slips, which are then baked in air at 1100° C. for 5 min. to activate the phosphor (Feldman & O'Hara, 1957). The evaporated surface is optically polished with tin oxide as abrasive. The layer of fluorescent material is very thin ($\sim 1\mu$), so that the region emitting light will all come within the depth of focus of the high-power optical microscope used to view the image. Being so thin, the screen will appreciably absorb only X-rays of long wavelength, such as aluminium K-radiation (8·3 Å), so that the image is formed with radiation that is also readily absorbed by a biological specimen, resulting in good contrast. The harder radiation from the X-ray source passes through both specimen and screen with little effect.

The sensitivity of this type of phosphor is very low and it must be used with a point source X-ray tube of high specific loading. The screen is placed about 0·2 mm. from a 10μ focal spot loaded to 10 kW/mm.^{2} (10 kV, 80μA) (Fig. 10.6). With a section of tissue 4μ thick this gives a geometrical blurring in the image of $0·2\mu$, approximately equal to the resolution of the light microscope used. The width of the illuminated region on the fluorescent screen is approximately equal to the source–screen distance, that is, about 200μ, very much smaller than the area exposed in the usual conditions of contact microradiography. However, the field of view of a light microscope is little more than 100μ at × 1000 magnification, so that nothing is lost by this restriction of illuminated area; in contact microradiography the initial micronegative must similarly be searched with a high-powered microscope.

The gain in visual intensity, as compared with normal contact conditions, is due to the fact that the target current has to be reduced only in proportion to the radius of the focal spot, not to its square, as spot size is decreased (§ 3.6.2). At the same time the spot can be brought closer to the specimen, so long as its angular size remains the same (that is, the geometrical blurring is constant), and as this is done the intensity on the screen increases by the inverse square law. For instance, when the diameter of focal spot is reduced from 1 mm. to 10 μ, the spot–specimen distance can be reduced from 2 cm. to 0·2 mm., whilst the beam current has to be reduced from 8 mA to 80 μA, to maintain thermal equilibrium. Thus the intensity at the screen increases by $100^2/100 = 100$ times, which more than offsets the reduced sensitivity of the fine-grained

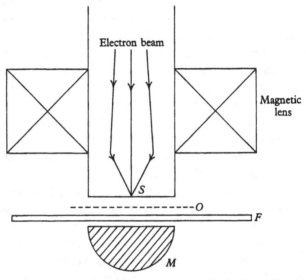

Fig. 10.6. Schematic arrangement of X-ray source S, fine-grain screen F, specimen O and optical microscope M, in the microfluoroscope (after Pattee).

as compared with the normal screen. In fact, direct viewing of the fluorescent image is found to be possible, without dark adaptation, at $\times 500 – \times 1000$ through the optical microscope.

In speed of working, microfluoroscopy gains over the normal contact method by eliminating the time-consuming procedure of photographing the X-ray image with a high-resolution emulsion. It also has the advantage of direct viewing of the X-ray image, which may be of particular value for observing living material. The dose rate, on the other hand, is correspondingly high.

In building the microfluoroscope (Fig. 10.6), Pattee adapted a Metropolitan-Vickers point source X-ray tube with single lens (§ 14.1) by adding at the bottom of the column a mechanism for rapidly changing targets. An inverted optical microscope, such as is used for viewing wet biological specimens, is placed below the X-ray source. The fluorescent screen is placed above the objective lens of the microscope, on a movable stage, and the thin section is laid on top of the screen without the use of mounting medium. When a permanent record is needed, the fluorescent screen image may be photographed through the optical microscope.

Microphotometry can be carried out directly on the image by placing a sensitive photometer, with a small aperture, in the focal plane of the microscope camera. The amount of light at a given image point is converted to a meter-reading and compared with other areas of the specimen. Stereoscopic methods can also be employed, by taking successive photographs of the same specimen in two different positions. Again, by using an eyepiece micrometer and noting the shift of a particular point in the image for a known shift of the calibrated specimen stage, the thickness of the specimen can be found, as well as the relative position of internal details. Preliminary quantitative measurements, using an aluminium target, have recently been described by Pattee (1960c).

Additional reference added in proof

To §10.1.3: Fournier (1960).

CHAPTER 11

TECHNIQUES OF PROJECTION MICROSCOPY

Special techniques are required in the practical application of the principles of point projection microscopy described in chapter 3. The electron source (*G*, Fig. 11.1) needs careful design and one or more lenses (L_1, L_2) are required for focusing an image of it on to the target *T*. Stabilized voltage and current supplies and high vacuum equipment have to be provided, together with adequate means for microscope alinement, focusing and photographic

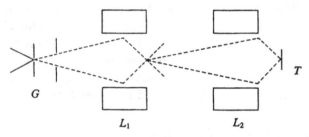

Fig. 11.1. Two lens projection X-ray microscope. *G*, electron gun; L_1–L_2 electron lenses; *T*, target.

recording. These experimental problems are discussed below, as well as methods of stereographic representation, target production and instrument maintenance.

11.1. The electron gun

The size of the focal spot and the electron intensity delivered into it principally determine the performance of the projection microscope as regards image resolution and exposure-time. As explained in § 8.2, heat dissipation in the target becomes a secondary factor at a spot size of less than a few microns. The main problem is to obtain the maximum beam current from the electron gun and to focus it with minimum loss by means of electron lenses.

The electron gun normally consists of an anode at ground potential, a tungsten hairpin filament heated to almost 3000° K.

and kept at high negative potential, and a Wehnelt control grid
(Fig. 11.2). Its properties have been studied in detail by Haine &
Einstein (1952) for electron microscopy, at 50 kV, and much of
their conclusions can be applied at the lower voltages of the X-ray
microscope. Details of how to make and mount the filament are
given in § 11.12. The grid is held at a negative potential with
respect to the cathode filament and acts as a beam-forming electrode, drawing no current. The bias voltage is generated by passing
the beam current through a fixed resistor in series with the filament

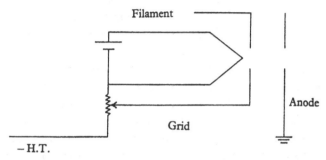

Fig. 11.2. Schematic diagram of triode electron gun showing the hot filament,
bias resistor for control grid voltage and anode (at ground potential). In practice
the filament is bent to a narrow point and passes part way through the grid
opening.

and so dropping the cathode potential. The value of the fixed
resistor is hard to determine, as the filament to grid spacing varies
from filament to filament. A variable potentiometer or rheostat,
controlled by an insulated shaft, allows some adjustment of the
bias voltage at any particular beam current. A more independent
type of bias voltage can be produced by using a high-resistance
chain to earth and placing a variable resistance at the high voltage
end. The grid is connected to the centre terminal and the cathode
to the lower voltage end, giving an independent control, so long as
the beam current is small compared with the chain current.

The bias voltage can be generated completely independently by
using batteries at high voltage or by a separate oscillator. As there
is no current drain, small 300 V portable Geiger counter batteries
can be used, such as Burgess U200 (U.S.A.) or EverReady
B1489 (U.K.). Alternatively an R.F. oscillator can be used, in

which several hundred volts can be generated, transformed, rectified and smoothed with very small components and at high voltage. With both these methods the controls must be insulated for the full gun voltage, and a bias voltmeter as well as beam current-meter must be mounted at high voltage. An advantage of a completely independent bias voltage is that it can be applied to the grid to switch off the electron beam while changing specimens or photographic plates, or to act as an X-ray switch for short exposures. Switching off the high voltage, the lens current or the filament heater, instead of the beam, makes it difficult to return to exactly the same focusing conditions. Deflecting the electron beam from the target is a practicable alternative, so long as there is no residual deflexion when the field is removed.

The final element in the electron gun is the anode, a flat plate with a central hole through which the electrons pass before reaching the electron lenses. It can be made movable in the plane parallel to the grid by means of two push rods operating through vacuum seals and against a retaining spring (Haine & Einstein, 1952). The filament should be pre-centred in the grid aperture and the anode shift employed to remove the effects of any subsequent filament movement. The filament height must also be accurately set with respect to the Wehnelt aperture; this should be done during operation so as to give the best beam brightness at any chosen kilovoltage. The filament should first be set high enough to give a beam current of several hundred microamperes with zero bias voltage, and the beam current then reduced by increasing the bias towards cut-off, which will normally be in the order of hundreds of volts (up to 500 V for 5–25 kV). A small Faraday cup that can be moved into the electron beam, and carrying an earthed aperture of known size, will help in finding the conditions of maximum brightness and in placing the beam on the axis of the instrument.

11.2. The focusing system

The electron gun delivers a narrow beam into an electron lens which focuses it into the focal spot. Theoretical and practical details of the various types of electrostatic and magnetic lenses are given in the standard texts (Cosslett, 1951; Klemperer, 1953; Hall,

1953). The modified forms used in the projection X-ray microscope are briefly described in § 11.3 and further details of their design will be found in the papers cited there. Fundamental data regarding the dependence of focal length and other lens parameters on excitation and geometrical form are given by Liebmann (1955) for magnetic lenses, and for electrostatic lenses by Liebmann (1949) and by Grivet (1950). In this section we discuss some of the properties of the focusing system.

11.2.1. Single-stage demagnification. A single lens will provide enough demagnification for many purposes, though the flexibility of a two lens system is usually more desirable. The maximum demagnification, in the presence of spherical aberration, is by (3.21) and (3.22):

$$r_c/r_2 = 1/M_x = \alpha_2 (V_1/V_c)^{\frac{1}{2}} = (4r_2/C_s)^{\frac{1}{3}} (V_1/V_c)^{\frac{1}{2}}. \quad (11.1)$$

For $V_1 = 10$ kV, $V_c = 0.25$ V, $C_s = 1$ mm. and $r_c = 50\mu$, this gives $1/M_x = 35$, so that the minimum spot-radius for full current utilization is 1.4μ. A smaller spot could only be obtained by moving the cathode further from the lens (unless a finer filament wire were used, which would shorten filament life). To obtain a larger spot the lens strength can be reduced, but this has the disadvantage both of increasing C_s and of requiring a longer image distance, so that the target would have to be movable with respect to the lens. A small physical aperture ($\sim$ 0.2–0.3 mm. dia.) must be fixed in the lens in order to obtain the minimum spot size, and this prevents more current being focused into the spot as the lens is reduced in power, unless interchangeable apertures are fitted. For these reasons a two lens arrangement is normally preferred. A single lens projection microscope is mainly of value for applications in which the spot size is kept constant or only rarely changed.

11.2.2. Two-stage demagnification. In a two lens system (Fig. 11.3) the final lens is maintained at maximum power so as to obtain minimum spherical aberration; in analogy with the corresponding element of an electron microscope, it has come to be called the objective lens. The first lens forms an image of the cathode, variable in size and position, which is further demagnified by the objective. As it controls the electron illumination reaching the latter, it is termed the condenser. It should be noted that this arrangement differs from that of the double condenser of the modern electron microscope, in which the roles of the two elements are reversed: the first lens forms a highly demagnified cathode image which is then projected into the specimen plane by the second condenser at a magnification of between one and two. Such a system is much more sensitive to alinement errors than is the two lens projection microscope (Haine & Agar, 1957).

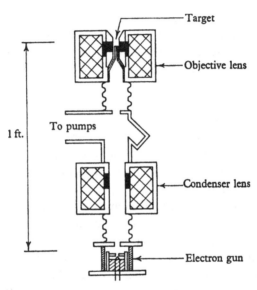

Fig. 11.3. Cross-section of a projection X-ray microscope, with 1 ft. scale. (Cosslett & Nixon, 1953.)

The overall reduction in the two lens system is given by

$$R = p_1(1 - t/f_2)/f_1 + (p_1 - t)/f_2 + 1 \qquad (11.2)$$

(Nixon, 1952), where R is the total reduction ($= 1/M$).

p_1 is the first object distance, that is, the distance from the filament to the centre of the condenser lens.

t is the distance between the centres of the two lenses.

f_1 is the focal length of the first lens (condenser).

f_2 is the focal length of the second lens (objective), usually fixed.

This reduces to

$$R = A/f_1 - B \qquad (11.3)$$

and, since $1/f \propto I^2$, this may finally be written

$$R = AI_c^2 - B, \qquad (11.4)$$

where A and B are constants and I_c is the condenser lens current. A plot of R against I_c is a parabola, with the intercept $R = B$ when I_c is zero. This graph is shown in Fig. 11.4, with the negative values of R replaced by positive values for convenience; since the image is normally round, only the absolute value of R is of interest.

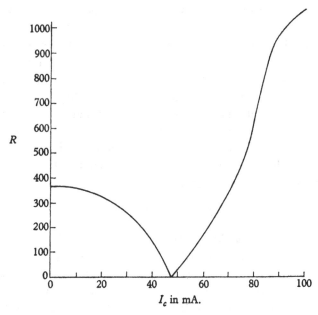

Fig. 11.4. Overall electron beam reduction in a two lens system plotted against variation in condenser lens current, with the objective lens focal length constant. The exact values of R and I_c depend on the separation of the two lenses and the position of the electron gun.

For large values of f_1, the image is inverted and R is negative. As f_1 becomes smaller R goes through zero and the image is erect, since an inverted intermediate image is formed between the two lenses.

At high values of I_c the focal length does not follow the simple formula exactly and the curve falls off. Values of $1/M$ of 5000 or more can be reached by using small bore pole-pieces in the condenser lens, to give a short focal length. A first object distance of 20 cm. and first image distance of 4 mm. would give a first reduction of × 50; a second object distance of 20 cm. and second image distance of 2 mm., gives the required second reduction of × 100. Such a combination would reduce a 50-μ electron source to a spot of 100 Å at the X-ray target.

Equation (11.4) is not exact for values of R close to zero, where there is a discontinuity. At $R = 0$ the condenser lens focuses the electron beam into the back focal plane of the objective lens, so that a parallel beam emerges. For values of $R > 10$ the formula is approximately correct and this is about the lowest value used even for X-ray microdiffraction or high definition medical diagnostics. It is difficult in practice to find the value of the condenser lens strength at which the X-ray image loses all definition, as the slope of the curve is very steep near $R = 0$ and a change of only 1 mA in lens current will change the value of R drastically.

The final image plane of the electron beam remains close to the plane of the target for wide variations of reduction, as shown by Smith (1956). For f_1 small compared to p_1 and t, a change Δf_1 in the condenser lens focal length produces a corresponding change Δq_2 in the objective lens image distance, given by

$$\Delta q_2 = (q_2/t)^2 \Delta f_1. \qquad (11.5)$$

That is, the change in final image distance is smaller than the change in the focal length of the first lens by a factor approximately equal to the square of the reduction of the final lens. After a large change of condenser lens strength a slight change of the objective lens current will bring the electron spot back to the target plane; this requires very fine control of the current (cf. § 11.8).

11.2.3. *Astigmatism and its correction.* The values of the spot size deduced above assume an aberration-free imaging system. In practice the spot may be enlarged by a number of effects, the most important of which are chromatic aberration, spherical aberration and astigmatism. The first of these may be kept negligible by stabilizing the electrical supplies (§§ 11.4 and 11.5). The effect of the second is made smaller than the desired spot size by inserting an appropriate aperture in the final lens (§ 3.5.2). The third, astigmatism, arises in two forms: true field aberration of an off-axis point, and 'mechanical' astigmatism due to inadequate care in machining the lens elements, resulting in ellipticity or mis-alinement of their bore and consequent lack of rotational sym-metry in the focusing field.

Astigmatism of the field is more serious than in electron micro-scopy, where the field of view at the specimen is at most a few microns at high magnification. In forming a point source the virtual cathode is imaged by one or more lenses; if the filament tip is off centre, the electron beam enters the lens at an angle and there will be off-axis astigmatism in the final spot. The amount of astigmatism depends on the square of the off-axis distance, so that for a 50μ source, off centre by its own diameter, it will be much greater than in the electron microscope. The final lens will produce two line foci at right-angles to each other and separated axially by several microns. Between these two lines there will be a disc of least confusion considerably larger than the smallest spot possible in the absence of this aberration. The effect on the X-ray image of a square grid pattern is to bring each of the two sets of wires into focus one after the other as the lens current is changed. The inter-mediate position will give poor resolution. The remedy is found in careful alinement of the electron optical column, but sometimes the filament may be bent so far that the available adjustments are not sufficient for realinement. In this case a new filament is needed, since a great deal of time can be wasted trying to aline on an impossible axis.

Elliptical astigmatism, which is normally an order of magnitude smaller than field astigmatism, arises from lack of perfection in manufacture of the lens. The electrodes or pole pieces may not have smooth, properly round bores or the several parts may be tilted and translated with respect to the ideal axis. In the magnetic lens, defects below the surface of the metal may disturb the field. All of these deficiencies produce an elliptical instead of circular field, so that the lens strength varies around the axis between a maximum and minimum. Instead of a point focus two line foci are formed, separated in space along the lens axis by a few microns and at 90° to each other as in geometrical astigmatism. This type of astigmatism has been studied in the conditions of electron microscopy by several authors and is found to vary as the first power of the angular aperture (Sturrock, 1951; Liebmann, 1952 and Archard, 1953). The radius of the astigmatic disk of least confusion is given by

$$r_a = 2\alpha f(\Delta R/R),$$ (11.6)

where R is the radius of the lens bore, f is the focal length and α is the semi-angular aperture. The effect of this astigmatism in the projection X-ray microscope is shown in Fig. 11.5 (Nixon, 1955*b*), in which the semi-angular aperture is plotted against the radius of

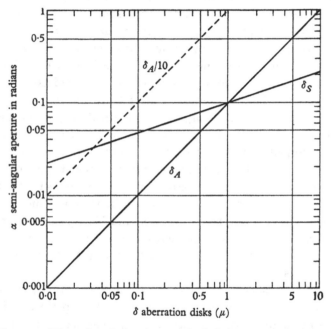

Fig. 11.5. Disks of confusion due to spherical aberration and astigmatism for various values of the semi-angular aperture of the objective lens. (Nixon, 1955*b*.)

the spherical and astigmatic disks of confusion. For severe errors in R, these effects are equal at $1\,\mu$ spot size; for a lens with ten times better mechanical finish, the effects are equal at 300 Å spot size.

For very high resolution X-ray microscopy it will be necessary to correct the astigmatism as in electron microscopy. Correcting fields, using either electrostatic or magnetic elements, can be arranged so as to introduce an astigmatic effect of equal magnitude and opposite sign to that of the lens field. Such a stigmator must be variable in azimuth and strength. A comparison of the various types has been made by von Borries (1956); they utilize magnetic elements with mechanical rotation, electrostatic elements with electrical rotation or electrical elements with mechanical rotation.

11.2.4. Comparison of electrostatic and magnetic lenses. A point focus of electrons can be formed by either type of electron lens and each has specific advantages. The electrostatic type does not need the lens current supplies essential for a magnetic lens and the high voltage may be of lower stability, especially if mechanical movement of the specimen is used for focusing. With electrical focusing the stability requirements are more severe (§ 11.5). At the lower voltages of X-ray microscopy (less than 20 kV) electrical breakdown is much less likely to occur than in electrostatic electron microscopes, although any leakage through a thin target could lead to pressures permitting an electrical discharge. The main disadvantage compared with the magnetic lens lies in the larger values of the spherical aberration coefficient, C_s, and of the minimum attainable focal length, which result in an important reduction in intensity at given focal spot size (r_s) or in resolution at given exposure-time.

For any lens we have $r_s = \frac{1}{4}C_s\alpha^3$. The current in the electron spot varies as the square of the aperture and also as the square of the spot radius:

$$I = kr_s^2\alpha^2. \tag{11.7}$$

Combining these two equations and eliminating α we have

$$I = k_1 r_s^{\frac{8}{3}}/C_s^{\frac{2}{3}}. \tag{11.8}$$

In practice, the current which can be used continuously, without danger of melting the target, may be less than this value because the rate of heat dissipation is limited (§ 8.2). The energy delivered by the beam is plotted against the source diameter d ($= 2r_s$), in Fig. 11.6, for typical values of C_s for magnetic and electrostatic lenses: for the magnetic lens $C_s = 1$ mm. and for the electrostatic, $C_s = 200$ mm. There is an important difference between comparing the lenses at constant I, along a horizontal line, or at constant d, along a vertical line. At constant d the intensity difference is $1/(200)^{\frac{2}{3}}$ or $1/34$, so that at 2μ spot size the beam current is

10 μA for the magnetic lens and 0·29 μA for the electrostatic lens, giving a difference in exposure-time of thirty-four times. However, at constant I, d varies as $C_s^{\frac{1}{4}}$ so that the ratio is 3·85. Therefore, for 10 μA beam current in both lenses, the magnetic spot size is 2 μ and the electrostatic spot size is 7·7 μ. Similarly if C_s for the electrostatic lens is taken as 32 mm., then $I_{\text{mag.}} = 10 I_{es}$ and $d_{es} = 2·4 d_{\text{mag.}}$.

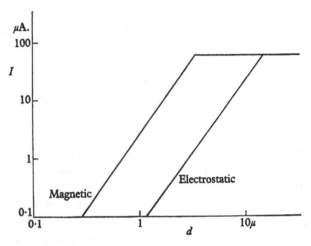

Fig. 11.6. Comparison of the total current available at given electron beam diameter in magnetic and electrostatic lenses; the common initial level represents the maximum beam current from the gun.

This means that by relaxing the resolution slightly in the electrostatic X-ray microscope the intensity can be brought up to that of the magnetic lens. Intensity in terms of target current or exposure-time can be accurately measured but the actual spot size cannot. Hence the performance of two instruments is best compared by arranging identical exposure-times and then estimating the difference in resolution.

Similar considerations apply to the choice of lens for other electron optical instruments and the practical outcome has been the predominance of the magnetic lens. The original proposal for a projection X-ray microscope (von Ardenne, 1939, 1940) envisaged the use of electrostatic lenses but it appears that an early change was made to magnetic lenses (von Ardenne, 1956). Similarly the microanalyser of Castaing (cf. § 7.2) was first used with electro-static lenses but later was changed to magnetic lenses (Castaing, 1956) in order to obtain better performance. In general the slightly greater complexity of the magnetic lens is found to be more than offset by its greater reliability, lower aberrations and shorter focal length.

11.3. Constructional details

11.3.1. *Magnetic lens model.* A brief outline will be given of the more important constructional details of the projection X-ray microscope with magnetic lenses. Details have been given by Cosslett & Pearson (1954) and Cosslett, Nixon & Pearson (1957). A diametrical section taken vertically through the instrument (Fig. 11.7) shows how the electron gun, which is of the type described in § 11.1, is mounted on the first electron lens, the condenser. The electron gun anode is earthed and is part of the main column of the tube. The anode itself can be a flat disk or a long tube with an aperture at one end. In either case it should be movable in a plane perpendicular to the electron beam so that it can be used for beam alinement as suggested by Haine & Einstein (1952).

The magnetic lenses may be designed on the basis of the data given by Liebmann & Grad (1951). Values of the focal length f and the spherical aberration coefficient C_s are as follows, for a beam voltage of 10 and 20 kV in symmetrical lenses (bore = gap) of three sizes.

Bore and gap (mm.)	10 kV		20 kV	
	f (mm.)	C_s (mm.)	f (mm.)	C_s (mm.)
3	0·9	0·6	1·2	0·7
6	1·8	1·2	2·4	1·5
10	3·0	2·0	4·0	2·5

Le Poole & Ong (1957) have used a rather shorter focal length (0·75 mm.) for very high resolution operation (cf. § 15.4).

The most suitable pole-piece shape is that proposed by Mulvey (1953) and Liebmann (1953) for avoiding iron saturation effects, but the beam voltage is normally so low that saturation does not occur. The above values correspond to 3000 ampere-turns excitation, and are those for the 'first' focal length, that is, for the first crossing of the lens axis by the beam. This is the condition obtaining in X-ray microscopy, since the target is immersed in the

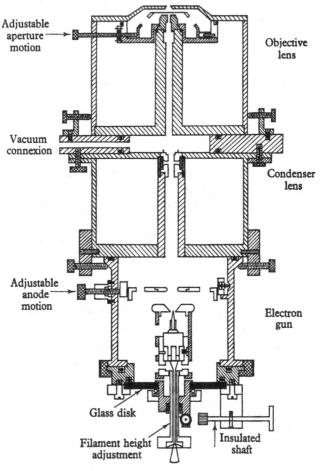

Fig. 11.7. Detailed diagram of the magnetic lens projection X-ray microscope.

field of the objective lens and the electrons are stopped at the target.

The bore and the opposing surfaces of the pole-pieces must be finished to high accuracy to avoid any limitation due to elliptical astigmatism. After machining, the pole-pieces may be joined together by shrinking on to a brass spacing ring or by using a thin layer of soft solder between the brass and the two iron pieces. A screw thread is not advisable as it might produce more severe astigmatism. The material used for the pole-piece should be free of inclusions or voids that might lie just below the surface and upset the magnetic field. The iron is best selected from a long billet, after testing to find short pieces which are magnetically uniform. Alternatively the type of lens described by Haine (1956) with integral pole-pieces can be used.

After the pole-piece bore, gap and shape have been decided, the pole-piece as a whole must be fitted to the lens coil and to the rest of the magnetic circuit. The choice of dimensions for electron lenses has been discussed fully by Zworykin, Morton, Ramberg, Hillier & Vance (1945, p. 145) and Hall (1953, p. 111). In practice the focal length determines the number of ampere-turns and the lens coil is designed to produce this at the lowest wattage. An example will illustrate the values used.

A convenient wire for the coil is British S.W.G. no. 26 copper wire, 0·018 in. (0·45 mm.) dia., which will carry 300 mA and when closely wound gives 390 wires/cm.² of coil cross-section. The required 3000 ampere-turns can be conveniently obtained with either 20,000 or 15,000 turns and with a variety of different shapes of coil; in general a long narrow coil will have a lower power dissipation than a short flat coil, since the outer turns add considerably to the total resistance. Some typical data are set out in Table 11.1. No allowance has been made for the insulation between layers, which is necessary even though the wire is enamelled. Owing to the method of layer winding, the full voltage drop in two layers will appear across two adjacent wires (in depth) at the end of a layer. Large induced voltages will be generated, if the coils are switched off from full load, which might short circuit several layers of wire. An interleaved layer of insulating paper should be used with enamelled wire, or the paper can be omitted if double silk covered enamelled or plastic coated wire is used. The insulation takes up some of the space allowed for windings and the effect is to raise the current and the wattage slightly over the values given in the table.

TABLE 11.1. *Alternative designs for a coil to produce 3000 ampere-turns excitation.* R_1 = *internal radius,* R_2 = *external radius, and* L = *axial length of the winding;* N = *number of turns;* I = *current*

NI	N	I (mA)	R_1 (cm.)	R_2 (cm.)	L (cm.)	Area (cm.²)	Length ($\times 10^5$ cm.)	Resistance (Ω)	Voltage (V)	Power (W)
3000	20,000	150	1	8		7·35	51·4	5·65	581	87	13
3000	20,000	150	1	6·14	10		51·4	4·5	464	69·5	10·4
3000	15,000	200	1	8		5·5	38·4	4·24	436	87	17·4
3000	15,000	200	1	4·84	10		38·4	2·74	282	56·4	11·3

11.3.2. *Electrostatic lenses.* Few details have been published of the electrostatic lenses used for projection X-ray microscopy. The field emission X-ray microscope of Marton, Schrack & Placious

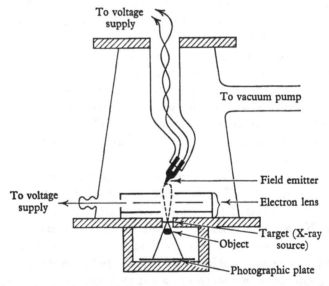

Fig. 11.8. An electrostatic lens projection X-ray microscope with field emission source. (Marton, Schrack & Placious, 1957.)

(1957) used a single unipotential lens to focus the electrons from a point filament onto the target (Fig. 11.8). In this type of lens the central electrode is at cathode potential and so must be well insulated from the outer electrodes, which are earthed. The lens was designed for a focal length of 1 cm. following the theory of Regenstreif (1951), the central electrode aperture being 5 mm. and the two outer apertures 2·5 mm. The electrode spacing was about 3–4 mm. at 10 kV. The chromatic aberration coefficient $C_c =$ 25 mm.; the spherical aberration coefficient was taken to be 342 cm. The central holes were round to about $\pm 1\mu$ in diameter and the aberration to be expected from this elliptical astigmatism would be 0·1μ. The final electron spot from the combination of field emitter and electrostatic lens was 5μ, well above that expected from theory. Possible causes are insufficient phase compensation (see below) or a larger spherical aberration than assumed.

The electrostatic X-ray projection microscope of Newberry and Summers (1956) incorporates two lenses of the unipotential type with the central electrode at high voltage; the target fits into the aperture of the final electrode. The condenser lens electrodes have a larger bore than those of the objective, so that it acts as a weak lens. The spherical aberration coefficient of the objective should be less than that of the normal type of unipotential lens, owing to the target closing the last electrode (Liebmann, 1949). The commercial version of this instrument is described in § 11.13.

11.4. Mechanical and electrical stability

Vibration and movement of the X-ray microscope specimen can be caused by the rotary mechanical backing-pump, by the liquid boiling in the diffusion-pump, or even by the slamming of a door. Although no vibration of the test specimen may be visible when focusing the microscope, the long exposures (5 min. or more) needed at high resolution offer the chance of movement of the specimen by distances of the order of the resolution sought, for example, 0·1μ (4 μin.). In the electron microscope the resolution is some 100 times better, but the exposure is only a few seconds and long-term drift does not affect the image. During high resolution operation, the backing-pump should be switched off, the diffusion-pump heater reduced below the bumping temperature and the pump outlet connected to a reservoir. A solid specimen stage and mount, securely fastened to the objective lens, will help to ensure that these parts move together so that the position of the specimen with respect to the focal spot does not change even if there is some slight movement of the whole microscope.

The electrical stability required for magnetic lenses can be calculated from the focal length relationship, as for electron microscopes (Drummond, 1950). For a focal length f, voltage V, and lens current I,

$$f = KV/I^2, \qquad (11.9)$$

where K is a constant depending partly on the design of the lens. On differentiating,

$$\Delta f/f = \Delta V/V - 2\Delta I/I. \qquad (11.10)$$

A small change in focal length, Δf, will produce a disk of confusion of radius δ_c in the target plane. If α is the semi-aperture of the beam at the image,

$$\delta_c = \alpha \Delta f. \qquad (11.11)$$

The spot size, δ_s, as limited by spherical aberration is:

$$\delta_s = \tfrac{1}{4} C_s \alpha^3 \qquad (11.12)$$

and for the lenses used f and C_s are approximately equal. If we set δ_s equal to δ_c, eliminate α and use (11.10) to find f in terms of V at constant I, then the resolution δ is

$$\delta = 2f(\Delta V/V)^{\frac{2}{3}}. \qquad (11.13)$$

Thus at constant focal length of the final lens the change in accelerating voltage and the resolution of the magnetic lens are related by a 3/2 power law. In other words $\Delta V/V$ falls more slowly than the resolution: for a change of δ by 8, $\Delta V/V$ changes by $8^{\frac{2}{3}}$ or 4. The change in lens current, also given by (11.10) is twice that for the high voltage. The negative sign shows that for a positive change in current the focal length decreases whereas an increase in voltage brings an increase in focal length. These relationships are plotted in Fig. 11.9. With typical values of $\delta = 1\mu$, $f = C_s = 1$ mm. and $\alpha = 1\cdot6 \times 10^{-1}$, then

$$\Delta V/V = 1/160;$$

and for $\delta = 100$ Å, $f = C_s = 1$ mm. and $\alpha = 3\cdot5 \times 10^{-2}$

$$\Delta V/V = 1/3400.$$

So that, even at the best resolution expected with the X-ray projection microscope, the high voltage and lens current stabilities are much less than those needed in the electron microscope at a few Ångströms resolution (1/100,000 and 1/200,000 respectively).

However, since the X-ray microscope stability must be maintained over much longer exposures, long-term drift is more dangerous than short-term fluctuations.

The unipotential type of electrostatic lens has adequate constancy of focal length with voltage in the non-relativistic range, but between 20 and 40 kV the chromatic error is of order $1/1000$ only. Focusing can be carried out mechanically or electrically, the latter having an effect on the stability requirements. The lens

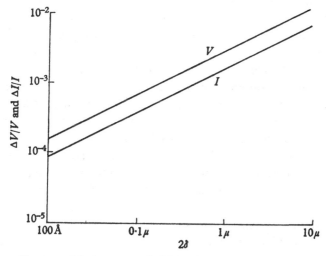

Fig. 11.9. Maximum permissible variation in high voltage and lens current for electron beams from 100 Å to 10 μ in diameter.

circuit must be phase-compensated by placing a variable condenser across the focusing resistance and adjusting its value whenever the focusing resistance is altered. Marton, Schrack & Placious (1957) have used this system in an electrostatic lens projection microscope with pulsed high voltage. In the absence of compensation the electron spot size was as much as 20 μ, completely spoiling the resolution.

The microanalyser of Castaing (1951) had two electrostatic lenses but used a dropping resistor only on the first lens. The final focusing was done with a 500 V battery of low internal resistance, so that the phase shift is negligible.

11.5. Electrical supplies

In general, the supplies for the high voltage, the lens current, the filament and grid bias of the electron gun all follow modern electron microscope practice. Some minor differences occur due to the lower kilovoltage and different stability requirements. The high voltage supply, which must provide from 3 or 5 kV up to 30 or 50 kV and a current up to 500 μA, is best produced by a voltage doubler circuit fed by a radio frequency oscillator. Air insulation is sufficient even at the upper limit of kilovoltage, but oil insulation may be preferred for compactness and shock-proofing. The required stability is obtained by using a high resistance chain between high voltage and ground potential and tapping off a low voltage near the bottom of the chain. This sample voltage is compared with a standard reference voltage, the difference being fed back to the control grid of a two or three stage D.C. amplifier and then to the screen grid of the oscillator valve. The reference voltage can be supplied by batteries or by a conducting neon tube; the latter is preferred for the X-ray microscope where long-term stability is more important than the short-term fluctuations detrimental in the electron microscope. The type 85A2 neon tube is stable within 1 part in 1000 to 1 part in 5000 after it has been used for 100 hr. and has a useful life of about 1000 hr. Any instability in the high resistance chain itself is not corrected by this system since, although the high voltage may be stable, the detecting system may be receiving signals due to change of resistance in the chain caused by internal heating or variations in the ambient temperature. If commercial high-quality resistors are used, this effect will not be noticeable but might occur with a chain of standard ± 20 % carbon resistors.

The lens current supplies can also be stabilized by the negative feed-back method, modified to maintain the current constant. The lens coil resistance changes slowly as the coil heats up, particularly when operating at the shortest focal lengths for high-resolution work, at high wattage. If the lens is supplied by accumulators at constant voltage, this change produces a variation in the focal length during an X-ray exposure, but the ease of construction and compactness of modern current regulators has almost eliminated the use of batteries for the lens coils. The current drain is typically 150 mA at 100 V, with slight variations in the objective lens for focusing and for different beam voltages, and with wide variations in the condenser lens for producing a range of values of electron spot size. This value of 15 W is about the maximum for continuous running without over-heating of the objective lens coil.

The electron gun filament supply should also be produced at radio frequency, to avoid the broadening of the electron beam which would be caused by the low frequency magnetic field of a heating current of 2–3 A. High frequency generation is also more convenient in practice than the use of the 50 cycles mains supply. The wattage necessary can be supplied by a single oscillator valve, since only 2–3 V at 2–3 A is required, a maximum of less than 10 W. Stability here is not so important as with the other supplies. Alternatively, a 6-V accumulator can be used as filament heating supply but, being at high voltage, must be insulated.

11.6. Vacuum requirements

A high vacuum is required in the X-ray tube for three main reasons. In the first place the mean free path of the electrons from the gun to the target must be at least as long as the X-ray microscope column. Secondly the prevention of high voltage breakdown between the electrodes in the electron gun, or in electrostatic lenses, requires a pressure at least as low as 10^{-4} mm. Hg. Finally, a reasonably long filament life depends on good vacuum. Both air and water vapour attack a hot filament and, unless a mercury diffusion-pump with liquid air-trap is used, a phosphorous pentoxide drying chamber should be included in the vacuum system. The relation between filament life and gas pressure has been investigated by Bloomer (1957); a life of 5 hr. at 10^{-3} mm. Hg rises to 50 hr. at 10^{-4} mm. Hg. In general a vacuum of 10^{-4} mm. Hg is the minimum permissible for reliable operation and 10^{-5} mm. Hg should be reached with a well-tested system.

The vacuum system should have a diffusion-pump and mechanical backing-pump, although a high-speed molecular pump has been used. A by-pass pumping line with suitable valving allows the X-ray microscope column to be let down to atmospheric pressure for changing the filament or other components, without having to cool the diffusion-pump liquid. Vacuum gauges are necessary to indicate when the tube has been pumped down. These can be very simple, such as a mercury manometer and discharge tube, a Pirani type hot-wire system, or thermocouple meter, which will all indicate pressures down to 10^{-3} mm. Hg or less and are not harmed by a sudden vacuum failure such as a burnt-out X-ray target. The Philips cold cathode discharge gauge is perhaps the best type, measuring down to 10^{-5} mm. Hg and being unaffected by atmospheric pressure. The hot cathode ionization gauge, which burns out at quite moderate pressures, is not recommended.

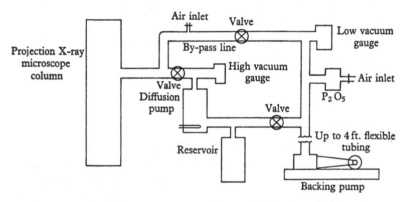

Fig. 11.10. A suitable vacuum system for a projection X-ray microscope.

The diffusion-pump fluid can be oil or mercury, but the inconvenience of using a liquid air trap with mercury more than offsets the advantage of reduced target contamination, and oil-pumps are usually fitted to X-ray microscopes. The oil diffusion-pump has a higher speed than the mercury-pump and so can deal with small leaks in the thin metal foil target. So long as small leaks are absent, either pump may be used with a vacuum reservoir as ballast so that the X-ray microscope can be operated for several hours at a time with the mechanical backing-pump turned off. This eliminates a source of vibration and noise and improves the mechanical stability and ease of operation. The diffusion-pump can be water or air cooled; as the tube itself requires no water cooling of the target, air cooling is convenient.

A suitable vacuum system for a projection X-ray microscope is shown in Fig. 11.10. The physical principles of high vacuum technique have been set out by Yarwood (1956) and Dushman (1949) and details of leak testing, pump maintenance, gauge calibration, valve types, etc., will be found there.

11.7. Microscope alinement

First the correct setting of the electron gun must be found and a beam obtained with the two lenses switched off. Rough alinement is then carried out with a fluorescent screen in place of the target; it is helpful for them to be interchangeable without breaking the vacuum. Either the condenser lens or objective lens may be taken as fixed reference, and the other components then brought onto a common axis with it. It is simplest to start with the condenser fixed and to aline the electron gun with it, the objective lens being switched off. The anode and gun lateral controls (or alinement coils) are used to bring the electron beam (of $0.1-1\mu A$) onto the condenser lens axis. The lens strength is varied and if the electron spot seen on the fluorescent screen swings away from the axis and out of view, the gun controls are operated so as to bring it back; beam tilt, by moving the anode, allows a more sensitive alinement control than gun traverse. This stage is complete when the electron beam opens out symmetrically on each side of a focused spot on the screen when the condenser lens current is varied. Then the objective lens is switched on and the same process repeated, now moving this lens so as to bring it on to the common axis of the condenser lens and electron gun. The amount of traverse needed in the objective lens is usually very small, since the fluorescent screen and target are almost at the focal plane of the lens, whereas any lack of alinement of the condenser is much more marked owing to the long distance between the lenses.

When rough alinement is finished, the fluorescent screen is turned aside and the target brought into position. For final alinement, the largest of the various adjustable apertures is brought into the beam; a beam current of a few microamperes is desirable. Test grids are placed in the specimen holder, preferably of two different gauges, for instance, a 200/in. electron microscope grid supporting a finer 1500/in. grid. A small fluorescent screen (a few millimetres square) close to the target should be used for detecting the first X-ray image. It may be very faint, owing to the rough alinement, but slight touches of the gun controls should increase the intensity. At this stage it is necessary to keep the total beam current low so that there is no chance of burning out the target when a slight realinement brings the full electron beam onto the target. The aim of the grid test is to produce conditions such that, when the condenser lens current is varied, the brightness of the grid image varies, but retains its sharpness and does not move. The external fluorescent screen is now placed at a few centimetres distance from the target, to obtain a highly magnified X-ray image. At × 1000 magnification, a sideways movement of 1 μ of the electron spot at the target will produce a shift of 1 mm. in the image of the 1500 mesh grid on the fluorescent screen, which is easily seen. This is a much more stringent test of alinement than viewing the electron spot on the internal fluorescent screen, and so the first stage should be accomplished quickly, leaving the final adjustment to the X-ray image test. When the objective lens current is changed, the grid image should stay at the same brightness, but its sharpness should vary on either side of focus. Since this behaviour is the opposite of that produced by the condenser lens, the two conditions can be tested separately.

As the filament ages it may shift, and this will result first in a reduction in X-ray intensity and then in a loss of alinement, producing astigmatism in the image. Realinement of the gun and lenses should cure this fault, unless the filament has twisted too far off the axis. If the misalinement cannot be corrected in a few minutes, it is better to remove the old filament and fit a new one. Some shifting is usually noticed with a new filament during the first ½–1 hr. of running, after which it should remain centred, except for minor adjustments when heated and cooled, until near the end of its useful life, which should be of the order of 50 hr.

11.8. Microscope focusing

In order to obtain the sharpest possible projection image (high resolution), the X-ray point source must be as small as the experimental conditions allow, that is, the electron beam must be focused on the target. The same procedure is followed as for alinement, the X-ray image of a test grid being viewed on a fluorescent screen with a × 10 magnifier whilst the lens currents are varied. Use of the eyepiece allows a lower X-ray magnification to be used, giving a brighter image on the fluorescent screen, since the screen may be placed closer to the target.

With the coarse grid in the specimen holder and the X-ray magnification set low by moving the holder away from the target, the objective lens current is varied about the focal value. As soon as an image is seen the grid can be moved closer to the target to increase the X-ray magnification. Finer adjustment of the objective lens strength will permit successively higher X-ray magnifications until finally the coarse grid will not be seen on the screen, when the issuing cone of X-rays passes through a single grid opening. At this stage the fine mesh should be visible and the adjustments can be continued, with successive increases of X-ray magnification; with a properly designed and well alined lens system, this should reach about × 1000, corresponding to a target –grid distance of 10μ with a target–screen distance of 10 mm. The final setting of the objective lens current must be accurate to the stability required by (11.10) and (11.13), as plotted in Fig. 11.9. For a resolution of $0\cdot1\mu$, $\Delta I/I = 1/1500$ and so a lens current of 150 mA would have to be adjusted to within $0\cdot1$ mA for exact focus.

This limiting depth of focus of the electron beam at the target may also be derived for a given resolution δ, by geometrical optics, from a knowledge of the angular aperture as fixed by the spherical aberration of the lens. The ray paths from the lens aperture to the target are shown in Fig. 11.11 with R as the radius of the physical aperture, α the semi-angular aperture, r the radius of the focal spot, f the focal length of the lens and $2\Delta f$ the depth of focus. By (11.12) we have $r \doteq \delta_s = \frac{1}{4}C_s\alpha^3$ and from Fig. 11.11, $\Delta f = r/\alpha$. Eliminating α, the depth of focus can be expressed as

$$\Delta f = \pm(C_s/4)^{\frac{1}{3}}r^{\frac{2}{3}}. \qquad (11.14)$$

For example, with a spherical aberration coefficient of 2 mm. (equal to the focal length), the depth of focus is $\pm 8\mu$ for $r = 1\mu$

($R = 250\mu$ and $\alpha = 1.25 \times 10^{-1}$) and $\pm 1.7\mu$ for $r = 0.1\mu$ ($R = 120\mu$ and $\alpha = 5.9 \times 10^{-2}$). If mechanical focusing is used to bring the target to the focal plane of a lens of fixed focal length, such as the electrostatic unipotential type, then this accuracy of adjustment must be achieved. However, if the spherical aberration of the lens is worse then the depth of focus will be greater, since the physical aperture must be reduced to keep the same spot size. For $C_s = 20$ mm. and the same focal length of 2 mm., $\Delta f = \pm 17\mu$ for $r = 1\mu$, and R has dropped to 120μ. The variation of Δf with respect to r, for a given value of C_s, is shown in Fig. 11.12.

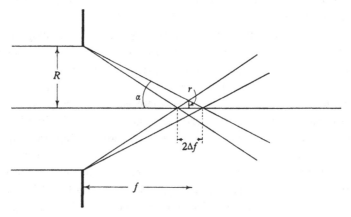

Fig. 11.11. Ray paths showing the relationship between depth of focus and angular aperture of the objective lens.

Adjustment of the condenser lens current is also necessary, if the smallest spot for a given beam intensity or X-ray exposure-time is to be attained. The variation of demagnification with condenser lens current I_c is plotted in Fig. 11.4, for given distance between the lenses and with the objective at maximum strength. From this the value of I_c can be found which will give the required size of electron spot, assuming the effective size of the electron source to be 50μ, the approximate value when filament wire of diameter 100μ is used. As the source size depends somewhat on the operating conditions of the electron gun (Haine & Einstein, 1952) and as the final size of the X-ray source is determined by electron scattering in the target (§ 3.6) as well as by the size of the electron spot itself, the value deduced from Fig. 11.4 for a given condenser current can be approximate only.

The final setting of the two lens strengths for the highest X-ray magnification should be carried out in complete darkness and with the eyes dark adapted so that the best focal setting can be found. When viewing at low light levels the visual acuity of the eye is worse than 0·1–0·2 mm., the value obtaining in good reading illumination, and so the maximum available magnification must be

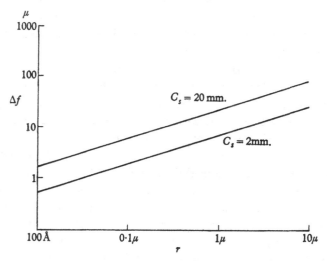

Fig. 11.12. Variation of depth of focus with respect to beam radius for two values of spherical aberration.

used, × 1000 by X-ray projection followed by × 10 optical. The final stage of fine focusing may be carried out by photography, when the X-ray image is too faint, successive exposures being made after slight changes of the objective lens control. Before commencing such a series it is advisable to switch the objective lens current off and on several times in order to reduce hysteresis effects in the iron and consequent magnetic drift during the exposure. A neon lamp and resistance overload should be provided to protect the current meter and the lens coil from high values of induced back e.m.f.

11.9. Photographic recording

When focusing is complete, the specimen holder with the test grid is removed and another carrying the actual specimen to be radiographed is inserted. Alternatively a test grid may be mounted alongside the specimen so that it is only necessary to change the magnification to a lower value to bring the specimen into view, the test grid remaining in one corner of the field to serve as a magnification check. This should be done with the X-ray beam switched off, but with the electron gun and lenses still operating, by using a beam switch as described earlier (§ 11.1). The part of the specimen to be photographed is positioned and brought to the required magnification by visual observation of the fluorescent screen, which is then replaced by a photographic plate at some standard distance, and the exposure made. The time of exposure will vary greatly with kilovoltage, beam current, thickness and atomic number of target and specimen. The exposure also depends on the camera length, the type of emulsion, and the development of the photographic plate. With a lantern-slide emulsion at the same position as the fluorescent screen, 5 min. is usually the longest exposure if the X-ray image can be seen on the fluorescent screen at all. The exposure can fall to one second or less, even with a very slow maximum resolution emulsion, if it is only 1 mm. from the X-ray source or to a small fraction of a second for high-speed ciné film at a distance of 1 cm. Photographic processing varies with the emulsion used, but in general the need is for a high contrast negative which is then printed onto a high contrast paper. Since X-rays are not scattered in the emulsion to the same extent as light or electrons, a larger optical enlargement is possible. With × 20 magnification of the negative very little grain is visible with a lantern-slide emulsion; it is thus possible to use a low X-ray magnification, giving a larger field of view and shorter exposure-time.

11.10. Stereographic projection

In the simple shadow projection method of image formation there is no unique image plane in space corresponding to a given plane in the object. Different object planes are registered on the photographic plate with different magnifications, but all planes are equally sharp when the negative is subsequently magnified. This 'infinite depth of field' of the projection microscope can be used to obtain stereographic views of the interior of the specimen; in combination with the penetrating power of X-rays it forms one of the most valuable methods of X-ray microscopy.

The use of stereographic methods in contact microradiography has been outlined in § 10.3. Since the X-rays form a parallel beam, the specimen must be tilted between two exposures so that two

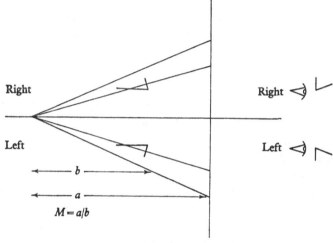

Fig. 11.13. Translation of the specimen across the cone of X-rays produces right and left views, which may be combined to give a stereographic effect.

different aspects of the specimen are recorded for stereographic viewing. The separation of the specimen from the photographic plate in projection X-ray microscopy makes it easier to change the emulsion between exposures than in the contact method. In the projection method the specimen can be tilted as before or, alternatively, it can be translated across the cone of illumination as shown in Fig. 11.13. The necessary amount of lateral motion x is related to the total magnification M and the distance between the eyes, taken as 65 mm., by

$$x = 65/M. \qquad (11.15)$$

When the magnification is greater than 65 times, the lateral shift will be less than 1 mm. and the normal stage motion controls for scanning the specimen may be used to move the specimen between the two exposures. The exact distance must be known for accurate measurements of depth, but any other distance will still produce a stereographic pair. A larger shift can be used to increase the effective base-line of the three-dimensional view, as is done in aerial photography, and this will give an exaggerated depth that can be measured more easily and accurately. The true depth can still be found if the exact shift is measured by using a micrometer stage advance or, indirectly, by the shift of a feature in the two micrographs if the magnification is found by calibration with 1500 mesh grid. The shift can be converted into the effective tilt-angle by dividing by the distance between target and photographic plate.

The change of magnification through the thickness of the specimen produces perspective in the stereographic view, the nearer of two similar features appearing larger and the farther smaller. This does not occur with parallel illumination, as in contact microradiography and electron microscopy. The amount of perspective depends on the thickness of the object with respect to its distance from the X-ray source b. With a specimen 0·1 mm. thick and 1 mm. from the source there will be 10% difference in magnification between the front and the back surfaces, and this is hardly noticeable. For higher magnifications where the source distance might be 0·1 mm., equal to the specimen thickness, the difference would be 50%. The stereographic view would be so deep that the eyes could not accommodate all of it, at any given separation of the two negatives. In this case the stereo-views should be studied without permanent mounting, so that any chosen section of depth can be brought into view by varying their separation.

For producing a stereographic effect it is important to have the X-ray micrographs as sharp as possible so that the eyes have a definite point in space to focus on. In practice it is sufficient if one of the pair is sharp with the other one just giving the third dimension. The two negatives should also be of the same density so that neither view predominates. Since the two negatives are seen overlapping, it is possible to take two stereo-views each with half the normal exposure, so that in combination they give the normal density. Thus the third dimension may be obtained without increasing the total exposure-time, which may be of importance with specimens that deteriorate in the microscope. The lateral shift or tilt-angle can vary within wide limits, but there must be no rotation of the specimen between the two exposures. At lower magnifications it is better to tilt the specimen rather than translate it, in order to keep the same field of view in each negative. A typical stereographic pair taken with the projection X-ray microscope by tilting the specimen is shown in Pl. IX A. Here the depth is so great that it is necessary to refocus the eyes between the front and back of the image.

Von Ardenne (1940) suggested an alternative method of producing a stereo-pair of images, by displacing the electron beam in the X-ray microscope. This is difficult in practice unless very high magnifications are used, since the amount of shift required would move the electron beam off the axis of the electron lens. Alternatively, double deflexion could be used with a long focus lens as in the X-ray emission scanning microscope (§ 7.3) so that the beam always passed through the lens on the axis. The X-ray absorption scanning microscope, using single deflexion, was designed for stereographic work by using two X-ray detectors set at an angle to each other, the outputs being displayed on two separate cathode-ray tubes of small size mounted 65 mm. apart (Pattee, 1953b).

It is also possible to follow the medical practice of 'Planigraphy', in which both plate and specimen are moved laterally during the exposure, thus producing a sharp image of only one plane of the specimen, that is, where the distance moved by the plate is equal to the distance moved by the specimen multiplied by the magnification of the desired plane. All other planes are blurred and add only to the background fog on the negative. In the projection method, owing to the large depth of field, it is possible to obtain both stereographic presentation, to show the true three-dimensional view of the specimen, and planigraphic presentation to study any one plane alone (Lindblom, 1954). In the optical microscope, the small depth of field does not permit stereoscopic viewing at high magnification and only the planigraphic presentation can be used.

A form of stereoscopic effect can be seen on the fluorescent screen when using a thick specimen of high contrast such as the aluminium–5 % tin alloy of Pl. XXII. When the magnification of the specimen varies by (say) 50 % from one surface to the other, the images of the two surfaces will move at different speeds on the fluorescent screen when the specimen is moved laterally. Thus the view of the specimen will appear stereoscopic as the stage is shifted back and forth.

This example indicates the value of taking more than two exposures in succession, continuing to move the specimen laterally between exposures. The negatives can then be combined in various pairs for stereo-viewing, using the first and second for the normal view, first and third for a view with exaggerated depth, and second and third for a normal view from a different direction to reveal any feature that might have been covered up in the first pair. Successive exposures can be made on the same plate, particularly on the 10 in. by 2 in. Kodak Lantern Plate. If five 2 in. by 2 in. exposures are made the stereo-views will be too close together to view directly, but by adjusting the plate advance mechanism four 2½ in. by 2 in. exposures may be made. As these are almost exactly 65 mm. apart, they allow direct stereo-viewing of adjacent pairs, without special mounting.

11.11. Transmission targets

The electron beam in the X-ray microscope is focused on to a thin target, in which the electron energy is converted mainly to heat, with a relatively small output of X-ray photons useful for image formation. The efficiency of X-ray production is greatest for elements of high atomic number, but thermal properties are also important since the target must be able to withstand considerable local heating while under atmospheric pressure. The relative importance of atomic number, melting-point, and thermal conductivity has been discussed in § 3.6.4. Usually the target will be in the form of a foil thick enough to stop all the electrons (a thick target in the classical X-ray sense), but for high resolution a thinner foil partially transparent to the beam may be used. Both types of target can serve as the vacuum window of the tube, without support or with a backing of beryllium or other light element. With very thin targets, and at low kilovoltages for soft X-ray production, it is desirable to evacuate the space on both sides of the target. The type of target used initially was rolled tungsten, $1\,\mu$ thick, supported over an aperture of $250\,\mu$ (Cosslett & Nixon, 1953). At 10 kV, electrons penetrate some $0\cdot2\,\mu$ into the tungsten; the rest of the target absorbs about 50 % of the X-rays generated in this layer. It was possible to burn a hole in the target only with an electron spot larger than $3\,\mu$ in diameter (Nixon, 1952).

A more useful target for routine work is $3\,\mu$ copper foil.* The increase in thickness gives better cooling and allows a slightly larger value of spot diameter before failure, thus guarding against accidental burn-out while focusing and alining the microscope. At voltages above 15 kV there is an appreciable output of the copper K lines, which helps to give better contrast than with tungsten. A disadvantage is that the density of copper is about half that of tungsten, so that electrons penetrate about twice as far into the target, giving poorer resolution. For convenience copper is preferable, but for resolution tungsten.

Both of these foils are normally vacuum tight, but will not support atmospheric pressure over apertures larger than about ½ mm. Over larger holes, such as are used in a system for interchanging targets and fluorescent screen while under vacuum, the target foil can be backed by a $50\,\mu$ beryllium layer, which transmits 50 % of the X-rays even at a wavelength of 6 Å. The beryllium foil will be imaged as well as the specimen itself, and so it must be free of impurities and imperfections that would show on the micrograph. A test exposure, with specimen absent, should be made whenever a new piece of beryllium is used. The target metal can be evaporated onto the beryllium, but the deposited layer must be thick enough to stop all the electrons at the kilovoltage used; otherwise X-rays will be generated within the beryllium and may spoil the resolution by causing background fog (Nixon, 1955b). The target foil, simple or composite, is sealed into the target holder with a vacuum sealing compound or a rubber washer. The total load on such a target is of the order of 1 W only (20 kV and 50 μA), and no water-cooling is necessary. This low power is still sufficient

* Obtainable in large sheets from S. Goodfellow, 10 Lime Street, London, E.C. 3, and the Chromium Corporation, Waterbury 20, Conn. U.S.A.

to melt some target sealing compounds, but lead washers, used once only, have been found satisfactory.

Of the other metals used as targets for special purposes, aluminium is the most common after tungsten and copper. With a foil thickness of the order of 5μ, the radiation from an aluminium target ($K\alpha = 8{\cdot}34$ Å) will still give enough contrast on a photographic plate. Mosley, Scott & Wyckoff (1956) used a 7μ Al target in an X-ray projection microscope operating at 10–15 kV and obtained 5-min. exposures.

In metallurgical applications the target metal must be chosen with care, because of the X-ray fluorescence which may be induced in the specimen (cf. § 10.1.2). The target element should have an emission line of longer wavelength than that of the elements to be detected in the sample, for example, for an iron specimen a chromium target is satisfactory. It can be prepared as a thin foil by vacuum evaporation and used with a beryllium window as vacuum seal.

In absorption microspectrometry the choice of target is determined by the element to be analysed (cf. § 6.3). If the composition of the specimen is completely unknown several micrographs should be taken at continually increasing kilovoltage: the contrast will change sharply as the wavelength of the absorption edge is passed. The successive use of targets of elements having emission lines on either side of the edge will then allow quantitative estimation of the element in question. Not all elements can be used as X-ray targets, but most metals can be evaporated on to a thin beryllium layer to make a composite target of the type described above.

For very high resolution, thin targets must be used unless the kilovoltage is lowered to 2–3 kV, which is undesirable because X-ray intensity is then so low that focusing is possible only by indirect methods. Commercial single-beaten gold foil will stand atmospheric pressure over an aperture of 100μ or less and is relatively vacuum tight. It is about $0{\cdot}1\mu$ thick and has been found to give an X-ray resolution of that order. Thinner foils of gold, aluminium or carbon, used for even higher resolution, need vacuum on both sides.

11.12. Maintenance of projection X-ray microscopes

The main servicing required in the projection microscope is the regular replacement of the filament and the target. The filament life is of the order of 10–50 hr., depending on the vacuum in the tube. Filaments may be purchased already mounted and pre-flashed, or the old mounts can be used again. Tungsten wire of diameter 0·004 in. to 0·006 in. is cut to length and bent into hairpin shape by pressing on it with a razor blade while the wire is resting on a rubber stopper. The filament is then spot welded to the support pins of the mount, preferably with nickel ribbon between and over the filament wire so that a good contact is made. The mounted filament is then placed in the electron gun and centred in the grid aperture; after pumping down to high vacuum, it should be run for a few minutes at operating temperature or slightly higher. This procedure releases the strains in the wire, and the filament may have to be recentred once or twice. Slight misalinement can be corrected with the gun-shift controls.

It is good practice to renew the target automatically whenever the filament is replaced, as this helps to ensure a run of many hours before target or filament fails. Copper targets tend to oxidize and can puncture more easily after long use. Target metals evaporated on beryllium will be slowly sputtered off by the electron beam, as the temperature can rise locally above the melting-point of the thin target metal without melting the mass of beryllium. The target will also collect carbonaceous contamination, as do electron microscope specimens (Ennos, 1953, 1954), which may deflect the electron beam and possibly reduce the total X-ray output, providing another reason for discarding the old target and replacing with a clean new foil.

The apertures which limit the electron beam also collect contamination and should be cleaned or replaced when the microscope is down at atmospheric pressure for filament replacement. Apertures of the platinum or molybdenum type can be brought to red heat to burn off the carbonaceous layer; the cheaper copper apertures can be thrown away each time and a new one used. Other parts of the column may need occasional cleaning, such as the anode plate and any other surfaces struck by the electron beam or even by scattered electrons. The grid shield should be cleaned each time the filament is replaced as it collects tungsten oxide from the filament, which may affect the high voltage stability, bias conditions and consequent beam forming action of the electron gun.

11.13. Commercial models of the projection microscope

The first X-ray microscope to be put into production (Pl. IX B) was constructed by the General Electric Co. (U.S.A.). It employs two electrostatic lenses of the unipotential type, the column being vertical, with the electron gun at the bottom (Newberry & Summers, 1956). The target, a thin layer of tungsten on a beryllium disk, rests on the top electrode of the final lens and acts as the vacuum seal. The aperture holder carries four holes of diameter ranging from 1500μ down to 100μ, any one of which can be selected without breaking the vacuum. The specimen stage has three-dimensional motion with 9 mm. horizontal traverses and vertical movement up to 6 mm. above the target. The fluorescent screen on a $\frac{1}{4}$ in. lead glass plate is viewed through a black hood or may be replaced with 4 in. by 5 in. plates or cut film for photography. The camera length is variable between 1 and 6 cm. Alternatively, a Polaroid-Land camera, without its glass lens, can be fitted.

This instrument was based upon the original projection X-ray microscope of Cosslett & Nixon (1951, 1952 a, b) with the substitution of electrostatic for magnetic lenses. The advantages of each type of electron focusing for projection microscopy were considered by Newberry & Nixon (1953) and it was decided that the electrostatic lens model was more economical to produce, although magnetic lenses would give better resolution and shorter exposure-times. The resolution of the General Electric model is about 1μ, with exposures of 20 min. for specimens of low absorption. A shorter focal length electrostatic lens has also been used (Newberry & Summers, 1957), giving shorter exposure at about the same resolution.

The Philips Electronics Company (New York) have modified the Philips EM 75 electron microscope (van Dorsten & Le Poole, 1955) as a projection X-ray microscope, with magnetic lenses throughout (Bessen, 1957a) (Pl. X). The electron gun and condenser lens are retained at the top of the column and a new objective lens is added, specially adapted for X-ray microscopy. The target assembly consists of seven different target materials mounted on a rotating disk in such a way that any one can be brought to the focal plane of the electron beam while the instrument is in use, without breaking the vacuum. The X-ray path, including the specimen and plate, can be at high vacuum or at atmospheric pressure. The photographic plate records five exposures each 2 in. by 2 in. or four successive stereographic views with $2\frac{1}{2}$ in. separation. For focusing, the X-ray image is viewed on a fluorescent screen through a binocular microscope at a total magnification by X-ray projection and optical enlargement of × 250. The useful magnification, after photographic enlargement, is × 1500.

The RCA electron microscope (type EMU 2B) was converted into an X-ray projection microscope by Siegel & Knowlton (1955, 1957), and a conversion kit based on their design is now marketed by Canalco of Washington, D.C. Similar adaptations of production models of electron microscopes have been made in Soviet Russia (Bagdikjanz, 1953, 1956) and in Japan, where specially designed X-ray projection microscopes have also been built, by the Shimadzu Company and by Kato, Yamanaka & Shinoda (1953).

Added in Proof

(a) The General Electric projection microscope has been modified to produce soft X-rays (Butler, Bahr, Taft & Jennings, 1960), by fitting an aluminium target $10\,\mu$ thick and operating at 4–5 kV. A vacuum camera was arranged for a direct magnification of × 6, higher magnifications requiring unduly long exposures. The resolution obtained in biological specimens was of the order of $1\,\mu$.

(b) The microdiffraction X-ray tube of Ehrenberg & Spear, described on p. 337, has been used for projection microscopy by von Batchelder (1959).

CHAPTER 12

APPLICATIONS OF X-RAY MICROSCOPY IN BIOLOGY AND MEDICINE

12.1. Living specimens

The differential penetration of X-rays through biological material forms the basis of medical diagnostic radiology, on the macroscopic scale. When the same methods are used for microscopy there is usually insufficient contrast between tissue and its water content to make any degree of magnification worth while. However, artificial non-toxic contrast media can be used to show vessels, tubes and ducts in living specimens. Air sacs within a body may be seen without injection.

12.1.1. *Rabbit.* The rabbit has been used extensively for X-ray microscopy both as a living subject and after death. The *peripheral circulation of the blood* has been studied by several workers, under the heading of micro-arteriography or micro-angiography (Barclay, 1951; Bellman, 1953; Saunders, Lawrence & MacIver, 1957). The rabbit ear, with the hair shaved, is easily penetrated by X-rays generated at 10–25 kV. Contrast is produced by injecting an opaque suspension into the blood vessels, as discussed in § 9.1.3. It is important to avoid vascular spasm and shock near the area under observation and the injection is made in an artery closer to the heart. Physiological pressures should not be exceeded. The anaesthetized rabbit is placed on the X-ray microscope with one ear over the point source, and is normally wrapped to reduce shock. An additional table is then brought up carrying the injection material and instruments. Exposures are made immediately after injection and continued until all of the contrast media has been eliminated from the system.

The X-ray image of the normal ear has been studied in detail to serve as a control for experiments in which the skin has been subjected to heat, cold, bruising, cutting and gas. The normal rabbit ear is shown in Pl. XI*a* and the same area 50 min. after freezing in Pl. XI*b*. The capillaries have become congested and the contrast media trapped. The full cycle of freezing and thawing under controlled temperatures can be followed in this way (Bellman & Adams, 1956; Saunders, 1957*a*, *b*).

The living rabbit ear can also be used to study *lymphatic drainage*. In this case the contrast medium is injected subcutaneously as shown by the large dark area in Pl. XII A, which the lymph vessels are draining. The non-return flow valves in these tubes are clearly seen (Bellman & Odén, 1957; Saunders, 1957*b*). The distribution of the lymph vessels in depth has been studied by stereo-microlymphangiography (Odén, Bellman & Fries, 1958). The thorotrast-filled lymph vessels could be seen around blood vessels, with much smaller lymph channels throughout the tissue of the living rabbit ear. The lymph trunks were shown to be wide sacs between the valves. The extremities of the rabbit were studied for lymph drainage patterns with both living and dead material. A double-exposure method was used to determine parallax on fixed tissue and then two single exposures to find the relative values of the movement of the specimen. These sequential exposures were too slow for recording the circulation of the blood with the conventional X-ray tube that was used. A suggestion for more rapid stereo X-ray microscopy using two point sources simultaneously has been made.

Lymph drainage has also been studied during inflammation of rabbit ears (Barer, 1952). Turpentine was injected to produce an abscess and hot water circulated to produce heat inflammation. Thorotrast was then injected at low pressure. Good lymph drainage could be seen in the control specimen but poor drainage in the inflamed area, with a medium reaction for mild abscesses. The effect of spreading of the thorotrast was compared with the efficiency of drainage. In some cases it was found that hyaluronidase would restore the spreading and give better drainage even in badly injured specimens.

In a similar study of lymph drainage and blood supply both umbradil and angiopac have been used to provide contrast (Collette, 1953).

Other parts of the living rabbit can also be studied, if the organ can be partially removed by operation. The *blood vessels of the living ovary* are shown in Pl. XIIB (Engström, 1956). It was prepared free by operation and then placed in contact with a fine-grained photographic emulsion protected and made light tight by a thin aluminium foil. After exposure the organ was replaced for further growth before the next exposure of the series, to show the development of the blood supply of the animal.

With the projection method the specimen does not need to be in contact with the photographic film and this may be an advantage for thick wet biological specimens in the living state. These are suitable objects, when injected, if imaged at 20 kV or above; for instance muscle bridges and trunk flaps can be prepared for living X-ray microscopy (Saunders, 1958).

12.1.2. *Insects.* Small complete isolated specimens such as insects can be examined while living, since contrast arises from the air spaces within the body. In some cases this contrast disappears when the specimen is dried for more detailed examination. An example is the tracheal system of a living anaesthetized beetle (*Feronia madida*), shown in Pl. XIIIA (Smith, 1957). These air-filled tubes are seen in the pro-, meso- and metathorax while the leg muscles of the thorax and the coxa are partially obscured by the water content of the tissues. In the freeze-dried comparison the muscles are well defined while the trachae are invisible; the fruit fly (*Drosophila melanogaster*), shown in Pl. XIIIB after freeze drying, may be compared with Pl. XIIIA. A thorough study of flightless insects and the development of flight muscles has been made by Smith (1957), by X-ray microscopy of undissected freeze-dried specimens.

Contrast media can also be administered by ingestion with food and the digestive channels studied (Le Poole & Ong, 1957).

12.1.3. *Ciné X-ray microscopy.* Slow processes such as the lymph drainage mentioned above can be successfully recorded with exposures of several minutes, without blurring due to movement, if the animal is anaesthetized. Many interesting medical problems occur at a faster rate and several attempts have been made to apply moving picture methods to X-ray microscopy. Sherwood (1937*b*) used soft X-rays on small biological specimens with ciné recording. More recently Barclay (1951) has discussed in detail ciné-radiography at very low or unit magnification, by both the indirect and direct methods, in a study of kidney circulation in the rabbit. In the former, the low intensity image on the fluorescent screen is photographed with a ciné camera containing fast, large-grain film. This is most useful where large fields of view, as in human radiography, must be compressed onto 35 mm. film, but little enlargement is possible. There is a great loss of intensity due to the conversion of X-rays to light, which is emitted in all directions, so that only a small fraction enters the lens of the ciné camera. The direct method is much more suitable for micro methods where the field of view is small. The X-rays fall directly onto 16 mm. or 35 mm. film in a ciné camera from which the lens has been removed. The X-rays are soft enough for the light shutter to be also opaque to X-rays. The intensity conditions are more favourable and finer grained film may be used, allowing enlargements up to several hundred times. The speed of transport of the film (number of frames per second) is chosen so that the movement of the specimen is not quite enough to cause blurring. Various speeds have been used from a time lapse of several seconds or minutes between frames up to 16 and 24 frames/sec., in a study of evolution of gas in the pupa of the blow-fly, *Calliphora erythrocephalia* (Thevenard, 1955, 1956) (see also Bellman, Frank, Lambert, Odén & Williams, 1960).

The projection X-ray microscope has been used by Newberry & Summers (1957) with normal speed ciné film at 5 frames/sec. at a camera distance of 9 mm., to image test grids and insects. Stereographic methods have been added by Newberry & Norton (1958). An example of a film-strip, taken at 16 frames/sec. with fast film at 10 mm. from a point source of X-rays and using test grids as object, has been published by Nixon (1957*a*).

12.1.4. *Radiation dosage to the specimen.* High X-ray intensities are needed for ciné recording and there may be some radiation damage to the living specimen under these conditions. Bellman (1953), using an X-ray tube with a total output of 1300 r/sec., calculated a dose rate of 8 r/sec. on the part of the ear examined. Exposures of 1 sec. produced no reddening of the skin, but one rabbit showed pigmentation after two weeks. Calculation of the output of a projection X-ray microscope shows that the dose rate to the specimen would be of the order of 1 r/sec. if the X-ray energy were absorbed within 1 g. of tissue. In fact, the specimen is usually much smaller and so the dose rate is much higher, reaching 10^3 to 10^6 r/sec. depending on the radiation used. Some reduction in the total dose to a specimen may be possible if X-ray image intensifiers can be used with ciné recording of the image (§ 15.2). Until that time any results of ciné work on living specimens must be interpreted in the light of possible radiation damage to the area studied.

Conversely, an interesting application of X-ray microscopy would be the study of the effects of X-rays on the specimen during continuous observation. Microbeam irradiation of parts of single cells using α-particles has been achieved by Davis (1956; Davis & Smith 1957a, b). One micron apertures were used to collimate the beam of particles and the specimen was viewed with a long working distance phase contrast microscope. In considering the choice of radiation for microbeam work they calculated that soft X-rays could be used equally well. At 5 kV a gold collimator 20μ thick and with a 1μ hole would reduce the intensity outside the direct beam to 1 % of that in the beam. The intensity at the specimen must be enough to produce 1 ion pair/μ^3/sec. or about 0·6 r/sec. An X-ray tube at 1 cm. distance would then need to produce 40 r/min., which could be obtained from a commercial X-ray tube with a focal spot of 1 mm.[2] Seed (1960) has continued experiments along these lines using 8–10 kV on the X-ray tube, lead glass capillary collimators and a phase contrast microscope for locating and studying the damaged area of tissue. A dosage of about 100 r is given in a 3 min. exposure (0·6 r/sec. as noted above) to the nucleolus of cells in mouse fibroblasts. A single ultra-violet exposure is made after the radiation experiment has been carried out.

A review of all types of partial cell irradiation, including microbeam and apertured macrobeam methods, has been given by Zirkle (1957).

12.2. Thick sections

With the living specimens discussed above relatively little detail is seen in the X-ray micrographs, but what is detectable relates directly to the living state. When the material is fixed, dried and sectioned much more information is gained from the microradiograph but this must be correlated in some way with observations on the living material.

Normal histology with the light microscope is concerned with this reconstruction process and many of the established techniques can be carried over into historadiography. One major difference is that the sections for X-ray microscopy can be much thicker, a large fraction of a millimetre or more, depending on the material, as compared with the usual histological section of $5-10\mu$. Both thick and thin sections may be used with X-radiation and in many investigations several different thicknesses will be cut or ground. The thicker sections, $> 100\mu$, will be discussed here. Many examples of this type have been reviewed by Mitchell (1951, 1954), including bone, teeth, nerves, stomach, kidney, lymph nodes, wood, grain, leaves, cloth and paper.

The *rabbit* has been used extensively for sectioned studies as well as in the living state. Almost all of the organs have been examined as well as many functions such as circulation of the blood, lymph drainage and urinary secretion. The normal rabbit has been throughly investigated by Tirman, Caylor, Banker & Caylor (1951) with the purpose of forming an atlas of the microradiographic appearance of each organ for reference by future workers. They used 300μ sections and magnifications from $\times 50$ to $\times 250$. Results were shown of the kidney, small intestine, gall bladder, liver, heart, spleen and lung. Many vascular channels and capillary anastomoses were noticed in the heart, whilst the circulation of the spleen was less organized. The action of various drugs on the kidney was also reported.

The *circulation of the blood* has been studied on the microscale in many regions of the body, using fixed material previously injected. The blood supply to the rabbit ear was investigated with a projection X-ray microscope by Saunders (1957b). Extremely fine capillary detail can be seen since there is more complete filling of the vessels when the blood is completely replaced, and also there is no blurring due to movement of the animal during the exposure, as may occur with a living specimen.

A most interesting aspect of the normal circulation is the role of *arteriovenous shunts* that serve to short circuit part of the capillary bed when another section of the tissue produces a greater metabolic demand. A shunt of this type is shown in Pl. XIV A, after thorotrast injection, at higher magnification than the general view in Pl. XI (Saunders, 1957b). A branch can be seen between the broad vein and the much narrower artery, the actual join appearing S-shaped. These anastomoses have been observed in many specimens under different conditions. Bellman (1953) found both shunts and arterial spasm after cold injury to the rabbit ear (see also Röhrl, 1951, 1952). A general discussion of circulatory shunts and microradiography is given by Barclay (1951), with demonstrations in the kidney and stomach. The role of shunts in gastric ulcers has been studied by Doran (1951). Silver iodide was injected after the specimen had been removed and 100μ sections were cut. Low vascular density in the gastric mucosa or stomach wall resulted in ulceration.

The *kidney and liver* have also been studied in detail by X-ray microscopy, in an effort to follow the extraction of waste products from the blood stream in both normal and pathological conditions. Thick sections and complete organs may be X-rayed after freeze-drying and injection. Micropaque injection of the rat kidney was used to obtain Pl. XIV B, showing the glomeruli and capillary connections (Saunders 1957b). Many similar examples are discussed by Barclay (1951) in connexion with his work on renal circulation. Both rabbit and rat kidneys sectioned to 200μ were used by Bellman & Engfeldt (1955) in a study of the production of kidney lesions due to overdoses of vitamin D. Heavy deposits of calcium salts occur, and kidney failure may be the cause of death in both animals and humans when very large doses of vitamin D have been given over long periods. The blood vessels, glomeruli, tubules and interstitial tissue all showed incomplete filling due to calcium and sometimes the deposits themselves could be seen.

The rabbit liver has been studied by microradiography by De Sousa & De Sousa (1956). The bile was removed, thorotrast injected and $400-500\mu$ sections cut. The biliary capillaries were $1-2\mu$ in diameter, considerably smaller than the blood vessels of the liver. The bile ducts and lymphatic vessels were also noticed at $\times 130$ magnification.

Excessive calcium will lead to the formation of *kidney stones* due to inadequate lymph drainage. Microradiography has been used during operation on living material to determine the extent of the stones, followed by detailed study of the excised freeze-dried specimen in an attempt to discover the position of origin within the kidney (Carr, 1954, 1957; Lagergren, 1956). A correct theory of formation of these stones is necessary to decide on the amount of kidney that must be removed to prevent a recurrence of deposits.

The *lung* does not need to be injected to provide X-ray contrast since the many air spaces serve to separate the different regions of tissue. A section of air-dried human lung (Pl. XV A) shows air sacs (alveoli) and septa (Saunders, 1957 b). Both normal and diseased human lung have been radiographed with relatively soft X-rays by Cunningham (1955; Cunningham & Miller, 1952), using 2–3 mm. sections. Bronchial abnormalities could be detected including fibrosis and bronchial carcinoma. Pulmonary disease such as pneumoconiosis could be followed in depth by serial sections of the diseased tissue. Deposits of coal dust, with possibly some calcium content, create nodular lesions that eventually break down; similar results were observed from silicosis (see also § 12.5). Normal and diseased collagen (lupus erythematosis) have been studied at low magnification by Oderr (1957) using 100–300μ sections of dried lung. Arteries, veins, cysts and bronchial tubules could be detected by using the soft radiation from a beryllium window X-ray tube.

Thick sections of *bone* have been used in a few cases for microradiography but the absorption of soft X-rays by this material is so great that most work has been done with sections of less than 100μ (§ 12.3). Both normal and abnormal bone cut to 2 mm. have been studied at low magnification ($\times$ 10) by Sissons (1950). The progress of developing bone, resulting in organized bone trabeculae, could be followed. Stereographic methods were used with these thick sections and some thinner sections down to 100μ were also prepared.

Rabbit *bone marrow* has been investigated by Okawa & Trombka (1956), using 500μ sections and micropaque injection. A finer barium sulphate injectant was also used for the smallest blood vessels. Stereographic techniques and × 100 magnification were used for negatives obtained at 15 kV.

Sections of *dog skull* several millimetres thick have been examined with the projection X-ray microscope (Hewes, Nixon, Baez & Kampmeier, 1956). The blood was removed from the diploe veins of the bone and vinyl plastic injected by perfusion. The bone was decalcified and dried through a series of alcohol solutions. The three-dimensional arrangement of the veins was obtained by taking a few micrographs with an exposure of several minutes. Sectioning for light microscopy, and reconstruction from the optical micrographs, would have taken at least 25 hr. to obtain the same information.

The *human eye*, sectioned to 1 mm., has been studied by Pattee, Garron, McEwen & Feeney (1957) in an attempt to determine the drainage path of aqueous humour into Schlemm's canal. Thoro-trast was injected by different passages and the tissue freeze-dried, sectioned and radiographed stereographically. The traces of injection medium showed an open pathway between the anterior chamber and Schlemm's canal. Similar results were obtained with single microradiographs by Francois, Collette & Neetens (1955) using formalin fixation of the tissue.

12.3. Thin sections

The relatively thick sections ($> 100\mu$) discussed above cannot be used for very dense material or for very high resolution. Thinner sections of bone or similar material may be prepared by grinding or cutting (§ 10.1.2). *Bone and teeth*, in sections about 50μ thick, have been studied intensively by microradiography in order to determine the distribution of mineral salts in normal and pathological conditions (see especially Engström, Björnerstedt, Clemedson & Nelson, 1958). The great penetration of X-rays permits the use of normal bone, without the decalcification needed for optical microscopy. A microradiograph of normal bone is shown in Pl. XVB. The large dark areas are the Haversian canals and the small black dots the osteocytes—both are transparent to X-rays since they contain no minerals. The grey areas depict the local variation of mineralization within the bone. Normal bone has been compared with a number of diseased conditions such as osteogenic sarcoma, a type of malignant bone tumour. There is a very high mineral concentration in sclerotic tumours, partly due to the relative absence of collagen as compared to normal bone (Engfeldt, 1954).

In a separate study ectopic bone tissue has been experimentally produced by injecting alcohol into rabbit muscle. The resulting bone formation when studied by microradiography showed a similarity to the pathological condition of osteogenesis imperfecta, although X-ray diffraction indicated that the mineral content was normal. In this disease the bone remains immature and the canal system does not become fully developed (Engfeldt & Engström, 1954; Engfeldt, Engström & Zetterström, 1954b). Another pathological condition investigated by microradiography was osteopetrosis or marble bone disease. In this case the bone marrow becomes calcified and the loss of production of red blood cells leads to severe anaemia. The distribution of mineral salts is different from that of normal bone and shows characteristic patterns. The cause of the disease is due to some breakdown in the bone-forming cells (Engfeldt, Engström & Zetterström, 1954a).

The microradiographic appearance of a bone specimen has been compared with that of a stained optical section by treating the sections in two different ways (Bohatirchuk, 1957). The fresh material was embedded in plastic after dehydration and sectioned to 5–50μ thickness. First the microradiograph was taken and then the specimen was stained with safranin-fast green, and mounted next to the microradiographic negative in such a way that identical areas could be seen with the same orientation and within the same field of view at high magnification.

Fluorescent X-rays have been used in order to increase the contrast in bone specimens. The calcium absorption edge at 3·06 Å gives strong absorption on the short wavelength side, and scandium oxide, which has an emission line at 3·02 Å, was used as a secondary X-ray source by Dershem (1939). With 10μ sections of rat bone the exposure-time was 40 min. with 40 kV and 10 mA on the primary X-ray tube.

Many other aspects of the development of bone have been studied with microradiography and a long bibliography is given by Vincent (1956), in connexion with a comparison of the bone structure of dog and man (see also Holmstrand, 1957; Wallgren, 1957). Another large field of application has been the correlation of microradiographs with autoradiographs of bone in order to follow the uptake of radioactive isotopes of calcium, phosphorus, sulphur and strontium (Engfeldt & Hjertquist, 1954; Engfeldt, Engström & Boström, 1954; and Engfeldt, Björnerstedt, Clemedson & Engström, 1954).

The microradiography of *teeth* has developed in parallel to that of bone and many of the techniques may be used for both types of specimen. The original work of Applebaum, Hollander & Bödecker (1933) showed the structure of enamel lamellae, Schreger bands and Owens contour lines, and Thewlis (1940) demonstrated the differences between normal and carious teeth. More recently thin sections of dentine have been magnified to × 1075 and compared with polarized light observations (Miller, 1954). Most of the work on teeth between 1931 and 1958 has been summarized by Röckert (1958) in connexion with a quantitative study of the absolute amount of calcium in the cementum (§ 12.5).

Very thin sections, about 5μ thick, may be cut with a normal histological microtome and many biological and medical specimens are seen with the best resolution and contrast when prepared in this way. In order to obtain sufficient initial X-ray absorption and optimum differential contrast the tube voltage must be very low (cf. § 2.3.1). With a voltage of 1 kV and sections of 2μ, a resolution of 0.2μ was attained (Engström, Greulich, Henke & Lundberg, 1957). A hair follicle is shown in Pl. XVIA, a large field of view at low magnification (Engström, Lundberg & Bergendahl, 1957), and a 2μ section of onion root tip in Pl. XVII, at × 1500, showing dividing cells and detail within the chromosomes (Engström & Greulich, 1956). These photographs show the highest resolution obtained in contact microradiography of biological material.

Fitzgerald, Li and Yermakov (1957) studied normal and cancerous human uterine cervical cells in 5μ sections at 3 kV, finding a slight decrease in mass in the cancerous tissue. Lindström & Moberger (1954), using $5-7\mu$ sections, studied squamous epithelial cells in carcinogenesis of the vaginal wall of the mouse. The cancerous tissue showed a variation in dry weight while the normal tissue was uniform. Similar results were obtained with thin sections of skin cancer (Moberger & Engström, 1954). Contrast media may be used with very thin sections, such as 5μ specimens of rabbit kidney, but the amount of diodrast needed was sufficient to cause nephritis in the rabbit, that is, 8 ml./kg. body weight (Engström & Josephson, 1953). Mosley, Scott & Wyckoff (1956, 1957) used a projection X-ray microscope giving soft radiation (aluminium target) to examine 10μ sections of frozen-dried mouse kidney and 5μ sections of decalcified and embedded human dentine.

12.4. Botanical specimens

Plant structures may often be examined by X-ray microscopy with little or no preparation since leaves, stems, seeds, petals, pollen and other parts are very thin in the natural state. Staining, or removal of the water content, will increase contrast in the microradiograph. After the early work of Ranwez (1896) and Burch (1896), botanical applications were neglected in comparison with the medical uses of microradiography. Recently, however, there have been a number of investigations emphasizing the value of X-rays for studying the microstructure of plants.

Barclay & Leatherdale (1948; Barclay, 1951) studied rose leaves both in the natural state and after injection, at magnifications up to × 480. Small opaque bodies about 5μ in diameter were seen along the leaf veins and were still visible at 50 kV, indicating a material of high atomic number. Seasonal variations were noted as well as differences between species. A separate study involved injecting the growing plant with lead acetate through the roots, by natural absorption, and then radiographing the vascular system of the leaves. The growth of plant galls in infected leaves could be followed without the sectioning required in normal morphological studies, whereby the specimen is destroyed at a certain stage of its growth.

Other isolated examples have been given by Mitchell (1954) and Clark (1955), while detailed surveys in plant histology have been made by Legrand & Salmon (1954; Salmon, 1957a). Various types of crystals could be seen, including those noticed by Barclay &

Leatherdale in the ribs of the rose leaf, calcium oxalate in the skin cells of the bulb in *Allium*, calcium carbonate in the leaves of *Ficus*, and concentrations of iodine in the algae *Falkenbergia*. These discrete crystallites must be distinguished from local concentrations of mineral salts that may be formed during rapid drying down of the specimen. Comparison should be made between dried and living tissue to obtain information about the amount and distribution of water content, using optical microscopy in normal and polarized light as well as microradiography. An advantage of microradiography is that it shows simultaneously all the crystals present, whereas with polarized light it is necessary to rotate the analyser through the full circle to find their positions. Also, since lignin is birefringent, the polarizing microscope cannot distinguish it from the crystals within lignified tissue. On the other hand, lignin is transparent to X-rays of the wavelength used, and so the crystals along the ribs of *Rosa* leaves were differentiated (Salmon, 1954).

The reproducibility of the results obtained with normal plant tissue allowed the extension of the method to tumours and other cancerous growths in plants. Abnormal leaf growth in the neighbourhood of a radioactive hot spring has been correlated with a distinctive radiographic appearance of calcium oxalate crystals in the leaves, compared with the appearance of normal leaves (Salmon, 1957 b). Artificial induction of tumours in experimental plants is a more controlled way of following cancer formation. Distinctive radiopaque zones surrounded the tumour in *Pelargonium* after virulent injection and as these zones were not birefringent they could not be detected in any other way. Comparisons were also made with autoradiographs using strontium 90 and phosphorus 32, but of much poorer resolution (Manigault & Salmon, 1956; Salmon, 1957 b).

The projection X-ray microscope has been used to examine wood sections by Jackson (1957 a). Radial and transverse sections (1 mm. thick) of obeche showed the relationship of the large sap-conducting cells to the parenchyma cells and the calcium oxalate crystals. A 1 mm. section of pine (Pl. XVI B) shows the tree rings due to different rates of growth; the lighter, less dense areas are spring wood and the dark areas summer wood. There is some distortion of the cell structure away from the centre of the image, due to the large angle of incidence of the X-rays on the specimen at the edge of the field of view (Jackson, 1957b).

12.5. Quantitative applications

The physical basis of X-ray absorption microanalysis has been outlined in chapter 6. The main field of application has been that of medical and biological material prepared as thin sections. *Dry weight determination* (§ 6.2) may be used to detect masses as small as 10^{-14} g. with an error of 5–10 % (Engström, 1957). A comparison of mass/unit area in a 15μ section of human skin has shown that the outer layers have a mass of $7 \cdot 8 \times 10^{-10}$ g./100μ^2 while the inner layers show $5 \cdot 6 \times 10^{-10}$ g./100μ^2, or a mass ratio of 1·4:1; the ratio varies with the type of skin selected. Nerve cells and nerve fibres were also analysed at the same time (Engström & Lindström, 1950). In fibres 10μ thick, from the sciatic nerve of the dog, the outer layer of myelin sheath contained 5–8 times more material than the inner part or axon. In absolute value the outer layer had a mass of $0 \cdot 3 – 0 \cdot 4 \times 10^{-12}$ g./μ^3. Determinations of mass before and after the extraction of lipids allowed a determination of the mass involved (Engström & Luthy, 1949). A detailed study of nerve cells by microweighing by Brattgård & Hydén (1952) included estimations of lipids and nucleo-proteins. Deiters nucleus, Purkinje cells and spinal ganglion cells all showed a mass in the region of 10^{-13} g./μ^3. For Deiters cells the lipids, about $5 \cdot 6 \times 10^{-13}$ g./μ^3, equalled the pentose nucleoproteins and proteins taken together. On the other hand the nucleoproteins formed only 5 % of the total protein in Purkinje cells and 10 % in the spinal cells; the lipids fell to 30–50 % of the total protein. After freeze-drying selected areas of the spinal cells gave the following weights: cytoplasm 3, nucleus 1, and nucleolus 5 ($\times 10^{-13}$ g./μ^3).

The X-ray method of determining dry weight has been compared with that of optical interference microscopy, on the gastric mucosa cells of dog (Davies, Engström & Lindström, 1953) and on bone tissue (Davies & Engström, 1954). For light microscopy the specimen may be living, but must be immersed in a fluid of different refractive index. Despite these difficulties, good agreement was found in measurements of mass/unit area of different layers of human skin, aorta, gastric mucosa and ventral horn cells and thyroid follicles.

Microspectrometry of the X-ray image at two or more wavelengths allows a quantitative determination of the *chemical elements* present (§ 6.3). Phosphorus, sulphur and calcium occur naturally in biological material in sufficient concentration to be detected. Other elements may be present in diseased conditions or artificially injected as a form of staining for experimental purposes. A comparison of the amounts of calcium and phosphorus in growing cells from guinea-pig teeth has been made by Engström (1946, 1950). To determine calcium, the K-lines of calcium and scandium were used and for phosphorus the K-lines of phosphorus and sulphur. The amount of each element increased rapidly with distance from the apex of the tooth; at 1·5 mm. distance calcium was present in 3×10^{-10} g./100μ^2, and phosphorus in half this amount. The results were correlated with the growth and differentiation of the dentine cells.

Calcium and phosphorus in single Haversian systems of human bone have been determined with 3% accuracy in 10μ sections. In one such system the amount of calcium was $4·61 \pm 0·14\mu\mu$g./μ^2 and phosphorus $2·76 \pm 0·07\mu\mu$g./μ^2, a calcium/phosphorus ratio of 1·7. The same elements were determined in the arterial wall in arteriosclerosis, hardening of the arteries due to calcification of deposited material. In this case the amount of calcium had risen to $11·5 \pm 0·2\mu\mu$g./μ^2 and the Ca/P ratio was 3·3. Calcium is also deposited in the human kidney, forming stones or renal calculi. The absolute value of the amount of calcium within such a deposit was determined by Lindström (1955), the value being $5·3 \pm 0·2\mu\mu$g./μ^2 in a 10μ section. As a final example of this type of analysis, the amount of phosphorus in the nucleolus of sea-urchin eggs was found to be $0·37\mu\mu$g./μ^2, and correlation with the dry weight determination for the same material gave the true percentage of phosphorus by weight as 4·4%.

Sulphur was determined in skin, and was found to be present in a well-defined layer about 0·1 mm. below the surface, to the extent of 8×10^{-10} g./100μ^2 (Engström, 1950). In a similar determination Lindström (1955) found different amounts of sulphur in the two layers of skin: 4.5×10^{-10} g./$100\mu^2$ in the stratum corneum and 2.9×10^{-10} g./$100\mu^2$ in the stratum spinosum. These values are at variance with earlier histochemical determinations of sulphur (MacArthur, 1957).

Absorption analysis has been applied to a study of teeth by Röckert (1958), using the projection system of Long (1958) (§ 6.5). Some 236 measurements of the calcium content of deciduous and permanent teeth from eighteen healthy rhesus monkeys, from birth to seven years of age, were analysed statistically. There was no significant difference in calcium content between the two types of teeth, the absolute values in each varying from 0.41 ± 0.10 mg./mm.^{3} to 0.21 ± 0.03 mg./mm.^{3} Calcium content was also investigated as a function of distance from the visible apex of the tooth. No difference could be observed between the deciduous and permanent teeth despite the demineralization which occurs prior to shedding of the former.

Direct X-ray emission microanalysis, described in chapter 7, has so far been applied primarily to metallurgical specimens. The present range of elements covered does not include carbon, nitrogen and oxygen, and the concentration of heavier elements in biological material is usually very low. In addition, the direct impact of the electron beam on the specimen produces heating and consequent ashing of biological material. However, the scanning X-ray microanalyser described in § 7.3 has been used to examine a section of human lung containing tin. The localization of the tin within the lung matrix could be seen and a quantitative estimate of the amount present was made (Duncumb, 1957 a). The specimen was previously coated with a thin layer of metal by evaporation, in order to eliminate charging by the electron beam.

Additional references added in proof

To § 12.1.1: Bellman, Frank, Lambert, Odén & Williams (1960).
To § 12.2: Bohr & Dollerup (1960); Fournier (1960); Juster, Fischgold & Metzger (1960); Sissons, Jowsey & Stewart (1960 a, b).
To § 12.3: Blickman, Klopper & Recourt (1960 a, b); Greulich (1960); Leroux, Francois, Collette & Neetens (1960 a, b); Oderr (1960); Saunders (1960); Saunders & van der Zwan (1960).
To § 12.4: Dietrich (1960); Salmon (1960); Wylie (1960).
To § 12.5: Butler, Bahr, Taft & Jennings (1960); Lindström (1960); Müller & Sandritter (1960).

CHAPTER 13

INORGANIC APPLICATIONS OF X-RAY MICROSCOPY

13.1. Metallurgy

Possibly the first application of X-rays in microscopy was the study by Heycock & Neville (1898, 1900) of alloys of sodium and aluminium with gold. By magnifying a contact negative five times they could distinguish the effects of slow and quick cooling of an AuAl₂ alloy, confirming the evidence of normal photomicrography (Pl. XVIII). After this initial application, metallurgical microradiography was neglected until, in the late 1930's, fine-grained emulsions became generally available.

The preparation of metallurgical samples for microradiography has been described in chapter 9, and a review article by Sharpe (1957 *a*, *b*) discusses the defects to be avoided, such as tapering and non-uniform sections, cracking and separation of layers and loss of constituents from the surface. An early survey of applications was given by Maddigan & Zimmerman (1944) and more recently Homes & Gouzou (1951) have discussed examples in the steel industry. Clark (1955) described many metallurgical investigations, such as gas porosity in bronze, shrinkage and manganese segregation in magnesium castings, copper-lead bearing linings with dendritic structure and electroplating studies. Trillat (1956) has surveyed the microradiography of ferrous and light metal alloys, and the electron radiography (§ 2.7.3) of pitchblende, uranium oxide and silica. Some applications of the method of emission microanalysis in metallurgy have been mentioned in §§ 7.2 and 7.3; others have been described by Philibert & Crussard (1956), Birks & Brooks (1957 *a*) and Fisher (1959).

13.1.1. *Iron and steel—segregations and inclusions.* The most prevalent inclusion in steel is manganese sulphide and this can be detected by differential absorption, using the line radiation from cobalt and copper in turn to straddle the iron absorption edge, as shown in Pl. XIX (Andrews & Johnson, 1957). The graphite flakes are seen dark (X-ray transparent) in micrographs of cast iron, whereas the manganese sulphide inclusions show up white (X-ray absorbing) in that taken with cobalt radiation (left) but cannot be seen in that taken with copper radiation (right). For the cobalt line the iron matrix is more transparent than manganese, giving good contrast. For copper radiation both iron and manganese have roughly the same absorption and no contrast occurs. Similar inclusions have been studied by Betteridge & Sharpe (1948 a, b) in overheated ball race steels. A special steel used by Betteridge (1950) was homogenized at 950° C. and 1250° C., oil hardened, tempered and radiographed with manganese, iron, cobalt and zinc radiation. Manganese segregated at both temperatures but molybdenum and chromium segregation was reduced at the higher temperature.

Kirkby & Morley (1948) have used microradiography to study corrosion resistant chromium-nickel-molybdenum steels. A comparison was made between quenching at 1050° C. and 1150° C. and slow cooling at 850° C. Wolfe & Robinson (1950) used both characteristic and white radiation to study inclusions and segregation of manganese, chromium and tungsten. They recommend cobalt and iron radiation to straddle the manganese absorption edge and manganese radiation to observe tungsten segregation. They found that the addition of molybdenum reduced the tendency of manganese to segregate. Andrews & Johnson (1957) have shown by microradiography that these manganese sulphide inclusions occur along the lines followed by fatigue cracks. The inclusions have a deleterious effect on the machinability of steels (Wolfe & Robinson, 1951), rapidly abrading machine tools. The ease of machining was controlled by using microradiography to determine the amount of segregation. Five different types of segregation as well as micro-cracks have been detected in permanent magnet alloys (Goldschmidt, 1953). Cobalt radiation was used for the Alcomax anisotropic alloy containing cobalt, nickel, copper, aluminium, manganese, carbon and nickel as well as iron. Micrographs of 50μ thick specimens could be magnified × 80.

13.1.2. *Cast iron.* Segregation in cast iron has been studied by Cohen, Hall, Leonard & Ogilvie (1955) and cast iron in general by von Batchelder & Schaum (1947), Rehder (1952 *a*, *b*) and Krynitsky & Stern (1952). Graphitic carbon is very transparent to X-rays and excellent contrast is seen with respect to the iron matrix. In ordinary brittle cast iron the graphite appears as flakes, which can line up to produce a weakness across the casting. The addition of a small amount of manganese as a nucleating agent causes the graphite to collect into spheres or nodules which are not in contact, and the cast iron is then not as brittle as grey cast iron. X-rays can show the size, shape and distribution of the graphite nodules as shown in Pl. XX A. The specimen was a 75μ section (3 % carbon) and the spheres are about 25μ in diameter. In some places two spheres are in line and show more X-ray transmission than does one sphere, as can be seen in the stereographic view of this specimen (Nixon & Cosslett, 1955). Some detail can be seen within each nodule and between the spheres some graphite that has not been completely removed from the iron.

13.1.3. *Iron ores and sinters.* Sintered iron ore is well suited to X-ray microscopy since it contains voids, pores, slag pools and constituents of widely differing atomic number. Preparation of a thin slice is difficult unless the specimen is mounted in Canada balsam and then sectioned, as described by Andrews & Johnson (1957). An example of their work is shown in Pl. XX B, where the dendrites of iron oxide are seen growing in a slag pool. Selected area X-ray diffraction (see § 14.2.5) identified the presence of wüstite, magnetite or haematite, in different samples. Cohen (1950) has studied a fossil oolitic iron ore with an iron carbonate matrix and hydrous iron-alumina-silicate oolites. The negatives obtained at 20 kV and 20–25 mA, with 10-min. exposures, could be enlarged 150 times. Magnetite crystals in an iron-free matrix could be detected in a good sinter while the presence of free iron in slag indicated overheating. Further work was done on experimental sinters (Cohen, 1953) using a standard ore and varying the coke content from 2 % to 5 %; in between 2 % and 4 %, the sinter changed from good to poor. Variation in flue dust content and lime addition was also studied in three ores (Grangesberg and Sierra Leone concentrates and low-grade Northamptonshire ore). A detailed study of ironstone sinter has been made by McBriar,

Johnson, Andrews & Davies (1954). At ×250 they showed dendrites of wüstite in an iron oxide rich monticellite matrix and wüstite laths in a dark silicate matrix. Dendrites of magnetite and haematite occurred in various matrices as well as in slag pools, as shown in Pl. XXB.

13.1.4. *Other ores and minerals.* The method of electron microradiography (see § 2.7.3) has been used to study various types of minerals, as well as other metallurgical specimens (Trillat & Urbain, 1943; Trillat & Legrand, 1947; Trillat, 1948, 1949). Relatively hard X-rays (180 kV) pass through a photographic film lying on the surface of the specimen and produce secondary electrons, which are easily absorbed in the film. The production of electrons is greater from elements of higher atomic number, which thus give denser regions in the negative. A picture of one surface of the specimen only is produced, not of the whole interior as in transmission X-ray microscopy. Pitchblende, containing uranium, shows good contrast, the lead appearing grey and silicon dioxide white compared to the black uranium regions. Galena, containing nickel arsenide and silicon dioxide, also gives excellent contrast. Magnifications were restricted to ×6.

A similar type of ore has been examined by Jackson (1957c) with projection X-ray microscopy (Pl. XXIA). The sample was a 25μ section of thucolite ore containing pitchblende and hisingerite. The pitchblende appears dark, being more X-ray absorbing, the grey areas are hisingerite and the white lines are probably small cracks.

13.1.5. *Light metal alloys.* Alloys composed mainly of aluminium or magnesium with small amounts of copper, nickel, zinc, etc. are ideal specimens for X-ray microscopy. The steel samples discussed above have to be reduced to about 50μ in thickness whereas the aluminium alloys can easily be radiographed at 500μ or even thicker, owing to the low X-ray absorption of the matrix. For the same reason the contrast between aluminium and the other metals in the alloy is usually better than in the case of iron based alloys. As a result, microradiography has been much used in light alloy research and quality control. Pl. XXIB shows a cast magnesium alloy in cobalt radiation (Sharpe, 1957a), in which characteristic precipitates and grain boundary segregations are easily recognized.

Zinc segregation on quenching has been investigated in an aluminium alloy containing 8·5% zinc, 1·5% copper, 2·5% magnesium and traces of chromium, iron and silicon (Paic, 1944). Selected area microdiffraction (see § 14.2.5) with X-ray beams from 3μ to 40μ in width, using Cu Kα radiation and transmission flat-film technique, was employed for a study of grain boundaries in a CuAl$_2$ alloy with 12% copper (Fournier, 1949). At lower magnification, gas porosity and segregation could be detected in aluminium copper alloys with small amounts of magnesium and manganese; these were 'as cast' samples (Lohodny & Nonveiller, 1951). Other light metal alloys have been radiographed by Osswald (1949).

Segregation to the grain boundaries is one of the most striking features of light alloys, as has been shown by Berghezan, Lacombe & Chaudron (1950) for an aluminium–12% zinc alloy, at ×250 magnification; they were able to correlate the effect with temperature. A very detailed study of the angles between grains has been made by Williams & Smith (1952). They chose an aluminium–5% tin alloy prepared so as to have large aluminium grains, up to 100μ in width, most of the tin being drawn into the boundaries on cooling. The excellent contrast in the radiograph enabled accurate measurements to be made of the angles where three grains meet. The negatives were enlarged about ×35 and examined stereoscopically. The three-dimensional view is very striking and can be compared to a soap bubble froth. The authors in fact prepared a soap solution containing X-ray opaque material (potassium argentocyanide) and radiographed the bubbles stereoscopically to see if the angles were the same as in the solid metal. The agreement was very good and the results verified the theory of grain formation. This same type of specimen has been viewed with the projection X-ray microscope at higher magnification, as shown in Pl. XXII. The specimen was 0·5 mm. thick, so that the magnification varies from one surface to the other, as seen stereoscopically (Nixon, 1956a). Small flakes of tin can be seen within the grains of aluminium, as well as the dark streaks of tin that delineate the boundaries. Several light marks through the tin indicate small fractures starting at the boundary surface.

This type of specimen is ideally suited for following the initial stages of cold work, fatigue or fracture by taking successive X-ray images while the deformation is in progress. When a small piece of the same alloy was reduced in thickness to 175μ by beating, the original smooth grain boundaries seen in the 'as cast' condition were found to have been replaced by very jagged and torn surfaces between the distorted aluminium grains (Baez, 1954). Williams & Smith (1952) showed a similar result after cold rolling to 66% reduction of the alloy. It should be possible to carry out quantitative stress-strain tests in the X-ray microscope, to follow the behaviour of the metal underneath the surface while subject to normal and abnormal conditions of service. Eventually this could be coupled with high- and low-temperature treatment of the specimen (Nixon, 1957b; Goldschmidt, 1957), as would be readily possible in the projection X-ray microscope; some initial work has been carried out by Bessen (1958).

Other non-ferrous alloys can be studied successfully by X-ray microscopy, although the thickness must be less than with those of aluminium. Good contrast is obtained with the leaded-copper based alloys investigated by Kura, Eastwood & Doig (1951). A similar type of specimen is annealed brass, approximately 60% copper and 40% zinc, with a small amount of lead added to increase machinability. For this purpose the distribution of the lead should be as uniform as possible, and the X-ray microscope showed that it was present as scattered particles about 1μ in diameter (Nixon, 1956a). Such a control is difficult to carry out with the optical microscope under normal metallographic conditions.

Another specimen that gives good contrast in the X-ray beam is porous bronze, containing 95% copper and 5% tin, shown in Pl. XXIII A (Nixon, 1956a). The tin appears as dendritic segregates, dark (X-ray absorbing) in the positive print. The many small circular light areas within the tin are voids or gas porosities. These small holes tend to be filled up with polishing compound when the specimen is prepared for metallurgical examination and a better idea of their size, shape and distribution is given by X-ray microscopy.

A final example of this type is the zinc–5% aluminium alloy shown in Pl. XXIV (Nixon & Cosslett, 1955). The dark areas are zinc rich and the striped regions correspond to a eutectoid of the two constituents. In microradiography this phase is seen without etching the surface; the relationship of the two areas within the sample has been studied stereoscopically.

13.1.6. *Diffusion.* The diffusion of one metal into another can be followed by X-ray microscopic methods, the contrast depending on the relative distance apart of the two elements in the periodic table. Such a problem is the study of the diffusion of copper into aluminium in the presence of calcium (Paic, 1951 *a, b*). A careful study was made with test samples consisting of sheets of aluminium, containing 3 % copper, 1 % magnesium and less than 1 % silicon and manganese, and clad with aluminium containing variable amounts of calcium from zero to 0·6 %, together with some traces of iron and silicon. The sample was heated for 20 and 400 min. at 500° C. and then radiographed at 47 kV. In the shorter time there was no discernable diffusion of the copper through the outer aluminium coating. After 400 min. copper could be seen throughout the added aluminium layer and, most important, the diffusion rate was accelerated by the presence of calcium, not reduced as had been expected. For an average amount of calcium, the copper reached the surface (150 μ from the original position) after 80 min. at 500° C. The diffusion zone at an aluminium/brass interface is shown in Pl. XXIIIb (Sharpe, 1957 *a*).

Microradiography has been used in a detailed study of voids and diffusion effects by Barnes (1952). A copper-nickel layer was held for 68 hr. at 1047° C. in an argon atmosphere to produce diffusion. A 50 μ section of the sample was radiographed with cobalt $K\alpha$ radiation and enlarged to × 150. The most striking result was the polyhedral shape of the voids, which had been seen in the optical microscope but was thought to be due to the polishing process. The X-ray method showed that the same shape occurred below the surface of the material.

Diffusion couples have been analysed by Ogilvie (1958), using an X-ray absorption technique to study the concentration gradient across the diffusion zone. A 25 μ slit, parallel to the diffusion boundary, was moved across the material from one side of the boundary to the other while the X-ray transmission was measured.

Nickel-chromium-nickel couples were viewed by projection X-ray microscopy, with 10 min. exposures at 20 kV.

13.1.7. *Spot welding.* X-ray inspection of welds at unit magnification is a standard method of control during the fabrication of pipelines, welded ships and other large projects which use this method of joining metal instead of riveting. Microradiography has been called into use for inspection of spot welds at low magnification by Gross & Clark (1943) and McMaster & Grover (1947). A detailed analysis of spot welds in aluminium has been made by Boss (1948). Typical welds are prepared for microscopic examination by cutting thin metal sections. Subsurface holes and cracks are readily visible due to the good contrast between a void and the aluminium sheet. Another effect that can be seen is the deposition of copper from the welding electrode onto the aluminium when the time and temperature characteristics are not correct. A common defect is the occurrence of 'spits' of molten metal from the central slug, which shoot out sideways at irregular intervals around the weld, reducing the amount of metal in it and leading to weakness. The scattered metal shows clearly in the microradiograph, as well as any departure from circular symmetry, which will also weaken a weld. It has been possible to correlate the size of the welded joint, as seen on the microradiograph, with the known welded strength of the joint and so to specify the inspection conditions that the spot weld must pass.

13.1.8. *Castings.* Defects in various types of castings can be analysed by X-ray microscopy once a thin section has been prepared. Porosity, arising from various causes, produces good contrast and has been studied for a bronze casting—80% copper, 10% tin and 10% lead. With normal sand casting the voids due to gas porosity could be seen but centrifugal casting eliminated them (Clark, 1955). Sand casting and gravity die casting of a manganese bronze were studied by Ball (1945). In this case sponge shrinkage occurred with the gravity method. Stress segregation could be seen in an aluminium-4% copper alloy and both gas porosity and shrink porosity in aluminium-10% magnesium. By correlating the various types of defects with the subsequent performance of the casting it was possible to prepare standards for inspection. Typical microradiographs were prepared showing the maximum amount of microcavity segregation, gas porosity, cracks or dross inclusions that could be passed as acceptable, and subsequent batches of castings were sampled and compared with these standards.

The cast aluminium alloy in Pl. XXVa shows Al_3Ti platelets, with an intermetallic phase and segregation (Sharpe, 1957a).

13.1.9. Other metallurgical applications. Many other aspects of the study of metals have required X-ray microscopy as an additional method of investigation. Some isolated examples are given here to indicate the type of information gained, without attempting to make an exhaustive list of all the metallurgical results of X-ray microscopy.

Phase transformation has proved to be detectable at a very early stage by microradiography (Goldschmidt, 1957). It was found that segregates could be seen within a solid solution matrix before the precipitating compound had been formed. In an iron-chromium alloy, X-ray diffraction suggested that the sample was pure ferrite. However, the microradiograph showed clear areas of chromium and iron accumulations within the ferrite. Since the chromium-rich regions can form the nucleus for sigma-phase precipitation, this indicated that the compound was about to go through a phase transformation.

Distortion and strain can be detected in microradiographs of metals and alloys, since contrast arises from Bragg diffraction from crystalline regions inclined at the correct angle to the incident X-ray beam (Smoluchowski, Lucht & Mann, 1946; Lucht, Mann & Smoluchowski, 1946; see § 14.3.3). With pure copper samples 50μ thick, in copper radiation, the contrast could be changed locally by tilting the specimen within the X-ray beam between exposures. Distortion was observed after mechanical polishing during preparation of the thin section, but not after electropolishing. The method was then used to follow the strain distribution and influence of temperature during annealing and cold working. Strain discontinuities at twin boundaries and grain boundaries were also studied by this means. Further work along these lines has been carried out by Votava, Berghezan & Gillette (1957), as discussed in § 14.3.3.

Electroplating can also be followed by microradiography. Samples should be prepared both parallel and perpendicular to the plating interface: the uniformity of plating will be seen in the former and the thickness of the layer in the latter. Porosity due to the evolution of gas at the interface can also be detected (Clark, 1955; Ogburn & Hilkert, 1957).

Nickel alloys have been studied in detail by Osipov & Fedotov (1951), including compounds of nickel and tungsten, molybdenum, niobium, titanium and tantalum. Other Russian work includes a phase transformation study at high temperature by Rovinsky & Gambashidze (1951) as well as papers by Ivanov (1949), Gogoberidze (1953) and Baranov, Zhdanov & Diesenrot-Mysovskaya (1944). Rovinsky, Lutsau & Avdeyenko (1957) include a micrograph of an iron-ore agglomerate as an example of the use of the point anode projection X-ray microscope. The fayalite and magnetite were analysed by microphotometry of the negative and gave good agreement with the known concentration in the ore.

13.2. Industrial applications

13.2.1. *Paper.* Paper is an ideal specimen for X-ray microscopy as it is already formed into thin sheets, of low atomic number, which are partially transparent to X-rays although usually opaque to light. Heavier elements added to paper for 'loading' give good contrast with the cellulose fibres. With bisulphate papers the process of loading with 10% of titanium oxide and kaolin has been followed by microradiography. With Lippmann emulsion a 5-min. contact exposure at 25 kV and 15 mA gave a negative that could be enlarged 125 times (Lambot, 1947).

An extensive study of paper has been made with the projection X-ray microscope by Isings, Ong, Le Poole & van Nederveen (1957). The normal methods of serial sectioning for optical microscopy tend to disturb the relationship of the various fibres. Also the optical depth of focus is limited at high magnification, whereas the whole thickness of a piece of paper can be imaged at once in the X-ray microscope. The best results were obtained by stereographic viewing of the sample, where both sides of the specimen as well as the interior detail can be seen at once. Attention was directed to the felting of the fibres and fibrils, the role of fillers and the surface layers of coated papers, which are particularly difficult to section for optical microscopy. The fillers used were titanium oxide and calcium carbonate, which both produce good contrast. Untreated paper did not give sufficient contrast until it had been processed with some form of iodine solution. Gold shadowing of the fibres was also used to overcome the difficulties due to lack of contrast. Coating layers of calcium carbonate or kaolin were studied on art papers and other machine-coated samples. With metal laminations thin spots or small holes could be seen as bright pits against a uniform grey field. An example of this type of X-ray micrograph of paper is given in Pl. XXVB. Other paper studies have been made by Pelgroms (1951) and by Lange (1954), who used a stereoscopic technique.

X-ray methods are also useful after the paper has been made up into various finished products. *Postage stamps* show details of the original metallic inks and watermarks even though they have been visibly obscured by the intensely black carbon ink used for cancelling. Both electron radiography and microradiography (at 9 kV) have been used by Cheavin (1947, 1948) to reveal the nature of repairs made to a stamp but hidden by the cancelling ink. Two versions of electron radiography were used by Pollack & Bridgman (1954; Pollack, Bridgman & Splettstosser, 1955). At 200 kV and with a 7 mm. copper filter the heavy metallic inks, such as mercury and lead oxide, gave rise to photo-electrons which produced an image of the printing pattern only. When a lead sheet was placed on top of the stamp, a uniform number of high-energy photo-electrons were liberated from the lead and were absorbed by the heavy inks. This method gave a picture which was in essence the negative of the first view but showing some details of the paper structure as well. Microradiography was also employed, with a tube at 6 kV, to investigate the carbon cancelling inks.

13.2.2. *Fibres and fabrics.* Cloth made of natural or artificial fibres also lends itself to X-ray microscopy. The details of the weave can be studied stereoscopically without sectioning and the staining methods mentioned above apply equally to the dyeing of materials. There is sufficient contrast to reveal the distribution of the individual fibres within each thread and the relationship between the threads in respect of spacing, compression or holes (Baez, 1954) (see also Lindsley, Fisher & Brant, 1949).

13.2.3. *Geology.* The value of X-ray microscopy for investigating various ores and minerals has been discussed above in connexion with metallurgical applications. Fossils can also be studied by this method. The fossilized material is often embedded in a rock matrix and cannot be removed without danger of crumbling. Thin sections can be used to show the outline of the species without removal, but if the fossil can be removed whole, its internal structure can then be studied with X-rays without grinding and sectioning. The classification of foraminifera often depends on internal detail. Calcareous, siliceous and pyritic microfossils have been radio-graphed with normal X-ray diffraction equipment (Schmidt, 1952). The internal arrangement of the chambers, muscle scar patterns and the position of pores were all detected. With specimens of 500μ thickness, a copper target tube operated at 30 kV and 20 mA gave exposures of less than 20 min. at 20 cm. from the target. Stereoscopic viewing of these fossils is desirable and has been applied to larger specimens in a survey of radiographic methods in paleontology (Schmidt, 1948).

A general survey of foraminifera by microradiography has been made by Hooper (1958, 1960) on sufficient specimens for statistical conclusions to be valid (500–1000 specimens). Fifty or more specimens were radiographed at one time on maximum resolution emulsion. To get sufficient field of view the plate was placed 40 cm. from the X-ray tube, requiring a 7 hr. exposure at 30 kV, 18 mA. A single sample of the genus *Operculina*, species *Victoriensis*, tertiary age, from Australia is shown in Pl. XXVII A. The central spherical chamber or proloculus is clearly seen, together with the septa radiating out from the centre and the marginal cord with the canal system. Apertures are seen at the base of some septa. In the outer whorl there is a loss of symmetry of the septa due to old age.

In mineralogy, X-ray microanalytical methods are proving of considerable value. A quantitative determination of calcium in 10μ diameter areas of rock sections 20μ thick has been made by Long & McConnell (1959). Larnite-bearing rocks (βCa_2SiO_4)

possess sealed hydration cavities which contain both calcium hydroxide (portlandite) and natural hydrogel. The calcium content of the hydrogel is important for determining the natural processes involved in its formation and when comparing these with laboratory methods of producing calcium silicate hydrogels. The corrected experimental results all showed a calcium content above $1\cdot50$ mg./cm.^{2} and a calcium oxide : silica ratio of $1\cdot5:1$, consistent with laboratory results obtained from solutions saturated with calcium hydroxide. For a low-lime hydrogel formed in the presence of circulating water, the calcium content would have been $1\cdot0$ mg./cm.^{2} at maximum and the conclusion is that the natural cavities were isolated during the formation of the narrow zone of hydrated larnite which covers the complete surface of the cavity. In this work a projection X-ray microscope was used, but converted for electrical instead of photographic recording of the degree of transmission through thin sections (Long, 1959). The experimental procedure and method of evaluating the results have been described in § 6.5. Mosley & Wyckoff (1958) have used differential absorption for qualitative studies of geological specimens.

The method of emission micro-spectrometry (§ 7.1) has also been applied to rock sections of a variety of types, in the apparatus shown in Fig. 7.6. Deflexion plates were fitted to the final electron lens, to permit the scanning technique to be used. By feeding the counter output to the first beam of a double-beam oscilloscope, an immediate quantitative result was displayed of the concentration of a given element along a line in the specimen traced out by the second beam (Pl. XXVI). In addition to the determination of nickel and iron in a meteorite (Agrell & Long, 1960a) the distribution of iron, chromium and titanium in a spinel was mapped (Agrell & Long, 1960b). There appears to be a wide range of problems in the micromorphology and morphogenesis of minerals which only the microemission method seems able to solve.

13.2.4. *Miscellaneous.* Various unrelated applications of X-ray microscopy have been made in other industrial fields. The *internal diameter of metal capillary tubes* has been determined by microradiography (Collett, Hughes & Morey, 1950). In certain applications of flow through capillary tubes it is necessary to know the bore exactly and to be sure that it is uniform throughout the length of the tube. With glass capillaries a thread of mercury can be observed at different positions within the tube, its length revealing any deviations from uniformity of bore. With metal tubes the same method can be used if the mercury is observed with X-rays. In this particular case the tubes were 15 ft. long with a bore of 0·015 in. and wall thickness from 0·005 in. to 0·014 in. Mercury of length 0·25 in. to 0·40 in. was introduced into the bore and moved along the full 15 ft. One hundred and ninety X-ray micrographs were made on spectroscopic plates and magnified × 50 in order to measure the length of the mercury column to within 0·0002 in. The total exposure for each tube occupied 6 hr. and the temperature variation over this period was kept within ±0·5° C. The X-ray images were closely inspected to ensure that there were no voids in the mercury thread. The results were used to select tubes with a maximum bore deviation of no more than 2·5% for subsequent experiments.

Electron radiography has been applied to *gum rubber* containing 32% sulphur (Trillat, 1945a) and to *paint* films (Trillat, 1945b; Trillat & Legrand, 1948); air bubbles within the paint film could be observed. *Plastics* have been studied with soft X-rays (O'Connor & Eagleson, 1950) as well as *ceramics* (Clark, 1955). Devitrification and bonding of enamels was looked for in the ceramic specimens as well as lack of uniformity, porosity, inclusions and cracks.

Additional references added in proof

To §13.1.1, 1.2 and 1.5: von Batchelder (1959); Dodd (1960); Ruff & Kushner (1960).
To §13.1.4: Cohen & Schlogl (1960).
To §13.1.5 and 1.7: Shinoda, Amono, Tomura & Shimizu (1960).
To §13.1.6: Lublin (1960).
To §13.2.3: Goldsztaub & Schmitt (1960).
For recent metallurgical applications of the X-ray emission microanalyser, see p. 368.

CHAPTER 14

MICRODIFFRACTION

One of the advantages of X-ray and electron microscopy, as compared with optical methods, is the possibility of obtaining information about the specimen by diffraction. Electron diffraction is used as a supplement to electron microscopy, and it is possible to examine the crystalline nature of a selected area as small as $1\mu^2$. X-ray diffraction can similarly be used as a supplement to X-ray microscopy, providing information about the crystalline structure, texture and chemical composition of the specimen that cannot be found by the absorption, fluorescence or emission methods of microanalysis.

X-ray microbeam techniques developed in conjunction with conventional types of X-ray tube have been reviewed by Hirsch (1955). However, the equipment designed for X-ray microscopy is usually more suitable for microdiffraction than the conventional tube. The reflexion X-ray microscope of Kirkpatrick & Baez (§ 4.8), the gas discharge X-ray tube and focusing monochromator of Henke (§ 10.1.4), and the point source X-ray tube used for projection X-ray microscopy (§ 11.3) can all be used for microbeam diffraction. In the simplest case the normal X-ray tube used for contact microradiography can also serve as the X-ray source for microbeam diffraction with a microcamera.

Information on the general principles of X-ray diffraction and on the interpretation of diffraction patterns may be found in the standard works of Klug & Alexander (1954) and Peiser, Rooksby & Wilson (1955).

14.1. X-ray tubes

The first microbeam X-ray diffraction experiments combined normal X-ray tubes with special cameras using fine glass capillary or narrow slit collimators (Kratky, 1931; Chesley, 1947). The specimens were very small crystals, or selected areas of larger objects such as fibres or metals containing many small crystallites. It was realized that selecting a small area of the X-ray output from a comparatively large focal spot ($>$ 1 mm.^{2}) did not give the optimum conditions for resolution and exposure-time. The true brightness of an X-ray source, which determines the flux density through a small aperture, increases as the source size is reduced due to the superior heat dissipation from a small focal spot (§ 8.2). This fact was recognized by Huxley (1953) in discussing the position of the X-ray tube, specimen and film which will give shortest exposure-time for diffraction from biological materials. He concluded that a large reduction in exposure-time (10–50 times) would be obtainable with most specimens by reducing the camera length used with a normal X-ray tube. A further reduction is possible with a fine-focus tube and he recommended a spot of variable width down to 50μ, a limit determined partly by the conclusion that 'it will often be impossible to come within the optimum distance of the focus of the X-ray tube', that is, closer than 1 cm. As a result, he stated that 'a fine-focus tube with a working distance of a few millimetres or less would be most valuable'. The production of the Ehrenberg and Spear tube, having a 40μ focus, made these advantages generally available.

14.1.1. *Micro-focus X-ray tubes.* The micro-focus X-ray tube most widely used for microdiffraction is the commercial version of that designed by Ehrenberg & Spear (1951 a) (see also Ehrenberg, 1957). It embodies a simple electrostatic focusing system, which can be adjusted during operation by moving the filament assembly F (which is earthed) with respect to the anode A (Fig. 14.1). Two focal spot sizes are available, a 40μ circle with a loading of 13·5 kW/mm.^{2} and a 1·4 mm. × 0·1 mm. line with a loading of 1·4 kW/mm.^{2} Both foci are 'viewed' foreshortened with a reduction of 10 in one direction, that is, an effective size of 40μ × 4μ for the circle and 1·4 mm. × 10μ or 100μ × 140μ for the line. The 40μ spot takes 0·5 mA and the 100μ line 4 mA at 50 kV anode potential. A recent description of this tube and its many applications has been given by Brech & Stansfield (1957).

A similar type of micro-focus X-ray generator has been described by Gay, Hirsch, Thorp & Kellar (1951). Particular attention was

paid to the focusing action of the grid bias cup, as in the Ehrenberg
and Spear tube, and in addition a rotating anode was used. A line
filament produced a line focus of width variable from 1 mm. to
50μ depending on the grid bias. Normal operating conditions for
a 4 mm. × 0·4 mm. filament were 25 mA and 40 kV on a 1·5 mm. ×
0·15 mm. focus, or 4·4 kW/mm.[2] Both the fine focus and the
rotating anode contribute to the increased specific loading of the

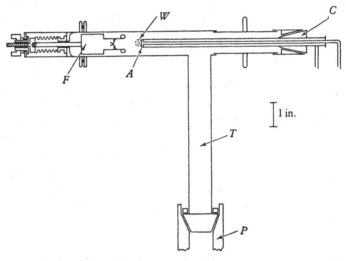

Fig. 14.1. Microfocus X-ray tube with electrostatic focusing. *F*, filament
assembly; *A*, anode; *W*, window; *C*, cooling system; *T*, supporting tube;
P, vacuum pump. (Ehrenberg & Spear, 1951 *a*.)

X-ray focal spot. The effect of rotation becomes less than that of
reduction in width for a spot below about 0·2 mm. In general, the
added complexity of a rotating anode is not warranted for spots of
100μ and below.

    These tubes relied upon electrostatic focusing to produce a fine
focus, but magnetic focusing has also been used to obtain fairly
small focal spots (Guinier & Devaux, 1943; Goldsztaub, 1947;
Poittevin, 1947; Drenck & Pepinsky, 1951). An example is the
single lens unit of Witty and Wood, described by DeBarr &
MacArthur (1950) and produced by Metropolitan-Vickers. A 50 kV
triode electron gun, as used in the electron microscope, carries a
line filament of 0·2 mm. tungsten wire and a 5 × 1 mm. grid slot

biased with a few hundred volts. The magnetic lens focuses the electron beam onto the water-cooled anode, which is several millimetres outside the lens gap. The focal spot width is 10μ, its length 50μ, and it carries a beam current of 500μA, giving a loading of 50 kW/mm.^{2} This tube formed the basis of the scanning X-ray microscope of Pattee (1957 b) and later of his microfluoroscope (§ 10.4).

Finally, the projection X-ray microscope described in chapters 3 and 11 can be used as a point source for X-ray diffraction, as suggested by Cosslett & Nixon (1951) and discussed in detail by Nixon (1957 b). This tube carries the point source principle to its logical conclusion, the spot width being as small as 1μ or less. In these conditions the total load is small, about 1 W, but the specific loading is very high, being 1000 kW/mm.^{2} when 50μA is focused on a 1μ spot at 20 kV. The self-cooling is so favourable that there is no need to use either a rotating anode or the liquid cooling of the target used in all other X-ray tubes.

These very small focal spots are used without foreshortening, since this method of increasing brightness (cf. § 8.6) can only be used when the focal spot is many times larger (in at least one direction) than the penetration distance of the electrons. For a square focal spot with depth equal to the length of a side, there is a reduction in energy when viewed at 45°. However, line foci can be profitably used in the point source tube by fitting a line filament and adjusting the lens strength so as to form a line focus of width equal to the penetration depth and length 10 times this depth. Viewing the source at a small angle is made possible by setting the target plane level with the top of the electron lens. The end-window target has an important advantage as compared with a solid target in that it is possible to place the specimen very close to the X-ray focal spot. Since the beam divergence must usually be limited to 10^{-2} or less, the specimen can be no closer than a few hundred microns to a focal spot of a few microns, but this distance is still appreciably better than the goal of a few millimetres set by Huxley (1953).

A final point of difference concerns the purity of the radiation emitted from a fine-focus tube, compared with that from the conventional type of tube. The large distance from the cathode to the anode, 30 cm. or more as compared with 2–3 cm., reduces the amount of evaporated tungsten on the anode by about a hundred-fold. In consequence the tungsten *L* lines are not excited, nor is the X-ray output reduced by absorption in the evaporated layer as in a normal tube. Also, since the electron spot strikes a carefully defined area of the anode, it is possible to be sure that only one type of radiation leaves the tube. In a normal X-ray diffraction tube, scattered electrons may excite X-rays from the many other types of metal used in its construction, giving unwanted effects in the diffraction pattern (or in fluorescence analysis).

14.1.2. *Reflexion types.* The microfocus tubes discussed above all form the point source within the tube itself. Various methods

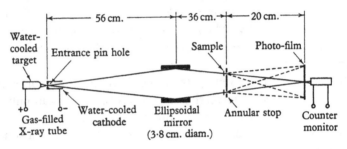

Fig. 14.2. A total-reflexion diffraction camera with ellipsoidal mirror and gas type X-ray tube. (Henke & DuMond, 1955.)

utilising X-ray reflexion have been proposed for forming a point source outside it. Such reflexion systems also give a high degree of monochromaticity to the X-ray beam and can be used to select any particular wavelength from the X-ray output of a tube operating at a high voltage.

A total reflexion diffraction camera using a gas type X-ray tube is shown in Fig. 14.2 (Henke & DuMond, 1955). The cold cathode gas discharge operates at 5 kV and 50 mA, a much higher beam current than in the point source tubes described above. Both cathode and anode are water-cooled. The radiation from the target is accepted in a direction opposite to that of the electron beam and passes through a small hole in the cathode to reach the diffraction camera. In this direction the continuous radiation is a minimum and the relative amount of the characteristic radiation therefore a maximum (cf. § 8.6). As there is no filament to burn out, this tube can run for several days unattended. The relatively poor vacuum is also easier to maintain than that in the filament type of tube.

The X-rays are focused by an ellipsoidal mirror (see § 4.7) placed half-way between source and photographic film. The sample is placed between the mirror and the film and an annular stop is used, to conform with the ring focusing of the mirror. This unit was designed primarily for long wavelength X-ray diffraction, not necessarily on the micro-scale, but it has also been used as a mono-chromatic source for ultra-soft quantitative microradiographic analysis (Henke, 1957).

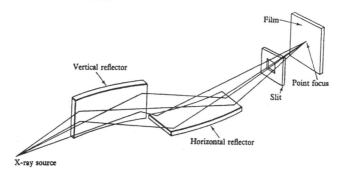

Fig. 14.3. Small angle diffraction camera using two cylindrical mirrors and a point source X-ray tube. (Franks, 1955.)

Another camera for small-angle diffraction combines the point source of the Ehrenberg and Spear tube with the twin cylindrical mirror system of Kirkpatrick and Baez so as to produce a point source outside the X-ray tube (Ehrenberg & Franks, 1952; Franks, 1955, 1958). A slit is placed close to the photographic film as shown in Fig. 14.3; the specimen is near this slit and 3 cm. in front of the optical image of the X-ray source. As in the Henke system the radiation used is highly monochromatic, because of the selectivity of the reflexion, and it is very intense due to the fine focus. Spacings of 1000 Å have been resolved with exposures of 10 min.

Kirkpatrick & Pattee (1953) have suggested a similar method for taking microdiffraction patterns from selected areas of an X-ray micrograph (Fig. 14.4). The specimen is exposed in a two mirror

X-ray focusing microscope (§ 4.8) and an enlarged image is recorded on the photographic plate. A small area of this image is selected for diffraction and a large X-ray-opaque plate is placed in the image plane with a small aperture at the region of interest. The original specimen remains in the X-ray microscope and the X-ray tube is moved from illuminating the *specimen* to illuminating what was the *image* plane. The X-rays passing through the small aperture are focused by the two mirrors and strike the corresponding area of the specimen. A photographic plate placed at the original

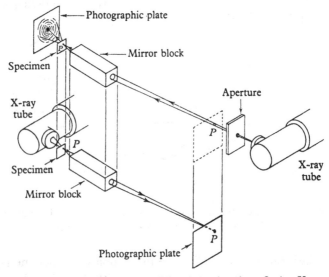

Fig. 14.4. Selected area microdiffraction using the reflexion X-ray microscope in reverse. (Kirkpatrick & Pattee, 1953.)

position of the X-ray tube records the focused source and the diffraction pattern from the small selected area of the specimen. This scheme, although elegant in principle, cannot be realized in practice until the performance of the reflexion method has been improved.

14.2. Cameras and methods

The standard types of X-ray diffraction cameras have all been modified for micro-beam operation, as shown in Fig. 14.5 (*F*, X-ray focal spot; *A*, aperture for collimation; *S*, specimen; *P*, photo-

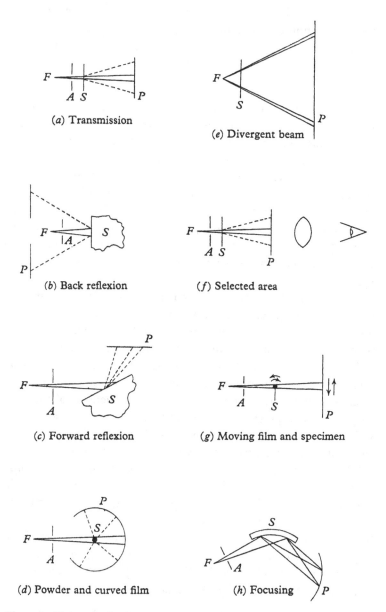

(a) Transmission

(e) Divergent beam

(b) Back reflexion

(f) Selected area

(c) Forward reflexion

(g) Moving film and specimen

(d) Powder and curved film

(h) Focusing

Fig. 14.5. Types of microbeam camera for use with a point source of X-rays.

graphic film). In several of these methods a Geiger counter can replace the film if automatic movement of the counter is provided (Thorp, 1951). The counter output actuates a pen recorder giving a permanent record in chart form.

14.2.1. *Transmission.* The simplest method (Fig. 14.5(a)) uses a fine glass capillary to collimate the output from a normal X-ray tube (Chesley (1947) and many others). Capillaries of a few microns internal diameter and a few millimetres long may be drawn from larger bore tubing or obtained commercially. X-rays incident at glancing-angle on the walls of the bore are totally reflected (chapter 4), so that if a capillary can be placed closer to the X-ray source than the specimen, there will be a gain in intensity at the specimen due to the larger collection angle. The corresponding decrease in angular resolution in the final diffraction pattern is a disadvantage which must be set against the increase in speed.

When a microfocus source is available, a pinhole in a lead sheet can be used in place of the capillary to define the width of the X-ray beam. A second aperture slightly larger than the pinhole may be needed between it and the specimen, to cut off the scattered radiation. The small platinum aperture-disks, used in electron microscopy, are suitable for microbeam X-ray diffraction and are available with various bores down to 5μ. Alinement of pinholes larger than 50μ presents little difficulty as the issuing radiation is visible on a fluorescent screen over a wide angle. Pinholes smaller than about 50μ are best alined with the aid of a Geiger counter.

The point projection X-ray microscope has been used for transmission diffraction by Nixon (1957b), an X-ray beam 10μ in diameter being obtained from a copper target, with beam current 50μA and voltage 20 kV. A pattern obtained from the silver bromide grains, less than 1μ in size, in a photographic emulsion required a 10 min. exposure on film 5 mm. from the specimen. A similar pattern, obtained by Hirsch (1955) with a commercial X-ray tube at 20 mA and 30 kV, took 9 hr.

14.2.2. *Back reflexion.* In this method the radiation diffracted back from the surface of a solid sample is utilized, the film being placed on the same side of the specimen as the X-ray source

(Fig. 14.5 (b)). In transmission a central hole is made in the film to allow the main X-ray beam to escape without causing photographic fogging. In back reflexion a similar hole allows the X-ray beam to pass through the film before striking the specimen. Since a large sample can be used, information may be obtained from the surface of a normal metallurgical specimen without the special preparation necessary for transmission. The method was used in a micro-beam study of the deformation, recovery and recrystallization of metals by Hirsch (1953) and by Gay, Hirsch & Kelly (1954).

The adaptation of the point projection X-ray microscope for back reflexion has been described by Nixon (1957b). The pole-pieces are removed and the electron beam reduction is shared equally by the condenser and objective lenses. The specimen is placed close to an aperture before the target and the film is within the magnetic field used to focus the electron beam. Transmission and back reflexion patterns obtained with the projection instrument are compared in Pls. XXVIIB and XXVIIc, from an area 20μ in diameter of a copper foil, 3μ thick (a diffracting volume of 10^{-9} cm.³). The exposure-time was 5 min. for the transmission pattern and 10 min. for the back reflexion pattern. The central punched hole in each case is 3 mm. in diameter.

14.2.3. *Forward reflexion.* If it is inconvenient to place the film close to the X-ray source for back reflexion, it is still possible to obtain a diffraction pattern from the surface of a solid by forward reflexion, as shown in Fig. 14.5 (c). In this case the diffraction pattern is not symmetrical, but appears as slightly less than one-half of the full circle recorded in back reflexion, which is still sufficient for measurement of ring diameters and for counting spots. Rooksby (1955) used the method to study the size and rate of growth of crystallites in oxide-coated cathodes at high temperatures, with the aid of a lead collimator 50μ in diameter.

Two other methods of forward reflexion, which have been used to obtain X-ray images of crystal surfaces, can be considered as X-ray microdiffraction methods. In the method developed by Berg (1931, 1934) and Barrett (1945), a parallel beam of monochromatic X-rays strikes the surface of a deformed crystal and is locally reflected wherever the orientation of the surface layers is suitable (cf. Fig. 1.5). The reflected X-ray image is recorded on fine-grained photographic film as in contact microradiography, and can be enlarged up to 100 times. The other method, also due to Berg, uses a divergent beam of polychromatic X-rays from a point source which forms a Laue type of diffraction pattern after striking the surface of a crystal.

Honeycombe (1951) has used the Berg–Barrett method to study the plastic deformation of metals, including the formation of kink bands and bands of secondary slip (Pl. XXVIII). Photographs were made on Kodak Maximum Resolution plates, with 2–6 hr. exposures, and subsequently enlarged up to 100 times. For magnifications up to × 20, faster plates were used, with 1–30 min. exposure. The X-ray images show small local changes in orientation, which are not revealed by optical microscopy until much heavier deformation has occurred.

The Berg–Barrett method has recently been used to investigate both dislocations and sub-grain boundaries (Ancker, Hazlett & Parker, 1956; Newkirk, 1958 a, b), and, in slightly different form, by Coyle, Marshall, Auld & McKinnon (1957) for studying the deformation of aluminium single crystals. In this case a divergent polychromatic beam of X-rays strikes the surface of the specimen, so that most parts of the surface reflect one wavelength or another in the primary beam. When monochromatic X-rays are used, many photographs have to be taken with the specimen tilted slightly by accurately controlled angles between exposures. The resolution depends on the focal spot size and on the specimen to source distance. Using a focal spot of 50μ, a specimen–film distance of 5 cm. and specimen–source distance of 10 cm., the resolution was 2 sec. of arc in one direction and 20 min. of arc at right angles, since a foreshortened focal spot was used; the exposure-time was 30 min. The results showed the formation of kink-bands and slip-lines, after elongation of the specimen, which could be related to the various crystallographic directions (see also Weissmann, 1956).

Similar Bragg diffraction effects are found in X-ray micrographs of transmission samples. When an X-ray beam passes through a thin crystalline sample some regions may be favourably orientated to reflect the X-rays out of the direct path to the photographic plate, so that a dark line is seen in the (positive) image of those regions. A corresponding light line is found in some other part of the plate, unless the angle of diffraction is such that the X-rays do not strike the plate at all. Such extinction lines were first reported in contact microradiography by Smoluchowski, Lucht & Mann (1946) and have been studied more recently by Votava, Berghezan & Gillette (1957). They suggest that it is a very sensitive method for studying the relative orientation of grains in a thin sample. A variant of it, in which a particular Bragg reflexion is isolated with an aperture, has been devised by Lang (1957, 1958, 1959) for investigating systematically the variations in orientation within a crystal, so that sub-grains and dislocations can be observed.

14.2.4. *Divergent beam.* If a crystalline specimen is placed very close to a micro-focus tube, and no aperture is used, a divergent beam X-ray diffraction pattern will be formed, the nature of which depends on the extent and the degree of perfection of the reflecting regions (Fig. 14.5(*e*)). From a single crystal the diffracted X-rays form conic sections at the film, known as Kossel lines. Both light and dark lines appear, but not necessarily together. The accuracy is such that the orientation of the crystal can be determined to within 2 sec. of arc and the lattice parameters measured to 1 part in 1000 (Lonsdale, 1947).

Kossel lines may be obtained with a conventional type of X-ray tube, but a micro-focus tube gives much shorter exposure-times, of the order of minutes, and also allows higher accuracy of measurement owing to the small angular width of the reflected beam (cf. Carlström & Lundberg, 1958; Long, 1959). A pattern obtained with a micro-focus tube is shown in Pl. XXXA. The specimen, a flake of lithium fluoride about 100μ thick, was placed 0·5 mm. from the X-ray source and the plate 3·3 cm. from the source. The exposure-time was 8 min. at a tube voltage of 20 kV and with less than 1μA target current (Long, 1959).

The same type of diffraction can occur in the X-ray target itself, in the absence of a specimen, and in fact the Kossel lines were first observed in these circumstances. Castaing (1951) has observed Kossel lines from the target of the X-ray microanalyser; with a 1μ electron probe striking a layer of copper 1μ thick, evaporated on aluminium, the exposure-time was 3 min.

Divergent beam diffraction will occur from any crystalline material examined by projection X-ray microscopy since this involves the same arrangement as that of Fig. 14.5(e); hence the possibility of wrongly interpreting the image may arise (Nixon, 1957b). So long as the Kossel lines originate from a single crystal, or from very large grains, they will appear as smooth conic sections which can be readily recognized. As the number of grains in the irradiated area becomes greater and their average extent smaller, or if only a few small crystalline regions are present, the possibility increases of confusing diffraction effects with differences of specimen thickness or density; that it is a diffraction effect can always be tested by tilting the specimen (or the beam).

The intermediate condition, in which short arcs are produced from small grains, has been the basis of an investigation into the behaviour of beryllium under stress (Sawkill, 1959). The specimen, in the form of a cylinder 1 cm. in diameter and 0·5 cm. thick, was placed with its near face 0·2 cm. from the micro-focus of an X-ray projection microscope. An iron target was used and the tube was operated at 25 kV. The photographic plate was 8 cm. from the source, so that the magnification varied from × 40 to × 11 through the specimen. A micrograph of the specimen in its original as-cast state shows a large number of Kossel lines of varying length (Pl. XXIX(a)); only deficiency lines (dark in the positive print) occur. The grain diameter was known to be about 0·1 cm., so that there would be on the average 5 grains in the path of the beam, the diffraction lines from which will be superimposed. The great length of some of the lines indicates that initially the local perfection of the lattice is high. After the specimen had been compressed to a strain of 2%, the Kossel lines were much shorter and more diffuse (Pl. XXIX(b)). Higher strain (13%) resulted in the almost complete disappearance of the lines, owing to deformation of the lattice (Pl. XXIX(c)); at the same time, cracks appeared (white streaks). After recrystallization the Kossel lines appeared again, but the cracks did not heal (Pl. XXIX(d)). The lines were shorter than obtained from the original metal, indicating a smaller grain size after recrystallization, but the lattice was more perfect, as shown by the sharpness of the lines; some lines were double, owing to the presence of two lines in the Fe K-radiation. The method is thus a powerful means of following crystallographic changes brought about in metals by mechanical or thermal treatment, such as grain growth or crack propagation (cf. Schwarzenberger, 1959).

14.2.5. *Selected area.* In the previous methods a sample of small dimensions, or a small part of a supposedly uniform large sample, is placed in the X-ray beam. With many specimens it is necessary to know the variation of crystallographic structure from point to point on the surface or to correlate the crystal habit with microanalysis of a small area carried out by some other method. A small area of the specimen selected under an optical microscope is placed accurately on the path of the X-ray beam (Fig. 14.5(f)). If the point source is small enough then the enlarged shadow projection image of the specimen can be used to locate the area of interest. It is brought to the centre of the field of view with the stage controls, an aperture is placed in the X-ray beam and the diffraction pattern recorded (Nixon, 1957b).

X-ray beams down to 25μ in diameter were used by Fankuchen & Mark (1944) to investigate the surface and interior of single nylon fibres. Variations of the method have been used to study kink bands in a deformed single crystal of aluminium (Gay & Honeycombe, 1951), to obtain Laue photographs from Neumann bands 5μ thick in crystals of iron (Kelly, 1953) and back reflexion Laue patterns from aluminium by Lewis (1955), with the Ehrenberg and Spear micro-focus tube.

14.2.6. *Powder cameras.* The normal powder camera (of 19 cm. radius) used with large focus X-ray tubes has been scaled down for microbeam work (Fig. 14.5(d)), in conjunction with the Hilger micro-focus unit (Ehrenberg, 1957). A general purpose cylindrical powder camera of diameter 57·3 mm., and a larger semi-circular camera of diameter 114·6 mm., are formed from the same unit by using different cassettes. A Brindley type of semi-focusing camera for the Ehrenberg and Spear tube has been described by Jeffery (1952). The camera is semi-cylindrical, of diameter 6 cm. and uses 16 mm. film. A fully focusing Seeman–Bohlin type is also used for powder work (Fig. 14.5(h)), in which the powder specimen, the source and the film are all arranged on the Rowland circle.

These powder cameras produce patterns in very short exposure-times in the Hilger micro-focus unit; for instance, 10 min. for zinc oxide as compared with 2½ hr. required with a normal 9 cm. powder camera. Some biological crystallites would require prohibitively long exposures with a normal diffraction system. The pattern from a bile acid fibre, shown in Pl. XXXB, required 10 days with the normal tube but only 20–30 hr. with a micro-focus unit (Rich & Davies, 1957).

14.2.7. *Moving specimen and film types.* With powder cameras the specimen is rotated during exposure to bring all reflecting planes into the X-ray beam. Other types of specimen movement are used for single crystals (Fig. 14.5(g)). A rotating crystal camera of 1 cm. radius for the Hilger tube has been described by Carlström (1954), and a similar type of wide angle rotating crystal camera for Hilger tubes by Finean (1956). Moving-film cameras of 1 cm. radius, having the same movements as the normal size Weissenberg and Buerger cameras, have been made by Mackay (1955).

14.3. Conclusion

Microdiffraction is well developed for X-ray tubes with foci of about 50 μ in diameter and the results are sufficiently encouraging to warrant a similar development for X-ray sources down to 5 μ in size. The need is for cameras specially adapted to these very small spots, so that the greater brightness of the X-ray source can be fully utilized. It is often impossible to grow crystals of biological origin larger than a few microns in size and in some cases it is necessary to use short exposures because of changes in the specimen. Both of these problems may be eased by the use of even finer and more intense X-ray sources. Similarly, with selected area work, as the need for smaller areas of identification arises so will the need for smaller X-ray focal spots.

Added in proof

Additional references on microdiffraction will be found on p. 370.

CHAPTER 15

SOME NEW EXPERIMENTAL METHODS

The main obstacle to improving the performance of the X-ray microscope, whether of the contact, projection or scanning type, is lack of brightness of the image. Insufficient intensity makes it difficult to focus the projection microscope, and requires unduly long exposure-time in high-resolution operation of all the methods. At the same time any great increase in X-ray inensity, even if it were attainable, would lead to a dangerously high dose rate for living material and especially in medical radiology. For these reasons a great deal of attention is being given to the development of image intensifiers for increasing the brightness of the final X-ray image, and to methods of focusing this image, in the absence of an intensifier, by observation of an electron instead of an X-ray image. Some of these new experimental devices are described in this chapter, together with a method of improving the resolving power in contact microradiography by using an electron microscope instead of an optical enlarger for magnifying a 'negative' formed in a plastic film.

15.1. The scanning system as an X-ray image intensifier

Various schemes have been suggested for intensification of the X-ray image in medical diagnostics, in an attempt to reduce the dose to the patient (Moon, 1950). The scanning X-ray microscope seemed to offer the possibility of greater image brightness and better resolution, as well as improvement in the width of field. In the arrangement as originally proposed (Pattee, 1953 b), the specimen was in contact with a target which was scanned with the electron beam. The transmitted X-rays were recorded with a scintillation counter the output of which modulated the grid of a cathode-ray tube scanned in synchronism with the main electron beam. Scanning images at a resolution of 8–10 μ, as determined by spot size, showed photon noise with about 10 photons per picture point, a signal-to-noise ratio of 3·2 ($= \sqrt{10}$). The beam current required for a 500 line picture was 4 μA, in agreement with theory (Pattee, 1957 *b*).

This type of X-ray scanning microscope, in which contrast is due to absorption, has been studied in more detail by Duncumb (1957 *a*). Taking into account the finite thickness of the specimen and the thermal limitations in the target, he was led to the conclusion that there is no advantage in the use of scanning for high-resolution absorption microscopy. Suppose that a specimen of

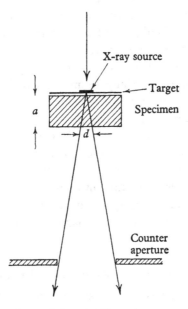

Fig. 15.1. Transmission scanning microscope geometry. (Duncumb, 1957 *a*.)

thickness *a* (Fig. 15.1) is placed in contact with a target on which is produced a minute X-ray source, moving with the scan. The target thickness and the distance between target and specimen are regarded as negligibly small. The size of the aperture that can be used in the X-ray counter is limited by the resolution *d* desired at the surface of the specimen away from the target, so that the number of X-ray photons collected per picture point is correspondingly limited. As in the scanning emission microscope, conclusions can be drawn as to the maximum number of lines per field compatible with a noise-free picture at the chosen resolution, assuming the maximum beam current obtainable from a hot cathode gun. Two other limits must also be considered: heating of the target and deflexion aberrations (§ 7.1). These limits are plotted in Fig. 15.2 in the same way as for the emission system (cf. Fig. 7.3). The deflexion limit is almost the same as for emission;

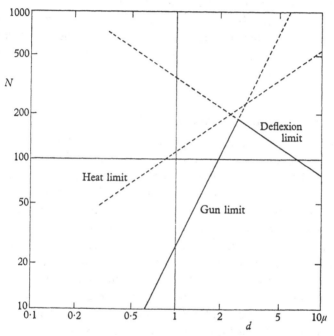

Fig. 15.2. Transmission scanning microscope limits, in terms of number of resolved lines in the field of view, in function of resolving power d (= spot size). (Duncumb, 1957 a.)

the heat limit is that for a copper target with thickness equal to the electron penetration. The gun limit is decided by choosing a specimen with thickness $a = 100\ d$. It is seen that a noise-free picture of 200 lines is just attainable at 2–3 μ resolution, by the absorption scanning system. This has been verified in practice with a specimen 200 μ thick and an electron beam of 25 kV and 25 μA, the target being a copper foil 3 μ thick.

It is seen that the gun limit becomes the most serious as improved resolution is sought. The field emitter cathode holds out some promise of improvement (§ 8.1.3), up to the point at which the heat limit will become the decisive factor. Alternatively, the number of X-ray photons collected could be increased by using a thinner specimen and a larger counter aperture. But a very thin specimen would be better examined by contact microradiography. An experimental comparison was made with the simple projection X-ray microscope to test if any increase in brightness or contrast could be obtained. No noticeable difference was observed in the photographic results, but for direct viewing on the fluorescent screen the scanning system was superior and this would seem to be its main advantage.

15.2. Other forms of image intensifier

Several other methods show better promise for image intensification than does the scanning X-ray microscope. Some of these may be adapted for use with the contact, projection or reflexion types of X-ray microscope so that the image intensity can be increased up to the limit set by quantum fluctuations. For applications involving living specimens this limit determines the minimum dose rate during an exposure and therefore the ultimate limit of resolution possible without causing radiation damage (Nixon & Pattee, 1957). It is necessary to be able to focus the X-ray microscope before making a photographic exposure and some form of intensifier will be essential for working at a resolution below $0.1\,\mu$. Theoretically, a 'blind' through focal series of exposures could be taken, but as each would need to be as long as the final exposure, the time required would be prohibitive.

One of the first useful types of intensifier was that described by Teves & Tol (1952). It consists of an evacuated glass tube some 10 in. in diameter, containing a photocathode at one end and a fluorescent screen at the other. X-rays enter the tube and strike a thin layer of fluorescent material over the photocathode. The emitted light quanta produce photoelectrons in the cathode which are accelerated by 30 kV before striking the final fluorescent screen, and the electron image is increased about 10 times in brightness by this means. In addition, the tube contains focusing electrodes that reduce the original image to a much smaller area, leading to a further increase in brightness by a factor of about 10. This final image is viewed by a binocular microscope with a magnification of about × 10, and with a large aperture so that it is enlarged without loss of brightness. The total useful gain is somewhere between 50 and 100 times over the original brightness of the fluorescent screen as viewed directly in medical fluoroscopy and at this level the quantum noise can be seen. Further gain up to × 1000 can be obtained by increasing the kilovoltage across the tube, and, although this does not permit a corresponding reduction in dose rate owing to quantum noise, it makes it possible to view the image in brighter external illumination. For X-ray microscopy a thin beryllium window would have to be used in place of the half-inch thick glass window of the tube developed for medical work, where hard radiation is demanded. In this way Lang (1954) has adapted the tube for X-ray diffraction with copper radiation of about 8 kV energy. Such an intensifier could best be used in conjunction with the projection and reflexion methods of X-ray microscopy, where the initial X-ray enlargement reduces the need for good resolution in the intensifier. With the contact method the intensifier resolution would have to be as good as the final resolution sought.

This type of intensifier is complicated and expensive; it is being used extensively for ciné medical work but not yet for microscopy. Simpler methods are available such as the gas discharge device of Lion & Vanderschmidt (1955), in which the photocathode and fluorescent screen are replaced by a parallel plate counter, with a gas at atmospheric pressure filling the narrow space between the plates. X-rays enter through one plate, which may be a thin sheet of beryllium, and cause ionization in the gas. The high voltage maintained across the plates produces a gas discharge and the electrons produced are accelerated to the second plate, which is made of a thin layer of aluminium on a fluorescent material, backed by a glass plate. The electrons penetrate the aluminium and a fluorescent image is seen. The plates are close enough together that the sideways spreading of the discharge does not appreciably reduce the resolution.

An even simpler approach is to replace the gas discharge space by a sheet of semi-conductor thin enough to have appreciable lateral resistance. Again a high voltage applied across the two faces of the sheet will accelerate electrons produced by absorption of X-rays in it, so that these electrons strike a fluorescent layer on one of the surfaces. In neither of these types of intensifier is there any reduction or enlargement of the image. However, if their behaviour with X-ray illumination is similar to that with ultra-violet (Williams & Cusano, 1955), an image could be held for long periods, as in a storage tube, thus allowing detailed examination of the specimen even though the X-ray exposure might only last a fraction of a second. A review of the different types of light amplifiers has been published by Kazan & Nicoll (1957), who have been responsible for much of the original work in this field.

Attempts have also been made to use the very sensitive image orthicon type of television camera to detect faint images on a fluorescent screen. The guiding principle here, as in the other types, is that at each stage in the image-forming process the number of quanta involved must be greater than the number in the original image (cf. Tol, Oosterkamp & Proper, 1955). It is not sufficient that the conversion processes from X-rays to light, to photo-electrons and back to light again, should each yield the same number of quanta. Although the statistical errors at each stage would be the same, the final fluctuations would be formed by summing the individual errors and so would be larger than the original noise. It is necessary to gain ten or more light photons from each X-ray photon and then to ensure that the subsequent conversion ratio is at least unity.

15.3. Image conversion X-ray microscope

In this method a contact X-ray image is formed on a grainless foil, which emits photo-electrons in proportion to the intensity of the X-rays at each image point. These photo-electrons are accelerated through several kilovolts before striking a fluorescent screen or photographic plate, thus giving an increase in intensity. At the same time the photo-electrons are imaged by a lens to give a magnified image of the distribution of X-ray intensity in the original image formed on the foil.

The principle of the method, due to Möllenstedt & Huang (1957), is shown schematically in Fig. 15.3. The X-rays originate at the anode of a high intensity fine-focus tube, in which either magnetic or electrostatic lenses may be used (Huang, 1957). The object is placed in contact with the emitting foil, usually of silver a few hundred Ångström units thick. In practice the energy spread

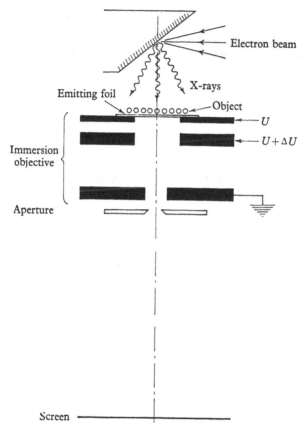

Fig. 15.3. An electrostatic immersion objective lens used as an X-ray image converter. (Möllenstedt & Huang, 1957.)

of the photo-electrons is as great as that of the primary X-ray beam, so that there is a serious loss of resolution owing to the chromatic aberration of the electron lens. This can be avoided if a foil of potassium chloride (see Fig. 15.4) is placed below the silver foil. The photo-electrons of varying energies strike the second foil and produce secondary electrons of almost uniform energy (Sternglass,

1955). These secondaries are then both accelerated and focused by the electrostatic immersion objective, so that an intensified and enlarged image is formed on the final screen, with contrast determined by the local degree of X-ray absorption in the object.

The process of X-ray image formation is essentially the same as in the contact method of microradiography (chapters 2 and 9), and has the same advantages and limitations. The geometrical unsharpness must be reduced below the required resolution by keeping the

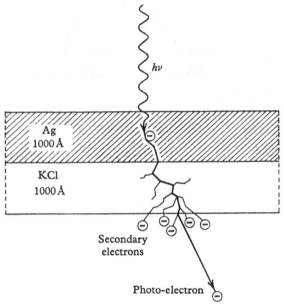

Fig. 15.4. The X-ray-electron conversion layer, in the apparatus shown in Fig. 15.3, is a composite film of silver and potassium chloride. (Möllenstedt & Huang, 1957.)

X-ray source and object far enough apart. The illumination is then almost parallel and stereographic X-ray images may be obtained, as in contact microradiography (cf. § 10.3). Fresnel diffraction will also occur, but for an object 40μ thick the first fringe is of width 0.14μ with X-rays of 5 Å wavelength, and therefore below the resolution of 0.3μ obtained by Huang.

The excellent contrast obtained and the large field of view (100μ) are illustrated by Pl. XXXI, for which the X-ray tube was operated at 22 kV and 500 μA. With the immersion objective at

40 kV and an electron magnification of ×400, the exposure-time was three minutes. Photographic enlargement brings the total magnification to ×1500.

The ultimate resolution is set primarily by the chromatic aberration of the electron lens, provided that the resolution of the original X-ray image is adequate and that the silver foil itself shows no structure. The limit set to the resolution by chromatic aberration depends on the electron energy spread and on the field strength at the secondary emitter. Huang has measured the energy spread for a composite foil of 2000 Å of collodion, 1000 Å of silver and 500 Å of potassium chloride, and found it to be 2 electron volts when a 50μ aperture was used in the lens. The field strength in the electrostatic lens was 20 kV/cm., corresponding to a theoretical resolution of 0.5μ, as compared with 0.3μ estimated from the micrograph. Improved resolution is to be expected from using a magnetic immersion objective 3 mm. from the foil and with 50 kV accelerating voltage (170 kV/cm.): 600 Å without an aperture, and 150 Å with a 20μ aperture in the lens. If these predictions can be realized, this may prove to be the most powerful method of X-ray microscopy.

15.4. Projection X-ray microscopy: focusing by back-scattered electrons

As explained earlier, lack of intensity is the main obstacle in the way of improving the resolution of the projection X-ray microscope. Theoretically it is possible to produce X-ray sources considerably smaller than 0.1μ. The difficulties arise in deciding when the smallest spot has been formed and in keeping this small spot stable during the exposure. With direct focusing by viewing the X-ray image of a test grid, a resolution limit of about 0.1μ can be reached (Nixon, 1955a, b), but beyond this an indirect method must be used, such as that due to Ong & Le Poole (1957).

The principle of the method is that electrons scattered back from the X-ray target are focused by means of the same electron lenses as are used to form the electron spot. When the electron beam strikes the target a small but finite number of electrons are elastically scattered back in the direction of incidence. These electrons, about 10^{-6} of the total beam current, retrace their initial paths through the lenses (Fig. 15.5) and are brought to a focus in the plane of the filament tip, in a spot of diameter determined in part by the size of the focal spot. In practice a fluorescent screen is placed in the anode plane and the objective lens is tilted or the electron beam slightly deflected with magnetic coils, so that the returning electrons strike the screen instead of passing through the anode aperture. This electron image, about $50–100\mu$ in diameter, is viewed with a low-power microscope through the side of the

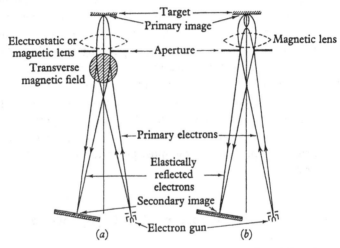

Fig. 15.5. Focusing by back-scattered electrons. Separation of the primary and return beams by (a) transverse magnetic field, (b) tilted magnetic lens. (Ong & Le Poole, 1958a.)

column and the lens controls are adjusted until the smallest spot is obtained. The return electron current, though small, is concentrated into a very small spot, which is thus visible even when the projected X-ray image is too faint for direct focusing to be possible. Although the return spot would be of minimum size in the plane of the filament and not at the anode, it is possible to calibrate the lens controls by taking a through focal series so as to allow for this in subsequent focusing.

The method has been used with an aluminium target and at voltages down to 6 kV, to reduce the spread due to electron penetration in the target; the exposure-time on high-resolution plates was 20 min. Images of 1500 mesh/in. silver grid show a 0·1μ Fresnel fringe (Pl. XXXII A) indicating resolution of the same order.

On energy considerations it is possible to compare the direct and indirect methods of focusing. Since the efficiency of direct back-scattering ($\sim$ 10^{-6}) is below that of soft X-ray production ($\sim$ 10^{-4}), the number of back-scattered electrons collected is about one hundred times smaller than the number of X-ray quanta. The electrons are focused onto an area about 50μ in diameter, or 6·25$\pi \times$ 10^{-6} cm.^{2} The X-rays will be spread over a much greater area, since it is necessary to provide sufficient X-ray magnification for direct focusing. For high resolution this should be of the order of $\times$ 1000, giving an overall magnification of $\times$ 10,000 if a $\times$ 10 viewing lens is used. Since the specimen cannot be closer than about 10μ to the X-ray source, the minimum source to screen distance will then be 1 cm., and the X-rays will be spread over an area of 4π cm.^{2} Assuming X-rays and electrons to be equally effective in causing fluorescence, the ratio of the electron to the X-ray energy density is given by the inverse ratio of the areas $\times$ 10^{-2}, that is, the intensity of the back-scattered electron image is 6·4 $\times$ 10^3 greater than that of the X-ray image.

With this large factor of brightness in hand, it would be possible to use the method to focus at a much higher resolution if no other difficulties were involved. The beam current that can be focused into the electron spot from a normal thermionic cathode varies as $r^{\frac{8}{3}}$ (cf. (3.24)), so that the beam current will be about 10^3 less at 100 Å resolution than at 1000 Å, the present limit at which it is still possible to focus the X-ray image directly. Since the indirect method gives a gain of well over 10^3, there should be sufficient intensity on the anode focusing screen for very high resolution indeed. So far this has not been obtained in practice, probably because the electron beam energy has to be lowered to reduce the spreading of the electrons in the target, when a smaller spot is sought, and at these lower voltages the output of the back-scattered electrons is not as great. However, it seems probable that, with some improvement in the electron gun and electron lenses, the method might prove to be practicable for a resolution in the range 100–500 Å.

15.5. Projection X-ray microscopy: focusing by forward-scattered electrons

For very high resolution X-ray microscopy it is necessary to use ultra-soft X-rays, in order to obtain sufficient contrast in a biological detail of dimensions comparable with the resolution; in consequence, at least a partial vacuum is required between specimen and photographic plate. Once the decision has been taken to evacuate the camera, it becomes practicable to consider other methods of focusing, which will in any case be necessary since the method described above will eventually fail as the target is made thinner in order to reduce spot broadening due to electron diffusion (cf. § 3.6.3.). With the X-ray path at atmospheric pressure, targets thinner than about $0.1\,\mu$ cannot be used (Nixon, 1955 a, b).

If a thin target is used and the X-ray path is *in vacuo*, then sufficient electrons will pass through and reach a viewing screen to give useful information about the focusing conditions (Fig. 15.6) (Nixon, 1960 a, b). The number of electrons penetrating a thin foil increases almost linearly as the thickness is reduced. For a foil of thickness equal to half the electron range, about 50 % of the incident electrons will penetrate through it as compared with 10^{-6} scattered back along the incoming electron path. The electrons passing through the target will mostly be scattered through large angles, so long as it is thick enough for multiple scattering to occur. With heavy metals the target thickness for single scattering is only 15 Å (gold), rising to 150 Å for carbon, for electrons of 10 kV incident energy (Cosslett, 1956). With a 3–4 kV beam and a target thickness of a few hundred Ångströms, plural and possibly multiple scattering will predominate and the overall scattering angle will be large. For 200 Å of tungsten at 4 kV the most probable angle of deviation in multiple scattering is over 90°. When a fluorescent screen is placed a few centimetres beyond such an electron transparent target a fairly uniform illumination is seen. Since these electrons come from a small focal spot, where the beam strikes the target, a test mesh can be placed close to the target so that an enlarged projection image is formed on a fluorescent screen by the scattered electrons. The imaging conditions are identical with those of X-ray projection microscopy, except that the electron wavelength is much shorter (0·38 Å at 1000 V) and therefore Fresnel diffraction is of less importance. High magnification by projection is possible since the test specimen can be placed a few microns from

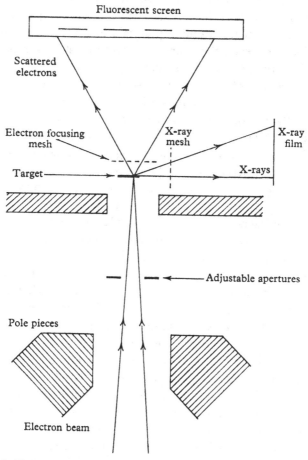

Fig. 15.6. Focusing by forward-scattered electrons. The electrons scattered in the thin target form a projection image of a test grid on the fluorescent screen; the X-ray image is recorded on a film placed to one side of the axis. (Nixon, 1960a.)

the target, and the field of view is large since a cone of electrons of angle one radian or more is available.

The projected electron image can be viewed with a ×10 eye-piece, and its sharpness will depend on the size of the electron source. When the final electron lens has been adjusted until the image is in best focus, the electron spot at the target will be of minimum size. The effectiveness of this type of focusing can be compared, in the same way as above, with that using back-scattered electrons. If 50 % of the electrons can be used, compared to 10^{-6} back-scattered, there is a numerical gain of 5×10^5 in the forward direction. However, the fluorescent screen must be at least 1 cm. from the target, to obtain a sufficiently high direct magnification, as in direct X-ray focusing. If the scattered beam issues in a cone of angle one radian the electrons will be spread over $25\pi \times 10^{-2}$ cm.^{2} whereas the return-focused electrons cover $6 \cdot 25\pi \times 10^{-6}$ cm.^{2}, that is, $2 \cdot 5 \times 10^{-5}$ less area. The energy density ratio will then be $5 \times 10^5 \times 2 \cdot 5 \times 10^{-5}$, so that image brightness is about ten times greater for the forward scattering as compared with the backward scattering method of focusing. Forward scattering may be used down to the lowest kilovoltages and thinnest targets, since the mean scattering angle will remain fairly large, so long as the target thickness is always appropriately related to the voltage.

Since the projected electron image is formed on the axis of the microscope, the specimen and film cannot be placed in the usual position for X-ray projection microscopy. Since the characteristic X-radiation is emitted uniformly in all directions, however, the specimen and film may be placed parallel to the axis and slightly to one side of the scattered electron beam used for focusing (Fig. 15.6). The specimen is a few microns, and the film from 1 to 4 mm., from the target. The specimen is placed in position by observing its electron shadow image on the fluorescent screen used for focusing. Very thin targets can be made by evaporating films of heavy metals onto thin layers of collodion supported on metal grids. When very soft X-rays are needed, thin carbon films may be used; they can be made by evaporation (Bradley, 1954) in any thickness down to about 50 Å, and prove to be very resistant to damage by the electron beam. Aluminium oxide films (Al_2O_3) can be prepared in appreciable areas with a thickness of a few hundred Ångströms, and will give both oxygen (24 Å) and aluminium (8·3 Å) radiation.

15.6. Contact microradiography with electron microscope enlargement

The use of an electron microscope to view a high-resolution micro-negative is attended by great technical difficulties (§ 2.5.2), and a more promising approach has been proposed by Ladd, Hess & Ladd (1956; Ladd & Ladd, 1957). They make use of the change in solubility produced in ammonium dichromate crystals by exposure to X-rays. An exposure is made as for contact microradiography, using the crystal in place of the emulsion, and its surface is then 'developed' for $2\frac{1}{2}$ hr. in methyl alcohol. The X-ray image is thereby transformed into surface relief on the crystal, with depth dependent on the initial X-ray intensity at each point and with remarkable absence of granularity. A thin replica of the crystal surface is then made in a plastic film, shadowcast with metal, and viewed at high magnification in an electron microscope. Even at enlargements of × 10,000 there is no visible background granularity. The image must be interpreted carefully since the relationship between X-ray intensity and surface relief is unknown, but this is a minor disadvantage if higher resolution is gained. A more serious difficulty is the very low sensitivity of the process, requiring exposure-times of 20 hr. at 20 kV and 20 mA. Other materials have been tried such as polyvinyl chloride acetate with development in acetone, but the exposure-time was not appreciably less.

The search for faster materials has been taken up by Pattee & Warnes (1957), but with negative results from plexiglas, mylar, collodion, ethylcellulose, polystyrene, polyethylene and nylon. Images could be obtained from a chloride polymer after an exposure of 1·7 hr. at 12·5 kV and 10 mA, but the resolution was spoiled by bubbles in the plastic possibly due to hydrogen released by X-ray action on the polymer bonds. The best results came from using freshly prepared thin sheets of plastic that had been heated to remove as much gas as possible before exposure. The exposures were still fairly long (7–9 hr.) but no granularity could be seen at × 645 magnification under a light microscope, after the surface had been shadowed with aluminium (Pl. XXXII B).

A more sensitive material employing a colour change instead of a change in solubility has been tried by Pattee (1960a): a grainless photosensitive dye, pararosanaline leuconitrile, which turns magenta on exposure to X-rays (Chalkley, 1952). The speed is much less than that of a Kodak 649 plate but faster than the plastics mentioned above. Using a 10μ point source of X-rays at a distance of a few hundred microns from the specimen and dye layer, 5-min. exposures were found to give sufficient effect, the density of the coloured layer being dependent on the X-ray intensity. Since the dye image has to be photographed through a light microscope, as with a normal contact negative, the resolution is still limited by the optical resolving-power; the dye layer itself seems to be grainless. At present there is no way of viewing it with the electron microscope, since the colour change produced by the X-rays gives no contrast in the electron image, but future developments may allow the inherent resolution of the dye layer to be exploited in this way.

Added in proof

The practical problems involved in the electron microscopy of such X-ray 'negatives' have been discussed by Pattee (1960b), and a number of possible processes have been investigated by Asunmaa (1960). Some of these are promising, but artefacts of processing need further investigation. The use of xerography also seems likely to have advantages over conventional photographic methods for X-rays. Auld & McNeil (1960) in preliminary trials with a technique involving a liquid developer, obtained a definition comparable with that of the high resolution emulsions and yet with no longer an exposure-time than needed with fast coarse-grained X-ray film.

COMMENTS ADDED IN PROOF

To Chapter 7: X-ray Emission Microanalysis

Increasing interest is being shown in microspectrometry of X-rays, whether with a static electron spot or by the scanning method. A large number of papers on the subject were contributed to the Stockholm Symposium on X-ray Microscopy and X-ray Microanalysis and also to the 8th Conference on Applications of X-ray Analysis, Denver 1959.

General principles (§ 7.1.1)

Survey articles have appeared by Duncumb (1960c) and Wittry (1960a), the latter dealing particularly with the relative performance of the various microanalysers described in the literature.

The discussion of the variation in K-line output with atomic number (p. 187) needs to be modified in the light of recent experimental results (see below, p. 369).

Corrections, and limits of accuracy (§ 7.1.2)

The corrections to be applied to the experimentally observed line intensities, on account of absorption and fluorescence, are fully discussed by Philibert & Bizouard (1959, 1960). They can be large when elements differing widely in atomic number are present.

The limits of accuracy and detection sensitivity have been further discussed by Duncumb (1959, 1960b) and Wittry (1958). The former has evaluated the optimum values of electron beam aperture and accelerating voltage at given count rate in the scanning microanalyser. The count rate is found to be proportional to the collection efficiency, to $C_s^{-\frac{3}{4}}$ and to $d^{4\cdot4}$, where C_s is the spherical aberration coefficient of the electron lens and d is the focal spot diameter. A finer spot can be used only if spherical aberration is reduced or if collection efficiency is increased, and Duncumb (1960b) describes an end-window counter designed for the latter purpose. The prospects of partially correcting spherical aberration by means of four- and eight-pole lens systems have been further considered by Archard (1960b). In the existing scanning microanalyser Duncumb was able to analyse particles as small as 0·3 μ diameter.

Discrimination between elements (§ 7.1.3)

A method proposed by Dolby (1959), for separating overlapping pulses produced in a proportional counter by neighbouring elements in the periodic table, has been experimentally tested (Dolby, 1960; Dolby & Cosslett, 1960). It was possible to discriminate between magnesium, aluminium and silicon by use of the filter network. The collection efficiency is then very much higher than when using a crystal spectrometer. Combined with the unexpectedly high K-output obtained from carbon (see p. 369), this result increases the prospects of analysing carbon, nitrogen and oxygen.

The static spot microanalyser (§ 7.2)

Apparatus. The Metropolitan-Vickers microanalyser, based on that of Mulvey (1959*a*, 1960*a*), has been described in detail by Page & Openshaw (1960). Mulvey (1960*b*) has described an improved version of his apparatus, and has now added a scanning system to it (Bernard, Bryson-Haynes & Mulvey 1959, 1960). Castaing & Descamps (1958) describe the vacuum spectrometer which has been fitted to the original Castaing microanalyser. It embodies a curved mica crystal and covers the wavelength range 4–12 Å, allowing elements down to sodium to be analysed.

Applications. The static spot microanalyser has been used mainly for metallurgical analysis: for interface segregation by Austin, Richard & Schwartz (1960); for concentration gradients in diffusion couples by Ogilvie & Lewis (1960); for a wide range of studies of ferrous and non-ferrous metals by Philibert & de Beaulieu (1959), Adda, Philibert & Faraggi (1957), and Philibert & Bizouard (1959, 1960); by Mulvey (1959*b*) on a variety of systems; by Reisdorf (1960) on precipitates extracted from steels; and by Wittry (1960*b*) on a number of problems in both ferrous and non-ferrous metallurgy. Some applications in mineralogy and geology, especially the variation in nickel content in meteorites, have been explored by Adler, Axelrod & Branco (1959), Maringer, Richard & Austin (1959) and Wittry (1960*b*). An indication of the wide variety of uses to which the Soviet microanalyser is being put has been given by Borowskii (1960).

The scanning microanalyser (§ 7.3)

Apparatus. Improvements in the original scanning microanalyser have been described by Duncumb (1959, 1960*a*). The details of a new instrument specifically designed for metallurgical use were described by Duncumb & Melford (1960). This is the instrument illustrated in Pl. VI*b*; it served as prototype for the commercial model now produced by the Cambridge Instrument Company. Mulvey has exhibited a version of his microanalyser to which a scanning system has been fitted (Bernard, Bryson-Haynes & Mulvey, 1959). Long (1959) has similarly incorporated scanning in the static spot instrument shown in Fig. 7.6.

Applications. Melford (1960) described some illustrative examples of the use of the scanning microanalyser in ferrous metallurgy. Agrell & Long (1960*a, b*) investigated a number of mineralogical applications with the instrument specially designed by Long (1959) for the purpose. It has to work with a very low beam current, because of the poor thermal conductivity of rock sections, but it has the advantage of allowing direct optical observation of the spot examined, since such sections can be viewed in transmission (cf. Fig. 7.6). The study of the nickel content of meteorites has been especially interesting.

Fluorescent microspectrometry (§ 7.4)

The reliability of trace determinations by the fluorescence method has been discussed by Liebhafsky, Pfeiffer & Zemany (1960). They pay particular attention to the specification of the smallest amount which can be recognized with a given degree of certainty against the background fluctuations in count rate, and propose a definition for the 'minimum amount guaranteed detectable'.

To Chapter 8: Production of X-rays

Electron range in the target (§ 8.3)

The data for electron penetration given on pp. 222–3 agree well with the quantum mechanical predictions for the rate of energy loss (Bethe, 1930), when account is taken of electrons lost by back-scattering (Holliday & Sternglass, 1959). Below a voltage of 10 kV, however, the exponent in the Thomson-Whiddington relation (8.15) appears to fall increasingly below 2. The data given in the first line of Table 8.1 for 2·5 kV may thus be in appreciable error.

Spectral distribution (§ 8.4)

Tables of the wavelengths of the characteristic line emission of the elements have been collected into book form by Sagel (1959).

Efficiency of line production (§ 8.5.2)

It must be emphasized that the relations 8.29 and 8.31 given on pp. 230–1 for the ratio of the efficiencies of line and continuous X-ray output apply to a copper target only. The way in which the ratio varies with atomic number (Z) is not yet clearly established. From the theoretical side it has been discussed by Worthington & Tomlin (1956), Archard (1960a) and Cosslett (1960). Although the quantum mechanical relation for the primary process of K-ionization is well confirmed, from experiments on very thin targets, it has to be corrected when thick targets are used for energy loss of the incoming electrons and for the loss of emitted X-ray quanta by absorption and fluorescence. The value of the fluorescent yield (w) in particular is poorly established for the lighter elements.

Experimental evidence is also meagre, but for light elements it gives rather higher values of output than expected. From the experiments of Dolby (1960) the ratio of K-quanta from carbon and copper (at given excitation ratio U) is about 1:20, or roughly proportional to Z^2. Taking account of Moseley's rule (p. 231), the *energy* of K-output is then approximately proportional to Z^4. On the other hand, the total energy emitted in the continuous spectrum is, by 8.21, approximately proportional to ZV^2 and thus to Z^5 at given U. It follows that the ratio of X-ray energy in the K-line to that in the whole of the continuous spectrum is roughly proportional at Z^{-1}, so that, although it is about 0·4 for copper, it may be considerably greater than 1 for carbon. Comparison of the few data in the literature suggests that the variation is steeper (Dyson, 1960), and may more nearly depend on Z^{-2}, giving a value approaching 10 for carbon. This conclusion agrees with the empirical observations of some authors that the K-lines of the lighter elements stand out better from the background.

370   COMMENTS

To Chapter 9: Specimen Preparation Techniques

A number of improvements in preparative techniques, especially for biological materials, were reported at the Stockholm Symposium on X-ray Microscopy and X-ray Microanalysis.

Freeze-drying (§ 9.1.1), and Embedding and Sectioning (§ 9.1.2)

Details of the preparation of a wide variety of tissues, including hair, sebaceous gland and kidney, were described by Saunders & van der Zwan (1960). Methods particularly applicable to bone were discussed by Sissons, Jowsey & Stewart (1960a).

Injection and staining (§ 9.1.3)

Bellman, Frank, Lambert, Odén & Williams (1960) gave details of their techniques for microangiography and microcinématography of the rabbit ear in vivo. Related microangiographical methods were described for rat liver by Blickman, Klopper & Recourt (1960a) and for the brain and spinal cord by Saunders (1960).

Mounting the specimen for contact microradiography (§ 9.1.4)

An improved version of the procedure given on p. 247 was described by Greulich (1960).

Metallurgical and mineral specimens (§ 9.2.1)

Techniques suitable for granular materials, such as mill products in mineral dressing processes, were described by Cohen & Schlogl (1960). Ruff & Kushner (1960) gave details of their preparative methods for aluminium-copper alloys.

To Chapter 14: Microdiffraction

A session of the Stockholm Symposium on X-ray Microscopy and X-ray Microanalysis was devoted to the techniques and applications of microdiffraction. A number of other advances have been reported to the Denver X-ray Conferences and elsewhere. For the most part they are direct developments of the topics discussed in chapter 14.

X-ray tubes (§ 14.1)

An improved design for a tube of the Ehrenberg and Spear type, operating at 50 kV, was described by Stansfield (1960); see also Franks (1960). A low-voltage tube (800 V to 3·5 kV) has been built by Fournier (1960) for investigating microcrystals in biological tissues. Biological applications were also described by Kennard (1960).

Cameras and methods (§ 14.2)

New designs for micro-cameras have been described by Bergmann (1960) and by Jeffery & Bullen (1960); the latter discusses a variety of applications of microdiffraction in metallurgy and mineralogy. Further details of the use of the Berg–Barrett method (§ 14.2.3), especially in the study of the deformation and recrystallization of aluminium, have been given by Weissmann (1959, 1960; Weissmann & Turner, 1959). Variant techniques, in which particular diffracted beams are selected, have been used by Newkirk (1959) and by Lang (1959) to produce striking micrographs of the distribution of dislocations and slip lines in single crystals. Similar transmission methods of X-ray imaging have been used by Barth & Hosemann (1958) and by Borrmann, Hartwig & Irmler (1958) to study imperfections in quartz, silicon and germanium; see also Gerold (1960). Applications of Kossel line diffraction (§ 14.2.4) in metallurgy have been investigated by von Batchelder (1959).

APPENDIX

ABSORPTION AND EMISSION DATA

A. I. *X-ray mass absorption coefficients of certain elements and of dry air for the wavelength range* 0.710–8.321 Å (*Dyson, 1956*)

The principal compilations of absorption coefficients in this energy region are those given by Compton & Allison (1935), and in the *Handbook of Chemistry and Physics* (of which the 36th edition has been used here). Compton and Allison give a table, with references, of the mass absorption coefficients for gases, and in the appendix there is a fuller table compiled by S. J. M. Allen. The table in the *Handbook* is also by Allen, and appears to follow closely the table in Compton & Allison.

In the Table which follows, data published or reviewed later than 1935 and taken directly from the literature are used where possible, but recourse is made if necessary to the tables of Allen. In this way it is believed that maximum reliability is achieved. Only experimental data are used. A review of the existing data for certain gases is given by Woo & Sun (1947), for comparison with theoretical values derived by them, and their collection of experimental values is made use of.

In the Table, parentheses denote interpolated or extrapolated values.

Mass absorption coefficients

λ (Å)	k eV	Be	C	N_2	O_2	Al	A	Cu	W	Au
0·710	17·46	—	—	—	1·22	—	13·00	51·0	104	—
0·746	16·62	—	—	—	—	—	—	54·48	—	—
0·829	14·96	—	—	—	—	—	—	73·3	—	—
0·880	14·09	—	0·99	1·5	2·20	—	(30)	87·1	—	159
0·925	13·41	—	—	—	—	11·88	—	99·04	—	137
0·949	13·06	—	1·20	—	—	—	—	—	—	—
1·000	12·40	0·55	1·36	2·10	3·15	—	35·00	—	260	—
1·039	11·93	—	—	—	—	—	—	145·2	—	—
1·104	11·26	—	—	—	—	—	—	—	—	87
1·175	10·55	—	—	—	—	23·48	—	—	—	102
1·235	10·03	0·95	2·42	3·95	5·7	—	62·5	—	95	—
1·253	9·89	—	—	—	—	—	—	—	—	120
1·337	9·27	—	—	—	—	—	—	—	—	141
1·389	8·93	1·25	3·35	5·50	8·1	(85)	—	—	—	—
1·433	8·65	—	—	—	—	—	—	41·54	130	168
1·539	8·06	1·60	4·52	7·45	11·16	51·15	116	51·37	176	216
1·656	7·48	—	—	—	—	—	—	62·73	—	251
1·787	6·94	—	—	—	—	76·34	—	77·70	—	—
1·934	6·40	3·05	8·75	14·0	22·0	—	214	96·24	300	387
2·099	5·90	—	—	—	—	122·35	—	120·1	—	—
2·285	5·42	—	15·0	—	—	152·1	—	155	—	563
2·50	4·95	6·1	17·8	(30)	45·5	196·3	436	198	—	720

Mass absorption coefficients (cont.)

λ (Å)	k eV	Be	C	N₂	O₂	Al	A	Cu	W	Au
2·74	4·52	—	25·0	(40)	60	255	(550)	256	—	925
2·91	4·26	—	—	(49)	(74)	—	664	—	—	—
3·03	4·09	—	35·0	(55)	84	333	(750)	319	—	1252
3·25	3·81	—	—	—	—	—	950	—	—	—
3·352	3·70	—	43·0	—	—	450	—	439	—	1470
3·38	3·66	—	46·0	79·5	117	—	(1000)	—	—	—
3·55	3·49	—	—	—	—	—	1180	—	—	—
3·57	3·47	(16.2)	—	96	138	—	(1200)	—	—	—
3·59	3·45	—	55·2	—	—	—	—	—	—	—
3·735	3·32	—	—	—	—	613	—	—	—	—
3·87	3·20	—	—	—	—	—	1460	—	—	—
—	—	—	—	—	—	—	148	—	—	—
3·93	3·16	—	71·0	121	189	—	(152)	673	—	2050
4·146	2·99	—	84·6	—	—	—	—	773	—	1910
4·36	2·84	(29·2)	97·8	166	258	—	202	—	—	2450
4·718	2·63	—	—	—	—	1167	—	—	—	2550
5·17	2·40	(49)	160	273	413	—	324	—	—	2910
5·36	2·31	—	—	—	—	1567	—	—	—	—
5·39	2·30	—	185	—	—	—	—	—	—	—
6·057	2·04	—	—	—	—	—	—	—	—	1210
6·97	1·78	—	390	645	976	—	748	—	—	—
7·111	1·74	—	—	—	—	3429	—	—	—	1730
8·321	1·49	—	656	1109	1589	396	1160	—	—	2450

X-ray absorption data for dry air, at 76 cm. Hg and 20° C.

λ (Å)	k eV	μ/ρ	Fraction transmitted				
			1	2	3	4	5 (cm.)
0·880	14·09	2·04	0·998	0·995	0·993	0·990	0·988
1·000	12·40	2·78	0·997	0·993	0·990	0·987	0·983
1·235	10·03	5·12	0·994	0·988	0·982	0·976	0·970
1·389	8·93	7·14	0·991	0·983	0·975	0·966	0·958
1·539	8·06	9·74	0·988	0·977	0·965	0·954	0·943
1·934	6·40	18·48	0·978	0·957	0·935	0·915	0·895
2·50	4·95	38·9	0·954	0·910	0·869	0·829	0·791
2·74	4·52	51·3	0·940	0·884	0·830	0·781	0·735
2·91	4·26	62·8	0·927	0·859	0·797	0·739	0·684
3·03	4·09	70·9	0·918	0·843	0·774	0·711	0·652
3·38	3·66	100·3	0·886	0·785	0·696	0·616	0·546
3·57	3·47	120·2	0·865	0·755	0·647	0·560	0·484
3·87	3·20	148	0·836	0·700	0·585	0·490	0·410
—	—	131	0·854	0·730	0·623	0·532	0·454
3·93	3·16	137	0·848	0·719	0·609	0·516	0·437
4·36	2·84	188	0·798	0·636	0·507	0·405	0·322
5·17	2·40	308	0·690	0·476	0·324	0·226	0·156
6·97	1·78	723	0·418	0·175	0·073	0·031	0·013
8·32	1·49	1220	0·229	0·053	0·012	0·003	0·001

Notes

Beryllium

Allen's data are used here, together with one or two extrapolated values given by Compton and Allison. Some measurements are given by Andrews (1938), but, as pointed out by Andrews, these are in error due to the presence of impurities. In some cases Andrew's values exceed Allen's by a factor of more than three times.

Carbon

Chipman (1955) has shown that there is a considerable amount of small-angle scattering from graphite and that the accepted values of μ/ρ for carbon are too high for application to plastic material and paraffin. Allen's data are used here, but it is clear that for accurate values of the absorption in, say, distrene, it is better to deduce them from, for example, Grosskurth's (1934) data for paraffin, rather than to use experimental values obtained with graphite.

For copper K_α radiation, Chipman recommends a value of 4·15 for carbon in compounds, compared with 4·87 (Andrews) and 4·52 (Allen).

Nitrogen

Values taken from the *Handbook*.

Oxygen

Values taken from Woo and Sun, along with some values from the *Handbook*.

Air

There appears to be no data later than that reviewed by Compton and Allison, so it was considered more satisfactory to calculate the required information from the mass absorption coefficients of the individual constituents, interpolating for one or more of the constituents where necessary. The composition of dry air is taken to be 75·5 % nitrogen, 23·2 % oxygen and 1·3 % argon. (See also Miller and Zingaro, 1960).

Aluminium

Biermann's (1936) data are used in the range 2–4 Å. For shorter wavelengths Grosskurth's data are used. From 2 to 4 Å, Allen's data is consistently lower, by a few per cent.

Argon

Values taken from the paper of Woo and Sun. The values follow Allen's data closely.

Copper

Data by Laubert (1941), Grosskurth, and Andrews have been used. Agreement between the data is very good indeed, but by comparison Allen's data are usually high by a few per cent, except for wavelengths greater than about 3 Å, when Allen's values are somewhat low.

At the upper limit of Andrews's measurements (4·146 Å for copper) the discrepancy with Allen's data is so great that an upward continuation of the table, using Allen's data, would lead to a reversal between 4·15 and 4·36 Å.

Tungsten

No data have been found for wavelengths greater than 2 Å. Allen's data are used up to 2 Å, and then extrapolated for the longer wavelengths.

Gold

The values are taken from the work of Laubert, Schulz (1936) and Andrews.

A. 2. *X-ray mass absorption coefficients of certain elements for the wavelength range 8·34–44 Å. (Henke, White & Lundberg, 1957)*

Absorber	Atomic No.	Al $K\alpha_{12}$ 8·34 Å	Cu $L\alpha_{12}$ 13·3 Å	Fe $L\alpha_{12}$ 17·6 Å	Cr $L\alpha_{12}$ 21·7 Å	O $K\alpha_{12}$ 23·7 Å	Ti $L\alpha_{12}$ 27·4 Å	N $K\alpha_{12}$ 31·6 Å	Ca $L\alpha_{12}$ 36·3 Å	C $K\alpha_{12}$ 44 Å
H	1	7·5	30	70	130	170	260	400	620	1,100
He	2	30	120	275	500	660	1,000	1,570	2,300	4,300
Li	3	78	280	640	1,200	1,450	2,300	3,500	5,300	9,400
Be	4	151·7	581	1,288	2,292	2,965	4,532	6,490	9,960	17,430
B	5	323·7	1,233	2,711	4,784	6,130	9,200	13,470	19,410	32,540
C	6	605	2,290	4,912	8,440	10,730	15,760	22,270	30,590	—
N	7	1,047	3,795	7,910	13,120	16,270	22,590	1,440	2,073	3,647
O	8	1,560	5,430	10,740	16,610	983	1,473	2,160	3,146	5,470
F	9	1,913	6,340	11,600	1,015	1,301	1,949	2,862	4,177	7,280
Ne	10	2,763	8,240	1,079	1,863	2,379	3,575	5,250	7,640	13,180
Na	11	3,129	661	1,402	2,429	3,100	4,651	6,810	9,830	16,650
Mg	12	3,797	981	2,085	3,601	4,592	6,830	9,880	14,000	22,850
Al	13	322·6	1,146	2,441	4,189	5,310	7,840	11,190	15,620	24,910
Si	14	510	1,813	3,812	6,420	8,040	11,510	15,940	21,600	33,840
P	15	640	2,259	4,661	7,670	9,470	13,280	18,090	24,520	38,610
S	16	814	2,839	5,710	9,160	11,190	15,520	21,200	28,730	45,230
Cl	17	990	3,364	6,530	10,210	12,450	17,330	23,720	32,080	50,100
A	18	1,163	3,795	7,110	11,070	13,540	18,820	25,650	34,670	—
K	19	1,429	4,504	8,310	12,960	15,820	22,030	28,650	40,520	—
Ca	20	1,706	5,150	9,450	14,800	18,030	24,910	—	—	—
Sc	21	1,819	5,280	9,750	15,210	18,480	—	—	—	—
Ti	22	2,002	5,680	10,480	16,300	19,820	—	—	—	—
V	23	2,168	6,100	11,230	—	—	—	—	—	—
Cr	24	2,409	6,740	12,360	—	—	—	—	—	—
Mn	25	2,556	7,160	—	—	—	—	—	—	—
Fe	26	2,799	7,850	—	—	—	—	—	—	—
Co	27	2,956	—	—	—	—	—	—	—	—
Ni	28	3,154	—	—	—	—	—	—	—	—
Cu	29	3,346	—	—	—	—	—	—	—	—
Zn	30	3,685	—	—	—	—	—	—	—	—
Zapon (H 5·3 %, C 46·7 %, N 6·6 %, O 41·4 %)		998	3,571	7,270	11,690	6,500	9,470	11,410	15,760	—
Parlodion (H 3·6 %, C 28·1 %, N 11·5 %, O 56·8 %)		1,177	4,167	8,390	13,320	5,450	7,870	7,660	10,640	—

A. 3. *Principal emission lines of the K-series (wavelengths in Å units) (Zingaro, 1954a)*

The Table lists the principal emission lines of the K-series of the elements, arranged in order of atomic number; the wavelengths of the K absorption edges are also given. The lines are indexed both by the usual Greek letter and subscript, and also by the transition levels from which electrons originate to fill vacancies in the K-level and thus give rise to the characteristic X-radiation. The relative intensities are approximate only. In practice they may be modified by absorption in media in the X-ray path, by reflexion from an analysing crystal and by variation in response of the detector with wavelength.

Table A. 3

Line		α^*	α_1 L_{III}	α_2 L_{II}	β_1 M_{III}	β_3 M_{II}	β_2 N_{II},N_{III}	β_4 N_{IV},N_V	β_5 M_{IV},M_V	O_{II},O_{III}	K absorption edge
Transition $K\to$		—									
Approximate intensity (Rel. $K\alpha_1$)...		150	100	50	15		5	<1	<1	<1	
Li	3	240	—	—	—	—	—	—	—	—	226·953
Be	4	113	—	—	—	—	—	—	—	—	—
B	5	67	—	—	—	—	—	—	—	—	—
C	6	44	—	—	—	—	—	—	—	—	43·767
N	7	31·603	—	—	—	—	—	—	—	—	31·052
O	8	23·707	—	—	—	—	—	—	—	—	23·367
F	9	18·307	—	—	—	—	—	—	—	—	—
Na	11	11·909	—	—	11·617	—	—	—	—	—	—
Mg	12	9·889	—	—	9·558	—	—	—	—	—	9·512
Al	13	8·339	8·338	8·341	7·981	—	—	—	—	—	7·951
Si	14	7·126	7·125	7·127	6·769	—	—	—	—	—	6·744
P	15	6·155	—	—	5·804	—	—	—	—	—	5·787
S	16	5·373	5·372	5·375	5·032	—	—	—	—	—	5·018
Cl	17	4·729	4·728	4·731	4·403	—	—	—	—	—	4·397
A	18	4·192	4·191	4·194	—	—	—	—	—	—	3·871
K	19	3·744	3·742	3·745	3·454	—	—	—	3·442	—	3·437
Ca	20	3·360	3·359	3·362	3·089	—	—	—	3·074	—	3·070
Sc	21	3·032	3·031	3·034	2·780	—	—	—	2·764	—	2·758
Ti	22	2·750	2·749	2·753	2·514	—	—	—	2·498	—	2·497
V	23	2·505	2·503	2·507	2·285	—	—	—	2·270	—	2·269
Cr	24	2·291	2·290	2·294	2·085	—	—	—	2·071	—	2·070
Mn	25	2·103	2·102	2·105	1·910	—	—	—	1·897	—	1·897
Fe	26	1·937	1·936	1·940	1·757	—	—	—	1·745	—	1·744
Co	27	1·791	1·789	1·793	1·621	—	—	—	1·609	—	1·608
Ni	28	1·659	1·658	1·661	1·500	—	1·489	—	1·489	—	1·488
Cu	29	1·542	1·540	1·544	1·392	1·393	—	1·381	1·382	—	1·381
Zn	30	1·437	1·435	1·439	1·296	—	1·284	—	1·285	—	1·284
Ga	31	1·341	1·340	1·344	1·207	1·208	1·196	—	1·197	—	1·195
Ge	32	1·256	1·255	1·258	1·129	1·129	1·117	—	1·119	—	1·116
As	33	1·177	1·175	1·179	1·057	1·058	1·045	—	1·049	—	1·045
Se	34	1·106	1·105	1·109	0·992	0·993	0·980	—	0·984	—	0·980
Br	35	1·041	1·040	1·044	0·933	0·933	0·921	—	0·926	—	0·920
Kr	36	0·981	0·980	0·984	0·879	0·879	0·866	0·866	0·871	—	0·866
Rb	37	0·927	0·926	0·930	0·829	0·830	0·817	—	—	—	0·816
Sr	38	0·877	0·875	0·880	0·783	0·784	0·771	—	—	—	0·770
Y	39	0·831	0·829	0·833	0·740	0·741	0·728	0·727	—	—	0·727
Zr	40	0·788	0·786	0·791	0·701	0·702	0·690	—	—	—	0·688
Cb	41	0·748	0·747	0·751	0·665	0·666	0·654	—	—	—	0·653
Mo	42	0·710	0·709	0·713	0·632	0·633	0·621	0·620	—	—	0·620
Tc	43	0·674	0·673	0·676	0·602		—	—	—	—	0·560
Ru	44	0·644	0·643	0·647	0·572	0·573	0·562	—	—	—	0·560
Rh	45	0·614	0·613	0·617	0·546	0·546	0·535	—	—	—	0·534
Pd	46	0·587	0·585	0·590	0·521	0·521	0·510	—	—	—	0·509
Ag	47	0·561	0·559	0·564	0·497	0·498	0·487	—	—	—	0·486
Cd	48	0·536	0·535	0·539	0·475	0·476	0·465	—	—	—	0·464
In	49	0·514	0·512	0·517	0·455	0·455	0·445	—	—	—	0·444
Sn	50	0·492	0·491	0·495	0·435	0·436	0·426	—	—	—	0·425
Sb	51	0·472	0·470	0·475	0·417	0·418	0·408	—	—	—	0·407
Te	52	0·453	0·451	0·456	0·400	0·401	0·391	—	—	—	0·390
I	53	0·435	0·433	0·438	0·384	0·385	0·376	—	—	—	0·374
X	54	0·418	0·416	0·421	0·369	—	0·360	—	—	—	0·359
Cs	55	0·402	0·401	0·405	0·355	0·355	0·346	—	—	—	0·345
Ba	56	0·387	0·385	0·390	0·341	0·342	0·333	—	—	—	0·332
La	57	0·373	0·371	0·376	0·328	0·329	0·320	—	—	—	0·319
Ce	58	0·359	0·357	0·362	0·316	0·317	0·309	—	—	—	0·307
Pr	59	0·346	0·344	0·349	0·305	0·305	0·297	—	—	—	0·296
Nd	60	0·334	0·332	0·337	0·294	0·294	0·287	—	—	—	0·285
Il	61	0·322	0·321	0·325	0·283		—	—	—	—	0·265
Sm	62	0·311	0·309	0·314	0·274	0·274	0·267	—	—	—	0·265
Eu	63	0·301	0·299	0·304	0·264	0·265	0·258	—	—	—	0·256
Gd	64	0·291	0·289	0·294	0·255	0·256	0·249	—	—	—	0·246
Tb	65	0·281	0·279	0·284	0·246	0·246	0·239	—	—	—	0·238
Dy	66	0·272	0·270	0·275	0·237	0·238	0·231	—	—	—	0·230
Ho	67	0·263	0·261	0·266	—	—	—	—	—	—	0·223
Er	68	0·255	0·253	0·258	0·222	0·223	0·217	—	—	—	0·215
Tu	69	0·246	0·244	0·250	0·215	0·216	—	—	—	—	0·209
Yb	70	0·238	0·236	0·241	0·208	0·209	0·203	—	—	—	0·202
Lu	71	0·231	0·229	0·234	0·202	0·203	0·197	—	—	—	0·195
Hf	72	0·224	0·222	0·227	0·195	0·196	0·190	—	—	—	0·189
Ta	73	0·217	0·215	0·220	0·190	0·191	0·185	—	—	—	0·184
W	74	0·211	0·209	0·213	0·184	0·185	0·179	—	—	—	0·178
Re	75	0·204	0·202	0·207	0·179	0·179	0·174	—	—	—	0·173
Os	76	0·198	0·196	0·201	0·173	0·174	0·169	—	—	—	0·168
Ir	77	0·103	0·101	0·196	0·168	0·169	0·164	0·163	0·167	0·163	0·163
Pt	78	0·187	0·185	0·190	0·163	0·164	0·159	0·159	0·162	0·158	0·158
Au	79	0·182	0·180	0·185	0·159	0·160	0·155	0·154	0·158	0·153	0·153
Hg	80	—	—	—	—	—	—	—	—	—	0·149
Tl	81	0·172	0·170	0·175	0·150	0·151	—	—	—	—	0·144
Pb	82	0·167	0·165	0·170	0·146	—	0·147	0·141	0·145	0·141	0·141
Bi	83	0·162	0·161	0·165	0·142	0·143	0·138	—	—	—	0·137
Th	90	0·135	0·133	0·138	0·117	0·118	0·114	—	0·116	0·113	0·113
U	92	0·128	0·126	0·131	0·111	0·112	0·108	—	—	—	0·107

A. 4. *Principal emission lines of the L-series (wavelengths in Å units) (Zingaro, 1954b)*

The absorption edges are listed as well as the emission lines, in the same way as for the K-series. The nomenclature used for the L-series is that in general use and due to Siegbahn (1925). Three general groups of lines are distinguished according to their hardness. The α group is the softest, the β medium, and the γ the hardest. The degree of hardness is relative and changes from element to element. The numerical subscripts to the lines within a group are assigned in accordance with their relative intensities, which again are only approximations.

Table A. 4

Line / Transition	α	α_1	α_d	β_1	β_2	β_3	β_4	β_5	β_6	β_7	β_9	β_{10}	β_{15}	β_{17}
$L_I\to$	—	—	—	—	—	M_{III}	M_{II}	—	—	—	M_V	M_{IV}	—	—
$L_{II}\to$	—	—	—	M_{IV}	—	—	—	—	—	—	—	—	—	M_{III}
$L_{III}\to$	—	M_V	M_{IV}	—	N_V	—	—	$O_{IV,v}$	N_I	O_I	—	—	N_{IV}	—
Approximate intensity (Rel. $L\alpha_1$)	110	100	10	80	60	30	20	30	<1	<1	<1	<1	<1	<1
Cl 17	—	—	—	—	—	—	—	—	—	—	—	—	—	—
A 18	—	—	—	—	—	—	—	—	—	—	—	—	—	—
K 19	—	—	—	—	—	—	—	—	—	—	—	—	—	—
Ca 20	36·393	—	—	36·022	—	—	—	—	—	—	—	—	—	—
Sc 21	31·393	—	—	31·072	—	—	—	—	—	—	—	—	—	—
Ti 22	27·445	—	—	27·074	—	—	—	—	—	—	—	—	—	—
V 23	24·309	—	—	23·898	—	—	—	—	—	—	—	—	—	—
Cr 24	21·713	—	—	21·323	—	—	—	—	—	—	—	—	—	—
Mn 25	19·489	—	—	19·158	—	—	—	—	—	—	—	—	—	—
Fe 26	17·602	—	—	17·290	—	—	—	—	—	—	—	—	—	—
Co 27	16·000	—	—	15·698	—	—	—	—	—	—	—	—	—	—
Ni 28	14·595	—	—	14·308	—	—	—	—	—	—	—	—	—	—
Cu 29	13·357	—	—	13·079	—	—	—	—	—	—	—	—	—	—
Zn 30	—	12·282	12·009	—	—	—	—	—	—	—	—	—	—	—
Ga 31	—	11·313	11·045	—	—	—	—	—	—	—	—	—	—	—
Ge 32	—	10·456	10·194	—	—	—	—	—	—	—	—	—	—	—
As 33	—	9·671	9·414	—	—	—	—	—	—	—	—	—	—	—
Se 34	—	8·990	8·735	—	—	—	—	—	—	—	—	—	—	—
Br 35	—	8·375	8·126	—	—	—	—	—	—	—	—	—	—	—
Kr 36	—	—	—	—	—	—	—	—	—	—	—	—	—	—
Rb 37	—	7·318	7·325	7·075	—	6·788	6·821	—	6·984	—	—	—	—	—
Sr 38	—	6·863	6·870	6·623	—	6·367	6·403	—	6·519	—	—	—	—	—
Y 39	—	6·449	6·456	6·211	—	5·983	6·018	—	6·094	—	—	—	—	—
Zr 40	—	6·070	6·077	5·836	5·586	5·632	5·668	—	5·710	—	—	—	—	—
Cb 41	—	5·725	5·732	5·492	5·238	5·310	5·346	—	5·361	—	—	—	—	—
Mo 42	—	5·406	5·414	5·176	4·923	5·013	5·048	—	5·048	—	—	—	—	—
Tc 43	—	—	—	—	—	—	—	—	—	—	—	—	—	—
Ru 44	—	4·846	4·854	4·620	4·372	4·487	4·523	—	4·487	—	—	—	—	—
Rh 45	—	4·597	4·605	4·374	4·130	4·253	4·289	—	4·242	—	—	—	—	—
Pd 46	—	4·368	4·376	4·146	3·909	4·034	4·071	—	4·016	—	—	3·792	3·799	—
Ag 47	—	4·154	4·162	3·935	3·703	3·834	3·870	—	3·808	—	—	3·605	3·611	—
Cd 48	—	3·956	3·965	3·739	3·514	3·644	3·681	—	3·614	—	—	3·430	3·437	—
In 49	—	3·752	3·781	3·555	3·339	3·470	3·507	—	3·436	—	—	3·268	3·274	—
Sn 50	—	3·600	3·609	3·385	3·175	3·306	3·344	—	3·270	3·155	3·155	—	3·121	—
Sb 51	—	3·439	3·448	3·226	3·023	3·152	3·190	—	3·115	3·005	2·973	—	2·979	—
Te 52	—	3·290	3·299	3·077	2·882	3·009	3·046	—	2·971	2·863	2·839	—	2·847	—
I 53	—	3·148	3·157	2·937	2·751	2·874	2·912	—	2·837	2·730	2·713	—	2·720	—
X 54	—	—	—	—	—	—	—	—	—	—	—	—	—	—
Cs 55	—	2·892	2·902	2·683	2·511	2·628	2·666	—	2·593	2·485	2·478	—	2·492	—
Ba 56	—	2·776	2·785	2·567	2·404	2·516	2·555	—	2·482	2·382	2·376	—	2·387	—
La 57	—	2·665	2·674	2·458	2·303	2·410	2·449	—	2·379	2·275	2·282	—	2·290	—
Ce 58	—	2·561	2·570	2·356	2·208	2·311	2·349	—	2·282	2·180	2·188	—	2·195	—
Pr 59	—	2·463	2·473	2·259	2·119	2·216	2·255	—	2·190	2·091	2·100	—	2·107	—
Nd 60	—	2·370	2·382	2·166	2·035	2·126	2·166	—	2·103	2·009	2·016	—	2·023	—
Il 61	—	2·283	—	2·081	—	—	—	—	—	—	—	—	—	—
Sm 62	—	2·199	2·210	1·998	1·882	1·962	2·000	1·779	1·946	1·856	1·862	—	1·870	—
Eu 63	—	2·120	2·131	1·920	1·812	1·887	1·926	—	1·875	1·788	1·792	—	1·800	—
Gd 64	—	2·046	2·057	1·847	1·746	1·815	1·853	—	1·807	1·723	—	—	1·731	—
Tb 65	—	1·976	1·986	1·777	1·682	1·747	1·785	1·577	1·742	1·659	—	—	1·667	—
Dy 66	—	1·909	1·920	1·710	1·623	1·681	1·720	—	1·681	1·599	—	—	—	—
Ho 67	—	1·845	1·856	1·647	1·567	1·619	1·658	—	1·622	—	—	—	—	—
Er 68	—	1·785	1·796	1·587	1·514	1·561	1·601	—	1·567	1·494	1·485	—	1·494	—
Tu 69	—	1·726	1·738	1·530	1·463	1·505	1·544	—	1·515	—	—	—	—	—
Yb 70	—	1·672	1·682	1·476	1·416	1·452	1·491	1·387	1·466	1·395	1·384	—	1·392	—
Lu 71	—	1·619	1·630	1·424	1·370	1·402	1·441	1·342	1·419	1·350	1·336	1·343	1·372	—
Hf 72	—	1·569	1·580	1·374	1·327	1·353	1·392	1·298	1·374	1·306	1·291	1·299	1·328	1·437
Ta 73	—	1·522	1·533	1·327	1·285	1·307	1·346	1·256	1·331	1·264	1·247	1·254	1·287	—
W 74	—	1·476	1·487	1·282	1·245	1·263	1·302	1·215	1·290	1·224	1·204	1·212	1·247	1·339
Re 75	—	1·433	1·444	1·238	1·206	1·220	1·260	1·177	1·252	1·186	1·165	1·172	1·208	1·293
Os 76	—	1·391	1·402	1·197	1·169	1·179	1·218	1·140	1·213	1·149	1·126	1·133	1·171	—
Ir 77	—	1·352	1·363	1·158	1·135	1·141	1·179	1·106	1·179	1·115	1·090	1·097	1·137	—
Pt 78	—	1·313	1·325	1·120	1·102	1·104	1·142	1·072	1·143	1·082	1·054	1·060	—	1·166
Au 79	—	1·277	1·288	1·083	1·070	1·068	1·106	1·040	1·111	1·050	1·021	1·028	1·072	1·128
Hg 80	—	1·242	1·253	1·049	1·040	1·034	1·072	1·010	1·080	1·019	0·986	0·990	1·041	1·090
Tl 81	—	1·207	1·218	1·015	1·010	1·001	1·039	0·981	1·050	0·990	0·957	0·964	1·012	1·056
Pb 82	—	1·175	1·186	0·982	0·983	0·969	1·007	0·953	1·021	0·962	0·927	0·934	0·984	1·022
Bi 83	—	1·144	1·155	0·952	0·955	0·939	0·977	0·926	0·993	0·935	0·898	0·905	0·957	0·989
Po 84	—	1·114	1·125	0·922	0·929	0·909	0·948	0·900	0·967	—	—	—	0·931	—
Fr 87	—	1·030	—	0·840	0·858	—	—	—	—	—	—	—	—	—
Ra 88	—	1·005	1·017	0·814	0·836	0·803	0·841	0·807	0·871	0·817	0·769	0·776	0·838	0·844
Th 90	—	0·956	0·968	0·766	0·794	0·755	0·793	0·765	0·828	0·775	0·723	0·730	—	—
Pa 91	—	0·933	0·945	0·742	0·774	0·732	0·770	0·746	0·803	0·755	0·701	0·708	—	—
U 92	—	0·911	0·923	0·720	0·755	0·710	0·748	0·726	0·789	0·736	0·681	0·687	—	—
Np 93	—	0·889	—	0·698	0·735	—	—	—	—	—	—	—	—	—

Line Transition	γ_1	γ_2	γ_3	γ_4	γ_5	γ_6	γ_8	l	η	s	t	L Absorption edges L_I	L_{II}	L_{III}
$L_I\rightarrow$	—	N_{II}	N_{III}	O_{III}	—	—	—	—	—	—	—			
$L_{II}\rightarrow$	N_{IV}	—	—	—	N_I	O_{IV}	O_I	—	M_I	—	—			
$L_{III}\rightarrow$	—	—	—	—	—	—	—	M_I	—	M_{III}	M_{II}	L_I	L_{II}	L_{III}
Approximate intensity (Rel. $L\alpha_1$)	40	10	10	10	<1	20	<1	30	10	<1	<1	—	—	—
Cl 17	—	—	—	—	—	—	—	67·84	67·25	—	—	—	—	—
A 18	—	—	—	—	—	—	—	56·212	56·813	—	—	—	—	—
K 19	—	—	—	—	—	—	—	47·835	47·325	—	—	—	—	—
Ca 20	—	—	—	—	—	—	—	41·042	40·542	—	—	42·184	—	—
Sc 21	—	—	—	—	—	—	—	35·671	35·200	—	—	—	35·200	35·561
Ti 22	—	—	—	—	—	—	—	31·423	30·942	—	—	—	—	—
V 23	—	—	—	—	—	—	—	27·826	27·375	—	—	—	27·29	—
Cr 24	—	—	—	—	—	—	—	24·840	24·339	—	—	16·7	17·9	20·7
Mn 25	—	—	—	—	—	—	—	22·315	21·864	—	—	—	—	—
Fe 26	—	—	—	—	—	—	—	20·201	19·73	—	—	—	—	—
Co 27	—	—	—	—	—	—	—	18·358	17·86	—	—	—	—	—
Ni 28	—	—	—	—	—	—	—	16·693	16·304	—	—	—	—	—
Cu 29	—	—	—	—	—	—	—	15·297	14·940	—	—	—	13·010	13·289
Zn 30	—	—	—	—	—	—	—	14·081	13·719	—	—	—	11·861	12·130
Ga 31	—	—	—	—	—	—	—	12·976	12·620	—	—	—	—	—
Ge 32	—	—	—	—	—	—	—	11·944	11·608	—	—	—	—	—
As 33	—	—	—	—	—	—	—	11·069	10·732	—	—	8·108	9·124	9·367
Se 34	—	—	—	—	—	—	—	10·293	9·959	—	—	7·505	8·417	8·645
Br 35	—	—	—	—	—	—	—	9·583	9·253	—	—	—	—	—
Kr 36	—	—	—	—	—	—	—	—	—	—	—	—	—	—
Rb 37	—	6·045	—	—	6·754	—	—	8·363	8·042	—	—	5·997	6·643	6·864
Sr 38	—	5·644	—	—	6·297	—	—	7·836	7·517	—	—	5·582	6·172	6·387
Y 39	—	5·283	—	—	5·875	—	—	7·356	7·040	—	—	5·233	5·756	5·962
Zr 40	5·384	4·953	—	—	5·497	—	—	6·918	6·606	—	—	4·867	5·378	5·583
Cb 41	5·036	4·654	—	—	5·151	—	—	6·517	6·210	—	—	4·581	—	5·223
Mo 42	4·726	4·380	—	—	4·837	—	—	6·150	5·847	—	—	4·299	4·719	4·913
Tc 43	—	—	—	—	—	—	—	—	—	—	—	—	—	—
Ru 44	4·182	3·897	—	—	4·288	—	—	5·503	5·204	—	—	—	4·179	4·369
Rh 45	3·944	3·685	—	—	4·045	—	—	5·217	4·922	—	—	3·626	3·942	4·129
Pd 46	3·725	3·489	—	—	3·822	—	—	4·952	4·660	—	—	3·428	3·724	3·908
Ag 47	3·523	3·307	—	—	3·616	—	—	4·707	4·418	—	—	3·254	3·514	3·698
Cd 48	3·336	3·137	—	—	3·426	—	—	4·480	4·193	—	—	3·084	3·326	3·504
In 49	3·162	2·980	2·926	3·249	—	—	—	4·269	3·983	—	—	2·926	3·147	3·325
Sn 50	3·001	2·835	2·778	3·085	—	—	—	4·071	3·789	—	—	2·778	2·982	3·156
Sb 51	2·852	2·695	2·639	2·932	—	—	—	3·888	3·607	—	—	2·639	2·830	3·000
Te 52	2·712	2·567	2·511	2·790	—	—	—	3·716	3·438	—	—	2·510	2·687	2·856
I 53	2·582	2·447	2·391	2·657	—	—	—	3·557	3·280	—	—	2·389	2·553	2·719
X 54	—	—	—	—	—	—	—	—	—	—	—	2·274	2·429	2·592
Cs 55	2·348	2·237	2·233	2·174	2·417	—	—	3·267	2·994	—	—	2·167	2·314	2·474
Ba 56	2·242	2·138	2·134	2·075	2·309	—	2·222	3·135	2·862	—	—	2·068	2·204	2·363
La 57	2·141	2·046	2·041	1·983	2·205	—	—	3·006	2·740	—	—	1·973	2·103	2·259
Ce 58	2·048	1·960	1·955	1·899	2·110	—	2·023	2·892	2·620	—	—	1·890	—	2·164
Pr 59	1·961	1·879	1·874	1·819	2·020	—	1·936	2·758	2·512	—	—	1·811	1·924	2·077
Nd 60	1·878	1·801	1·797	1·745	1·935	1·855	—	2·675	2·409	—	—	1·735	1·843	1·995
Il 61	—	—	—	—	—	—	—	—	—	—	—	—	—	—
Sm 62	1·726	1·659	1·655	1·606	—	—	—	2·482	2·218	—	—	1·598	1·702	1·845
Eu 63	1·657	1·597	1·591	1·544	1·708	—	1·632	2·395	—	—	—	1·536	1·626	1·776
Gd 64	1·592	1·534	1·520	1·485	—	—	—	2·312	2·049	—	—	1·477	1·561	1·709
Tb 65	1·530	1·477	1·471	1·427	—	—	—	2·234	—	—	—	1·421	1·501	1·648
Dy 66	1·473	1·423	1·417	1·374	1·518	—	—	2·158	1·898	—	—	1·365	1·438	1·579
Ho 67	1·417	1·371	1·364	1·323	1·462	—	—	2·086	1·826	—	—	1·318	1·390	1·535
Er 68	1·364	1·321	1·315	1·276	1·406	—	—	2·019	1·757	—	—	1·269	1·339	1·482
Tu 69	1·316	1·274	1·268	—	1·355	—	—	1·955	1·695	—	—	1·222	1·288	1·433
Yb 70	1·268	1·228	1·222	1·185	1·307	1·243	1·250	1·894	1·635	—	1·831	1·181	1·243	1·386
Lu 71	1·222	1·185	1·179	1·143	1·260	1·198	1·204	1·836	1·478	—	1·776	1·140	1·198	1·342
Hf 72	1·179	1·144	1·138	1·103	1·215	1·155	1·161	1·782	1·523	1·663	1·723	1·099	1·154	1·298
Ta 73	1·138	1·105	1·099	1·065	1·173	1·114	1·120	1·728	1·471	1·612	1·672	1·061	1·113	1·256
W 74	1·008	1·068	1·062	1·028	1·132	1·074	1·081	1·678	1·421	—	—	1·024	1·074	1·215
Re 75	1·061	1·032	1·026	0·993	1·094	1·037	1·044	1·630	1·374	—	—	0·990	1·037	1·177
Os 76	1·025	0·998	0·992	0·959	1·057	1·001	1·008	1·585	1·328	—	—	0·956	1·001	1·140
Ir 77	0·991	0·966	0·959	0·928	1·022	0·967	0·974	1·541	1·285	—	—	0·923	0·967	1·105
Pt 78	0·958	0·934	0·928	0·897	0·988	0·934	0·941	1·499	1·243	—	—	0·893	0·934	1·072
Au 79	0·927	0·905	0·898	0·867	0·956	0·903	0·910	1·460	1·202	1·352	1·414	0·864	0·903	1·040
Hg 80	0·897	0·876	0·869	0·839	0·925	0·873	0·880	1·422	1·164	—	—	0·836	0·872	1·009
Tl 81	0·868	0·848	0·842	0·812	0·895	0·845	0·852	1·385	1·127	1·279	1·342	0·808	0·844	0·979
Pb 82	0·840	0·822	0·815	—	0·867	0·817	0·822	1·350	1·092	1·244	1·308	0·782	0·815	0·950
Bi 83	0·814	0·796	0·790	0·761	0·840	0·791	0·799	1·317	1·058	1·210	—	0·757	0·789	0·924
Po 84	0·788	—	0·765	—	—	0·765	—	—	—	—	—	—	—	—
Fr 87	0·716	—	—	—	—	—	—	—	—	—	—	—	—	—
Ra 88	0·664	0·682	0·675	0·649	0·717	0·673	0·680	1·167	0·908	—	—	0·644	0·670	0·803
Th 90	0·653	0·642	0·635	0·611	0·675	0·632	0·640	1·115	0·855	1·011	1·080	0·606	0·630	0·761
Pa 91	0·634	0·624	0·617	0·594	0·655	0·613	—	1·091	0·830	—	—	—	—	—
U 92	0·615	0·605	0·598	0·577	0·635	0·595	0·601	1·067	0·806	0·964	1·035	0·569	0·592	0·722
Np 93	0·597	—	—	—	—	—	—	—	—	—	—	—	—	—

BIBLIOGRAPHY

The number(s) in brackets after each entry indicate the
page(s) on which it is quoted in the text.

GENERAL TEXTS AND SURVEY ARTICLES

CLARK, G. L. (1955). *Applied X-rays* (New York: McGraw-Hill); 4th
edition. [224, 256, 317, 322, 329, 331, 335]
COMPTON, A. H. & ALLISON, S. K. (1935). *X-rays in Theory and Experiment* (London: MacMillan); 2nd edition. [27, 89, 124, 125, 126, 208,
227, 228, 372]
COSSLETT, V. E., ENGSTRÖM, A. & PATTEE, H. H. (eds.) (1957). *X-ray
Microscopy and Microradiography: Proceedings of Cambridge Symposium*, 1956 (New York: Academic Press). [xii]
EASTMAN KODAK COMPANY (1955, 1957). *Bibliography of Microradiography and Soft X-ray Radiography*; and Supplement I (Eastman
Kodak Company, Rochester, N.Y.). [254]
ENGSTRÖM, A. (1956). *Physical Techniques in Biological Research*. Ed. by
G. Oster and A. W. Pollister (New York: Academic Press), vol. III,
chap. 10. [20, 24, 42, 250, 252, 260, 265, 308]
ENGSTROM, A., PATTEE, H. H. & COSSLETT, V. E. (eds.) (1960). *X-ray
Microscopy and Microanalysis: Proceedings of Stockholm Symposium*,
1959 (Amsterdam: Elsevier). [xii]
HILDENBRAND, G. (1958). *Ergebnisse der Exakten Naturwissenschaften*,
30, 1–133.
KIRKPATRICK, P. & PATTEE, H. H. (1957). *Handbuch der Physik*, XXX,
305–35. [105, 112]
LIEBHAFSKY, H. A. & WINSLOW, E. A. (1956). *Analyt. Chem.*, 28, 583.
LIEBHAFSKY, H. A. & WINSLOW, E. A. (1958). *Analyt. Chem.*, 30, 580.
SAGEL, K. (1959). *Tabellen zur Röntgen-Emissions- und Absorptions-
Analyse* (Springer, Berlin). [369]

REFERENCES

ADDA, Y., PHILIBERT, J. & FARAGGI, H. (1957). *Rev. Mét.* 54, 597. [368]
ADLER, I., AXELROD, J. & BRANCO, J. J. R. (1959). *Proc. 7th Conf. on X-ray
Analysis*, 1958 (Denver: The University), 167. [368]
AGRELL, S. O. & LONG, J. V. P. (1960a). X-ray Microscopy and X-ray
Microanalysis (Amsterdam: Elsevier), 391. [334, 368]
AGRELL, S. O. & LONG, J. V. P. (1960b). *Miner. Mag.* (in publication).
[334, 368]
ALLISON, S. K. (1932). *Phys. Rev.* 41, 13 and 688. [125]
AMPRINO, R. & ENGSTRÖM, A. (1952). *Acta Anat.* 15, 1. [159]
AMREHN, H. & KUHLENKAMPFF, H. (1955). *Z. Phys.* 140, 452. [223]
ANCKER, B., HAZLETT, T. H. & PARKER, E. R. (1956). *J. Appl. Phys.* 27,
333. [346]

ANDERSON, T. F. (1953). C.R. Premier Congrès International de Microscopie Électronique, Paris, 1950. (Rev. D'Optique, Paris), 567. [242]

ANDERSON, T. F. (1956). Proc. Internat. Conf. Electron Microscopy, London, 1954 (Royal Microscopical Society, London), 122. [242]

ANDREWS, C. L. (1938). Phys. Rev. 54, 994. [374]

ANDREWS, K. W. & JOHNSON, W. (1957). X-ray Microscopy and Microradiography (New York: Academic Press), 581. [322, 323, 324]

APPLEBAUM, E., HOLLANDER, F. & BÖDECKER, C. (1933). Dent. Cosmos, 75, 1097. [316]

ARCHARD, G. D. (1953). J. Sci. Instrum. 30, 352. [278]

ARCHARD, G. D. (1955). Proc. Phys. Soc. B, 68, 156. [65]

ARCHARD, G. D. (1960a). X-ray Microscopy and X-ray Microanalysis (Amsterdam: Elsevier), 331. [369]

ARCHARD, G. D. (1960b). X-ray Microscopy and X-ray Microanalysis (Amsterdam: Elsevier), 337. [367]

ARDENNE, M. VON (1939). Naturwissenschaften, 27, 485. [7, 49, 280]

ARDENNE, M. VON (1940). Elektronen-Übermikroskopie (Berlin: Springer), 72. [49, 280, 299]

ARDENNE, M. VON (1956). Tabellen der Elektronenphysik, Ionenphysik und Übermikroskopie (Berlin: Deutscher Verlag der Wissenschaften), 441. [49, 280]

ARNDT, U. W. (1955). Chap. 7 in X-ray Diffraction in Polycrystalline Materials, ed. by H. S. Peiser, H. P. Rooksby and A. J. C. Wilson (London: Institute of Physics). [176, 198]

ARNDT, U. W. & RILEY, D. P. (1952). Proc. Phys. Soc. A, 65, 74. [176, 230, 236]

ASSELMEYER, F. (1954). Z. angew. Phys. 6, 272. [138]

ASUNMAA, S. K. (1960). X-ray Microscopy and X-ray Microanalysis (Amsterdam: Elsevier), 66. [366]

AULD, J. H. & McNEIL, J. F. (1960). X-ray Microscopy and X-ray Microanalysis (Amsterdam: Elsevier), 5. [366]

AUSTIN, A. E., RICHARD, N. A. & SCHWARTZ, C. M. (1960). X-ray Microscopy and X-ray Microanalysis (Amsterdam: Elsevier), 401. [368]

BAEZ, A. V. (1949). Ph.D. Thesis, Stanford Univ., U.S.A. [117]

BAEZ, A. V. (1952a). Nature, Lond. 169, 963. [18]

BAEZ, A. V. (1952b). J. Opt. Soc. Amer. 42, 756. [18]

BAEZ, A. V. (1954). Private communication. [327, 333]

BAEZ, A. V. (1957). X-ray Microscopy and Microradiography (New York: Academic Press), 186. [111, 112]

BAEZ, A. V. & EL-SUM, H. (1957). X-ray Microscopy and Microradiography (New York: Academic Press), 347. [18, 58, 76]

BAEZ, A. V. & WEISSBLUTH, M. (1954). Phys. Rev. 93, 942. [111]

BAGDIKJANZ, G. O. (1953). Izvestia Akademee Nauk, U.S.S.R., Series Physica, 17, 255. [305]

BAGDIKJANZ, G. O. (1956). Akademee Nauk, U.S.S.R., Biophysica, 1, 341. [305]

BALL, L. W. (1945). Metal Ind., Lond. 67, 130 and 210. [329]

BARANOV, V. I., ZHDANOV, A. P. & DIESENROT-MYSOVSKAYA, M. Ju. (1944). Izvestia Acad. Nauk. Ser. Khem. 1, 20. [331]

BARCLAY, A. E. (1947). *Brit. J. Radiol.* N.S. **20**, 394. [20]
BARCLAY, A. E. (1951). *Micro-arteriography* (Oxford: Blackwell). [14, 20, 245, 246, 256, 264, 306, 308, 312, 317]
BARCLAY, A. E. & LEATHERDALE, D. (1948). *Brit. J. Radiol.* N.S. **21**, 544. [244, 317]
BARER, G. R. (1952). *Brit. J. Exp. Path.* **33**, 123. [307]
BARER, R. & SAUNDERS-SINGER, A. E. (1951). *J. Sci. Instrum.* **28**, 65. [249]
BARNARD, J. E. (1915). *J. R. Micr. Soc.* **35**, 1. [19]
BARNES, R. S. (1952). *Proc. Phys. Soc.* B, **65**, 512. [328]
BARRETT, C. S. (1945). *Trans. Amer. Inst. Min.* (*Metall*) *Engrs*, **161**, 15. [16, 345]
BARTH, H. & HOSEMANN, R. (1958). *Z. Naturforsch.* **13**a, 792. [371]
BATCHELDER, F. W. VON & SCHAUM, J. H. (1947). *Iron Age*, **160**, 94. [324]
BATCHELDER, F. W. VON (1959). *Proc. 7th Conf. on X-ray Analysis*, 1958 (Denver: The University), 283. [305, 335, 371]
BEARDEN, J. A. (1938). *Phys. Rev.* **54**, 698. [89]
BELL, G. E. (1936). *Brit. J. Radiol.* N.S. **9**, 578. [162]
BELLMAN, S. (1953). *Acta Radiol.*, *Stockh.*, Suppl. 102. [20, 163, 246, 265, 306, 309, 312]
BELLMAN, S. & ADAMS, R. J. (1956). *Angiology*, **7**, 339. [307]
BELLMAN, S. & ENGFELDT, B. (1955). *Amer. J. Roentgenol.* **74**, 288. [312]
BELLMAN, S. & ENGSTRÖM, A. (1952). *Acta Radiol.*, *Stockh.* **38**, 98. [41]
BELLMAN, S., FRANK, H. A., LAMBERT, P. B., ODÉN, B. & WILLIAMS, J. A. (1960). *X-ray Microscopy and X-ray Microanalysis* (Amsterdam: Elsevier), 257. [309, 321, 370]
BELLMAN, S. & ODÉN, B. (1957). *Acta Radiol.*, *Stockh.* **47**, 289. [307]
BENDIT, E. G. (1958). *Brit. J. Appl. Phys.* **9**, 312. [230]
BERG, W. F. (1930). *Wiss. Veröff. Siemens*, **9**, 119. [16]
BERG, W. F. (1931). *Naturwissenschaften*, **19**, 391. [345]
BERG, W. F. (1934). *Z. Kristallogr.* **89**, 286. [16, 345]
BERGMANN, M. E. (1960). *X-ray Microscopy and X-ray Microanalysis* (Amsterdam: Elsevier), 431. [370]
BERGHEZAN, A., LACOMBE, P. & CHAUDRON, G. (1950). *C.R. Acad. Sci.*, *Paris*, **231**, 576. [326]
BERNARD, A., BRYSON-HAYNES, D. & MULVEY, T. (1959). *J. Sci. Instrum.* **36**, 438. [368]
BESSEN, I. I. (1957a). *Norelco Reporter*, **4**, 119. [237, 238, 305]
BESSEN, I. I. (1957b). *J. Appl. Phys.* **28**, 1369 (abstract only). [200]
BESSEN, I. I. (1958). Private communication. [327]
BESSEN, I. I. (1959). *Proc. 7th Conf. on X-ray Analysis*, 1958 (Denver: The University), 79. [237].
BETHE, H. (1930). *Ann. Phys.*, *Lpz.* **5**, 325. [369]
BETTERIDGE, W. (1950). *Metal Ind.*, *Lond.* **76**, 5. [323]
BETTERIDGE, W. & SHARPE, R. S. (1948a). *J. Iron & St. Inst.* **158**, 185. [251, 323]
BETTERIDGE, W. & SHARPE, R. S. (1948b). *Iron & Steel*, *Lond.* **21**, 242. [323]
BIERMANN, H. H. (1936). *Ann. Phys.*, *Lpz.* **26**, 740. [374]
BIRKS, L. S. & BROOKS, E. J. (1957a). *Proc. 6th Conf. on X-ray Analysis*, 1957 (Denver: The University), 339. [198, 322]

BIRKS, L. S. & BROOKS, E. J. (1957 b). *Rev. Sci. Instrum.* **28**, 709. [198]
BISI, A. & ZAPPA, L. (1955). *Nuovo Cim.* **2**, 988. [195]
BLICKMAN, J. R., KLOPPER, P. J. & RECOURT, A. (1960 a). *X-ray Microscopy and X-ray Microanalysis* (Amsterdam: Elsevier), 224. [321, 370]
BLICKMAN, J. R., KLOPPER, P. J. & RECOURT, A. (1960 b). *X-ray Microscopy and X-ray Microanalysis* (Amsterdam: Elsevier), 226. [321]
BLODGETT, K. B. & LANGMUIR, I. (1937). *Phys. Rev.* **51**, 964. [137]
BLOIS, M. S. (1951). Ph.D. Thesis, Stanford Univ., U.S.A. [119]
BLOOMER, R. N. (1957). *Brit. J. Appl. Phys.* **8**, 83. [289]
BOHATIRCHUK, F. (1942). *Fortschr. Röntgenstr.* **65**, 253. [20, 256]
BOHATIRCHUK, F. (1944). *Acta Radiol., Stockh.* **25**, 351. [20]
BOHATIRCHUK, F. (1957). *X-ray Microscopy and Microradiography.* (New York: Academic Press), 473. [244, 315]
BOHR, H. & DOLLERUP, E. (1960). *X-ray Microscopy and X-ray Microanalysis* (Amsterdam: Elsevier), 184. [321]
BOROVSKII, I. B. (1953). *Collection of Problems in Metallurgy.* (NA. U.S.S.R.), 135. [183]
BOROVSKII, I. B. (1960). *X-ray Microscopy and X-ray Microanalysis* (Amsterdam: Elsevier), 344. [368]
BOROVSKII I. B. & IL'IN, N. P. (1956). *Dokl. Akad. Nauk, SSSR,* **106**, 655. [183]
BOROVSKII, I. B. & IL'IN, N. P. (1957). *Zavodskaya Laboratoriya,* **10**, 1234. [200]
BORRIES, B. VON (1956). *Proc. Internat. Conf. Electron Microscopy, London, 1954* (Royal Microscopical Society, London), 4. [279]
BORRMANN, G., HARTWIG, W. & IRMLER, H. (1958). *Z. Naturforsch.* **13 a**, 423. [371]
BOSS, G. H. (1948). *Metal Prog.,* **53**, 522. [329]
BOURGHARDT, S., BRATTGÅRD S. O., HYDÉN, H., KIEWERTZ, B. & LARSSON, S. (1953). *J. Sci. Instrum.* **30**, 464. [148]
BOURGHARDT, S., HYDÉN, H. & NYQVIST, B. (1955). *J. Sci. Instrum.* **32**, 186. [148]
BOUWERS, A. & DIEPENHORST, P. (1928). *Fortschr. Röntgenstr.* **38**, 894. [232]
BRADLEY, D. E. (1954). *Brit. J. Appl. Phys.* **5**, 65 and 96. [120, 249, 364]
BRAGG, W. L. (1939). *Nature, Lond.* **143**, 678. [17]
BRAGG, W. L. (1942). *Nature, Lond.* **149**, 470. [17]
BRATTGÅRD, S. O. (1952). *Acta Radiol., Stockh.,* Suppl. 96. [146]
BRATTGÅRD, S. O., HALLÉN, D. & HYDÉN, H. (1953). *Acta Radiol., Stockh.* **39**, 494. [147]
BRATTGÅRD, S. O. & HYDÉN, H. (1952). *Acta Radiol., Stockh.* Suppl. 94. [146, 149, 259, 264, 319]
BRECH, F. & STANSFIELD, J. R. (1957). *Proc. 6th Conf. on X-ray Analysis, 1957* (Denver: The University), 17. [337]
BROEK, S. L. VAN DEN (1957). *X-ray Microscopy and Microradiography* (New York: Academic Press), 64. [20, 259, 260]
BROMLEY, D. & HERZ, R. H. (1950). *Proc. Phys. Soc. B,* **63**, 90. [162]
BUERGER, M. J. (1950). *J. Appl. Phys.* **21**, 909. [17]
BURCH, G. J. (1896). *Nature, Lond.* **54**, 111. [1, 19, 317]
BURFOOT, J. (1953). *Proc. Phys. Soc. B,* **66**, 775. [65]

BUSCHMANN, E. C. & NORTON, J. F. (1957). *Proc. 6th Conf. on X-ray Analysis, 1957* (Denver: The University), 207, 219. [200]

BUTEUX, R. H. (1953). *J. Opt. Soc. Amer.* 43, 618. [120]

BUTLER, T. J., BAHR, G. F., TAFT, L. I. & JENNINGS, R. B. (1960). *X-ray Microscopy and X-ray Microanalysis* (Amsterdam: Elsevier), 268. [182, 305, 321]

CARLSON, L. (1957). *Exp. Cell Res., Suppl.* 4, 193. [152]

CARLSTRÖM, D. (1954). *J. Histochem. Cytochem.* 2, 149. [350]

CARLSTRÖM, D. & LUNDBERG, B. (1958). *J. Ultrastructure Res.* 2, 261. [347]

CARR, R. J. (1954). *Brit. J. Urol.* 26, 105. [313]

CARR, R. J. (1957). *X-ray Microscopy and Microradiography* (New York: Academic Press), 551. [247, 313]

CASPERSSON, T. (1955). *Experientia,* 9, 45. [158]

CASTAING, R. (1951). Ph.D. Thesis, Paris. [9, 183, 185, 197, 202, 287, 347]

CASTAING, R. (1954). *Electron Physics Symposium, Washington, 1951* (Nat. Bureau of Standards, Washington), 305. [197]

CASTAING, R. (1956). *Proc. Internat. Conf. Electron Microscopy, London, 1954* (Royal Microscopical Society, London), 300. [198, 281]

CASTAING, R. (1958). *Electron Probe Microanalyser Conf., Washington, February, 1958.* [188]

CASTAING, R. & DESCAMPS, J. (1955). *J. Phys. Radium,* 16, 304. [189]

CASTAING, R. & DESCAMPS, J. (1958). *Recherche Aéronaut.,* 63, 41. [368]

CASTAING, R. & GUINIER, A. (1950). *Proc. Conf. Elect. Microscopy, Delft, 1949.* (Martin Nijhoff, Delft), 60. [197]

CASTAING, R. & GUINIER, A. (1953). *Analyt. Chem.* 25, 724. [197]

CASTAING, R., PHILIBERT, J. & CRUSSARD, C. (1957). *J. Metals, N.Y.* 9, 389. [198]

CAUCHOIS, Y. (1946). *C.R. Acad. Sci., Paris,* 223, 82. [6, 132]

CAUCHOIS, Y. (1950). *Rev. d'Opt.* 29, 151. [132, 134]

CAUCHOIS, Y. (1952). French Patent no. 1,019,616. [134]

CAUCHOIS, Y. & HULUBEI, H. (1947). *Tables de constantes et données numériques,* 1 (Paris: Hermann and Cie). [27]

CHALKLEY, L. (1952). *J. Opt. Soc. Amer.* 42, 387. [366]

CHEAVIN, W. H. S. (1947). *Philately,* 1, 149. [332]

CHEAVIN, W. H. S. (1948). *Philately,* 1, 176. [332]

CHESLEY, F. C. (1947). *Rev. Sci. Instrum.* 6, 68. [336, 344]

CHIPMAN, D. R. (1955). *J. Appl. Phys.* 26, 1387. [374]

CLARK, G. L. (1939). *Photo Technique,* 1, 19. [20]

CLARK, G. L. (1955). See 'General Texts'.

CLARK, G. L. & EYLER, R. W. (1943). *Rev. Sci. Instrum,* 14, 277. [266]

CLARK, S. M. & IBALL, J. (1955). *J. Sci. Instrum.* 32, 366. [243]

CLARK, S. M. & IBALL, J. (1957). *Progress in Biophysics* (London: Pergamon), 7, 225. [243]

CLEMMONS, J. J. (1955). *Biochim. Biophys. Acta,* 17, 297. [149, 150]

CLEMMONS, J. J. (1957). *Exp. Cell Res.,* Suppl. 4, 172. [146, 150]

CLEMMONS, J. J. & APRISON, M. H. (1953). *Rev. Sci. Instrum.* 24, 444. [259]

CLEMMONS, J. J. & WEBSTER, T. C. (1953). *Biochim. Biophys. Acta*, **11**, 464. [146, 149]

COHEN, E. (1950). *Metallurgia, Manchr*, **41**, 227. [324]

COHEN, E. (1953). *J. Iron St. Inst.* **175**, 160. [324]

COHEN, E. & SCHLOGL, I. (1960). *X-ray Microscopy and X-ray Microanalysis* (Amsterdam: Elsevier), 133. [335, 370]

COHEN, J., HALL, E., LEONARD, L. & OGILVIE, R. (1955). *Non-destr. Test.* **13**, 33. [323]

COLLETT, C. T., HUGHES, J. C. & MOREY, F. C. (1950). *J. Res. Nat. Bur. Stand.* **45**, 283. [335]

COLLETTE, J. M. (1953). *J. Belge Radiol.* **36**, 293. [307]

COMBÉE, B. (1955). *Philips Tech. Rev.* **17**, 45. [20, 259]

COMBÉE, B. & ENGSTRÖM, A. (1954). *Biochim. Biophys. Acta*, **14**, 432. [151, 259]

COMBÉE, B. & RECOURT, A. (1957). *Philips Tech. Rev.* **19**, 221. [20, 39, 164, 260]

COMER, J. J. & SKIPPER, S. J. (1954). *Science*, **119**, 441. [40]

COMPTON, A. H. (1923). *Phil. Mag.* **45**, 1121. [87]

COMPTON, A. H. & ALLISON, S. K. (1935). See 'General Texts'.

COSSLETT, A. & COSSLETT, V. E. (1957). *Brit. J. Appl. Phys.* **8**, 374. [120]

COSSLETT, V. E. (1940). Unpublished. [7, 49]

COSSLETT, V. E. (1951). *Practical Electron Microscopy* (London: Butterworths). [272]

COSSLETT, V. E. (1952a). *Proc. Phys. Soc.* B, **65**, 782. [50, 64, 67]

COSSLETT, V. E. (1952b). *Internal Rep.* (MG/EC/227/52) *of the British Iron and Steel Research Association.* [9]

COSSLETT, V. E. (1956). *Proc. Internat. Conf. Elec. Microscopy, London, 1954* (Roy. Microscopical Soc. Lond.), 311. [50, 362]

COSSLETT, V. E. (1957). *X-ray Microscopy and Microradiography* (New York: Academic Press), 321. [61, 85]

COSSLETT, V. E. (1959). *Appl. Sci. Res.* B, **7**, 338. [86]

COSSLETT, V. E. (1960). *X-ray Microscopy and X-ray Microanalysis* (Amsterdam: Elsevier), 346. [369]

COSSLETT, V. E. & DUNCUMB, P. (1956). *Nature, Lond.* **177**, 1172. [9, 183, 202]

COSSLETT, V. E. & DUNCUMB, P. (1957). *Proc. Stockh. Conf. Elect. Microscopy, 1956* (Almqvist and Wiksell, Stockholm), 12. [9]

COSSLETT, V. E. & DYSON, N. A. (1957). *X-ray Microscopy and Microradiography* (New York: Academic Press), 405. [226, 232, 233, 234]

COSSLETT, V. E. & HAINE, M. E. (1956). *Proc. Internat. Conf. Elect. Microscopy, London, 1954* (Roy. Microscopical Soc. Lond.), 639. [68, 215, 216]

COSSLETT, V. E. & JONES, S. S. D. (1946). *Progress Report, Electrical Laboratory, Oxford.* [49]

COSSLETT, V. E. & NIXON, W. C. (1951). *Nature, Lond.* **168**, 24. [8, 49, 304, 339]

COSSLETT, V. E. & NIXON, W. C. (1952a). *Nature, Lond.* **170**, 436. [8, 49, 304]

COSSLETT, V. E. & NIXON, W. C. (1952b). *Proc. Roy. Soc.* B, **140**, 422. [8, 49, 304]

25

C & N

COSSLETT, V. E. & NIXON, W. C. (1953). *J. Appl. Phys.* **24**, 616. [8, 49, 274, 301]

COSSLETT, V. E. & NIXON, W. C. (1954). *Proc. Elect. Phys. Symp.*, *Washington, 1951* (Nat. Bureau of Standards, Washington), 257. [49]

COSSLETT, V. E., NIXON, W. C. & PEARSON, H. E. (1957). *X-ray Microscopy and Microradiography* (New York: Academic Press), 96. [281].

COSSLETT, V. E. & PEARSON, H. E. (1954). *J. Sci. Instrum.* **31**, 255. [281]

COSSLETT, V. E. & TAYLOR, D. (1948). *Progress Report, Cavendish Laboratory, Cambridge.* [8]

COYLE, R. A., MARSHALL, A. M., AULD, J. H. & McKINNON, N. A. (1957). *Brit. J. Appl. Phys.* **8**, 79. [16, 346]

CUNNINGHAM, G. J. (1955). *Amer. J. Clin. Path.* **25**, 253. [313]

CUNNINGHAM, G. J. & MILLER, J. W. (1952). *Thorax*, **7**, 170. [313]

CZERMAK, P. (1897). *Ann. Phys., Lpz.* **60**, 760. [7, 48]

DAUVILLIER, A. (1927). *C.R. Acad. Sci., Paris*, **185**, 1460. [19, 256, 261]

DAUVILLIER, A. (1930). *C.R. Acad. Sci., Paris*, **190**, 1287. [19]

DAVIES, H. G. & ENGSTRÖM, A. (1954). *Exp. Cell Res.* **7**, 243. [319]

DAVIES, H. G., ENGSTRÖM, A. & LINDSTRÖM, B. (1953). *Nature, Lond.* **172**, 1041. [151, 319]

DAVIS, B. & STEMPEL, W. M. (1921). *Phys. Rev.* **17**, 608. [126]

DAVIS, B. & STEMPEL, W. M. (1922). *Phys. Rev.* **19**, 504. [126]

DAVIS, M. I. (1956). *Research Film*, **2**, 172. [310]

DAVIS, M. I. & SMITH, C. L. (1957a). *J. Sci. Instrum.* **34**, 32. [310]

DAVIS, M. I. & SMITH, C. L. (1957b). *Exp. Cell Res.* **12**, 15. [310].

DEBARR, A. E. & MACARTHUR, I. (1950). *Brit. J. Appl. Phys.* **1**, 305. [338]

DERSHEM, E. (1929). *Phys. Rev.* **34**, 1015. [89]

DERSHEM, E. (1939). *Proc. Nat. Acad. Sci., Wash.* **25**, 6. [315]

DE SOUSA, A. & DE SOUSA, J. F. (1956). *Brit. J. Radiol. N.S.* **29**, 377. [312]

DESPUJOLS, J. (1953). *C.R. Acad. Sci., Paris*, **236**, 282. [134]

DIETRICH, J. (1960). *X-ray Microscopy and X-ray Microanalysis* (Amsterdam: Elsevier), 306. [321]

DODD, C. G. (1960). *Proc. 8th Conf. on X-ray Analysis, 1959* (Denver: The University), 11. [335].

DOLAN, W. W. & DYKE, W. P. (1954). *Phys. Rev.* **95**, 327. [215]

DOLBY, R. M. (1959). *Proc. Phys. Soc.* **73**, 81. [195, 367]

DOLBY, R. M. (1960). *Brit. J. Appl. Phys.* **11**, 64. [367, 369]

DOLBY, R. M. & COSSLETT, V. E. (1960). *X-ray Microscopy and X-ray Microanalysis* (Amsterdam: Elsevier), 351. [196, 367]

DORAN, F. S. A. (1951). *Lancet*, **260**, 199. [312]

DORN, E. (1896). *Abh. naturf. Ges. Halle*, **21**, 1. [48]

DORSTEN, A. C. VAN & LE POOLE, J. B. (1955). *Philips Tech. Rev.* **17**, 47. [304]

DRECHSLER, M., COSSLETT, V. E. & NIXON, W. C. (1960). *Proc. Internat. Conf. Elect. Mic., Berlin, 1958* (Berlin: Springer), 13. [68, 215, 216]

DRECHSLER, M. & HENKEL, E. (1954). *Z. angew Phys.* **6**, 341. [215]

DRENCK, K. & PEPINSKY, R. (1951). *Rev. Sci. Instrum.* **22**, 539. [59, 338]

DRUMMOND, D. G. (1950). *J. R. Micr. Soc.* **70**, 132. [286]

DUDLEY, L. (1951). *Stereoptics* (London: Macdonald). [266]

DUNCUMB, P. (1957a). Ph.D. Thesis, Cambridge. [185, 189, 191, 192, 194, 195, 202, 203, 225, 228, 230, 321, 352, 353]

DUNCUMB, P. (1957b). *X-ray Microscopy and Microradiography* (New York: Academic Press), 617. [202, 206]

DUNCUMB, P. (1959). *Brit. J. Appl. Phys.* **10**, 420. [367, 368]

DUNCUMB, P. (1960a). *Proc. Internat. Conf. Elect. Mic., Berlin, 1958* (Berlin: Springer), 267. [185, 205, 206, 368]

DUNCUMB, P. (1960b). *X-ray Microscopy and X-ray Microanalysis* (Amsterdam: Elsevier), 365. [367]

DUNCUMB, P. (1960c). *Brit. J. Appl. Phys.* **11**, 169. [367]

DUNCUMB, P. & COSSLETT, V. E. (1957). *X-ray Microscopy and Microradiography* (New York: Academic Press), 374. [202]

DUNCUMB, P. & MELFORD, D. A. (1960). *X-ray Microscopy and X-ray Microanalysis* (Amsterdam: Elsevier), 358. [206, 368]

DUSHMAN, S. (1949). *Vacuum Technique* (New York: Wiley and Sons). [291]

DYKE, W. P. & DOLAN, W. W. (1956). *Advanc. Electron.* **8**, 89. [215]

DYSON, J. (1952). *Proc. Phys. Soc.* B, **65**, 580. [4, 98, 99, 101, 104, 109, 110, 111]

DYSON, N. A. (1956). Ph.D. Thesis, Cambridge. [72, 226, 227, 229, 230, 231, 232, 372]

DYSON, N. A. (1957). *X-ray Microscopy and Microradiography* (New York: Academic Press), 310. [53, 166, 167, 168, 169, 170, 227]

DYSON, N. A. (1959a). *Proc. Phys. Soc.* **73**, 924. [226, 227, 232]

DYSON, N. A. (1959b). *Brit. J. Appl. Phys.* **10**, 505. [231]

DYSON, N. A. (1960). Private communication. [369]

EHRENBERG, W. (1947). *Nature, Lond.* **160**, 330. [95, 117, 119, 120]

EHRENBERG, W. (1949a). *J. Opt. Soc. Amer.* **39**, 741. [95, 117]

EHRENBERG, W. (1949b). *J. Opt. Soc. Amer.* **39**, 746. [119, 120]

EHRENBERG, W. (1957). *X-ray Microscopy and Microradiography* (New York: Academic Press), 331. [337, 349]

EHRENBERG, W. & FRANKS, A. (1952). *Nature, Lond.* **170**, 1076. [341]

EHRENBERG, W. & FRANKS, A. (1953). *Proc. Phys. Soc.* B, **66**, 1057. [69]

EHRENBERG, W. & JENTZSCH, F. (1929). *Z. Phys.* **54**, 227. [94]

EHRENBERG, W. & SPEAR, W. E. (1951a). *Proc. Phys. Soc.* B, **64**, 67. [60, 213, 257, 259, 337, 338]

EHRENBERG, W. & SPEAR, W. E. (1951b). *Nature, Lond.* **168**, 513. [238]

EHRENBERG, W. & WHITE, M. (1957). *X-ray Microscopy and Microradiography* (New York: Academic Press), 213. [39, 61, 164, 239, 259]

ELY, R. V. (1957). *X-ray Microscopy and Microradiography* (New York: Academic Press), 59. [257]

ELY, R. V. (1960). *X-ray Microscopy and X-ray Microanalysis* (Amsterdam: Elsevier), 47. [257]

ENGFELDT, B. (1954). *Cancer,* **7**, 815. [315]

ENGFELDT, B., BJÖRNERSTEDT, R., CLEMEDSON, C.-J. & ENGSTRÖM, A. (1954). *Acta Orthopaed. Scand.* **24**, 101. [316]

ENGFELDT, B. & ENGSTRÖM, A. (1954). *Acta Orthopaed. Scand.* **24**, 85. [315]

ENGFELDT, B., ENGSTRÖM, A. & BOSTRÖM, H. (1954). *Exp. Cell Res.* **6**, 251. [316]

ENGFELDT, B., ENGSTRÖM, A. & ZETTERSTRÖM, R. (1954*a*). *Acta Paediat., Stockh.* **43**, 152. [315]

ENGFELDT, B., ENGSTRÖM, A. & ZETTERSTRÖM, R. (1954*b*). *J. Bone Jt. Surg.* **36**B, 654. [315]

ENGFELDT, B. & HJERTQUIST, S. O. (1954), *Acta Path. Microbiol. Scand.* **35**, 205. [316]

ENGSTRÖM, A. (1946). *Acta Radiol., Stockh.*, Suppl. **63**. [20, 144, 153, 154, 155, 257, 261, 320]

ENGSTRÖM, A. (1947*a*). *Experientia*, **3**, 208. [266]

ENGSTRÖM, A. (1947*b*). *Biochim. Biophys. Acta*, **1**, 428. [29]

ENGSTRÖM, A. (1950). *Progress in Biophysics and Biophysical Chemistry*, **1**, chap. 7 (London: Butterworths). [320]

ENGSTRÖM, A. (1951*a*). *Acta Radiol., Stockh.* **36**, 305. [265]

ENGSTRÖM, A. (1951*b*). *Acta Radiol., Stockh.* **36**, 393. [155]

ENGSTRÖM, A. (1955*a*). *Analytical Cytology* (R. Mellors, ed.). chap. 8 (New York: McGraw-Hill). [20]

ENGSTRÖM, A. (1955*b*). *Exp. Cell Res.*, Suppl. **3**, 117. [29, 259]

ENGSTRÖM, A. (1956). See 'General Texts'.

ENGSTRÖM, A. (1957). *X-ray Microscopy and Microradiography* (New York: Academic Press), 24. [32, 319]

ENGSTRÖM, A., BERGENDAHL, G., BJÖRNERSTEDT, R. & LUNDBERG, B. (1957). *Exp. Cell Res.* **12**, 440. [45]

ENGSTRÖM, A., BJÖRNERSTEDT, R., CLEMEDSON, C-J. & NELSON, A. (1958). *Bone and Radiostrontium* (Stockholm: Almqvist and Wiksell). [314]

ENGSTRÖM, A. & GLICK, D. (1950). *Science*, **111**, 379. [150]

ENGSTRÖM, A. & GLICK, D. (1956). *Science*, **124**, 27. [152]

ENGSTRÖM, A. & GREULICH, R. C. (1956). *J. Appl. Phys.* **27**, 758. [29, 32, 243, 316]

ENGSTRÖM, A., GREULICH, R. C., HENKE, B. L. & LUNDBERG, B. (1957). *X-ray Microscopy and Microradiography* (New York: Academic Press), 218. [39, 261, 316]

ENGSTRÖM, A. & JOSEPHSON, B. (1953). *Amer. J. Physiol.* **174**, 61. [317]

ENGSTRÖM, A., LAGERGREN, C. & LUNDBERG, B. (1957). *Exp. Cell Res.* **12**, 592. [32, 159, 175]

ENGSTRÖM, A. & LINDSTRÖM, B. (1949). *Nature, Lond.* **163**, 563. [146, 147]

ENGSTRÖM, A. & LINDSTRÖM, B. (1950). *Biochim. Biophys. Acta*, **4**, 351. [146, 147, 247, 259, 261, 319]

ENGSTRÖM, A. & LINDSTRÖM, B. (1951). *Acta Radiol., Stockh.* **35**, 33. [38, 39, 164, 264]

ENGSTRÖM, A. & LUNDBERG, B. (1957). *Exp. Cell Res.* **12**, 198. [261, 262]

ENGSTRÖM, A., LUNDBERG, B. & BERGENDAHL, G. (1957). *J. Ultrastructure Res.* **1**, 147. [316]

ENGSTRÖM, A. & LUTHY, H. (1949). *Experientia*, **5**, 244. [319]

REFERENCES 389

ENGSTRÖM, A. & WEGSTEDT, L. (1951). *Acta Radiol., Stockh.* **35**, 345. [148, 256]

ENGSTRÖM, A., WEGSTEDT, L. & WELIN, S. (1948). *Acta Radiol., Stockh.* **30**, 440. [147]

ENGSTRÖM, A. & WEISSBLUTH, M. (1951). *Exp. Cell Res.* **2**, 711. [158]

ENNOS, A. E. (1953). *Brit. J. Appl. Phys.* **4**, 101. [303]

ENNOS, A. E. (1954). *Brit. J. Appl. Phys.* **5**, 27. [303]

FANKUCHEN, I. & MARK, H. (1944). *J. Appl. Phys.* **15**, 364. [349]

FARRANT, J. L. (1950). *J. Appl. Phys.* **21**, 63. [90, 120]

FAUST, R. (1950). *Phil. Mag.* **41**, 1238. [119]

FELDMAN, C. & O'HARA, M. (1957). *J. Opt. Soc. Amer.* **47**, 300. [267]

FINEAN, J. B. (1956). *J. Sci. Instrum.* **33**, 161. [350]

FISHER, R. M. (1959). *Electron Probe Microanalyser Conf.*, Washington, *1958*. [322]

FISHER, R. M. & KNECHTEL, H. E. (1955). *J. Appl. Phys.* **26**, 1395 (abstract only). [198]

FITZGERALD, P. J. (1956). *Ann. N.Y. Acad. Sci.* **63**, 1141. [261]

FITZGERALD, P. J. (1957). *X-ray Microscopy and Microradiography* (New York: Academic Press), 49. [261]

FITZGERALD, P. J., LI, T. G. & YERMAKOV, V. (1957). *X-ray Microscopy and Microradiography* (New York: Academic Press), 520. [316]

FOURNIER, F. (1938). *Rev. Métall.* **35**, 349. [19]

FOURNIER, F. (1949). *Rev. Métall.* **46**, 360. [326]

FOURNIER, F. (1960). *X-ray Microscopy and X-ray Microanalysis* (Amsterdam: Elsevier), 518. [269, 321, 370]

FRANCOIS, J., COLLETTE, J. M. & NEETENS, A. (1955). *J. belge Radiol.* **38**, 1. [314]

FRANKS, A. (1955). *Proc. Phys. Soc.* B, **68**, 1054. [341]

FRANKS, A. (1958). *Brit. J. Appl. Phys.* **9**, 349. [341]

FRANKS, A. (1960). *Proc. 8th Conf. on X-ray Analysis*, 1959 (Denver: The University), 69. [370]

FURNAS, T. C. (1957). *Rev. Sci. Instrum.* **28**, 1042. [240]

GABOR, D. (1949). *Proc. Roy. Soc.* A, **197**, 454. [17]

GAY, P., HIRSCH, P. B. & KELLY, A. (1954). *Acta Cryst.*, Camb. **7**, 41. [345]

GAY, P., HIRSCH, P. B., THORP, J. S. & KELLAR, J. N. (1951). *Proc. Phys. Soc.* B, **64**, 374. [337]

GAY, P. & HONEYCOMBE, R. W. K. (1951). *Proc. Phys. Soc.* A, **64**, 844. [349]

GEROLD, V. (1960). *Proc. 8th Conf. on X-ray Analysis*, 1959 (Denver: The University), 289. [371]

GERSH, I. (1932). *Anat. Rec.* **53**, 309. [241]

GERSH, I. (1948). *Bull. Int. Ass. Med. Mus.* **28**, 179. [242]

GLICK, D., ENGSTRÖM, A. & MALMSTRÖM, B. (1951). *Science*, **114**, 253. [158]

GLOCKER, R. (1958). *Materialprüfung mit Röntgenstrahlen* (Berlin: Springer). [166, 236]

GLOCKER, R. & FROHNMAYER, W. (1925). *Ann. Phys., Lpz.* **76**, 369. [142]

GLOCKER, R. & SCHAABER, O. (1939). *Z. Tech. Phys.* **20**, 286. [19]

GLOCKER, R. & SCHREIBER, H. (1928). *Ann. Phys., Lpz.* **85**, 1089. [207]

GOBY, P. (1913a). C.R. Acad. Sci., Paris, 156, 686. [2, 19, 256]
GOBY, P. (1913b). Bull. Soc. franç. photogr. 4, 310. [19, 256]
GOBY, P. (1914). Bull. Soc. franç. photogr. 5, 196. [19]
GOBY, P. (1925). C.R. Acad. Sci., Paris, 180, 735. [19, 264]
GOGOBERIDZE, D. B. (1953). Uspekhi Fizicheskikh Nauk, 50, 577. [331]
GOLDSCHMIDT, H. J. (1953). Metallurgia, 47, 215. [323]
GOLDSCHMIDT, H. J. (1957). X-ray Microscopy and Microradiography (New York: Academic Press), 600. [327, 330]
GOLDSZTAUB, S. (1947). C.R. Acad. Sci., Paris, 224, 458. [338]
GOLDSZTAUB, S. & SCHMITT, J. (1960). X-ray Microscopy and X-ray Microanalysis (Amsterdam: Elsevier), 149. [335]
GOOD, R. H. & MÜLLER, E. W. (1956). Handbuch der Physik, XXI, 176–231. [215]
GOUY, G. (1916). Ann. Phys., Paris, 5, 241. [127]
GRAAF, J. E. DE & OOSTERKAMP, W. J. (1938). Philips Tech. Rev. 3, 259. [235]
GREULICH, R. (1960). X-ray Microscopy and X-ray Microanalysis Amsterdam: Elsevier), 273. [321, 370]
GREULICH, R. C., BERGENDAHL, G. & EKBLAD, H. G. (1957). Private communication. [247]
GREULICH, R. C. & ENGSTRÖM, A. (1956). Exp. Cell Res. 10, 251. [247]
GRIVET, P. (1950). Advances in Electronics, 11, 48–101. [273]
GROSS, S. T. & CLARK, G. L. (1943). Iron Age, 152, 44. [329]
GROSSKURTH, K. (1934). Ann. Phys., Lpz. 20, 197. [374]
GUINIER, A. (1952). X-ray Crystallographic Technology (London: Hilger and Watts), 6. [230]
GUINIER, A. (1956). Theorie et technique de la Radiocristallographie (Paris: Dunod). [240]
GUINIER, A. & DEVAUX, J. (1943). Rev. Sci., Paris, 81, 341. [338]
HAINE, M. E. (1956). Proc. Internat. Conf. Electron Microscopy, Lond., 1954 (Roy. Microscopical Soc. Lond.), 92. [283]
HAINE, M. E. (1957). J. Brit. Instn Radio Engrs, 17, 211. [62]
HAINE, M. E. & AGAR, A. (1957). Brit. J. Appl. Phys. 8, 259. (Summary only.) [274]
HAINE, M. E. & EINSTEIN, P. (1952). Brit. J. Appl. Phys. 3, 40. [214, 271, 272, 281, 294]
HAINE, M. E. & MULVEY, T. (1952). Nature, Lond. 170, 202. [18]
HALL, C. E. (1953). Introduction to Electron Microscopy (New York: McGraw-Hill). [272, 283]
HALL, C. E. (1955). J. Biophys. Biochim. Cytology, 1, 1. [45]
HALL, C. E. & INOUE, T. (1957). J. Appl. Phys. 28, 1346. [45]
HALLÉN, O. (1956). Acta Anat. 26, Suppl. 25. [151]
HALLÉN, O. & HYDÉN, H. (1957a). Exp. Cell Res., Suppl. 4, 197. [150]
HALLÉN, O. & HYDÉN, H. (1957b). X-ray Microscopy and Microradiography (New York: Academic Press), 249. [148, 150]
HALLÉN, O & RÖCKERT, H. (1958). Nature, Lond. 182, 1225. [244]
HALLÉN, O. & RÖCKERT, H. (1960). X-ray Microscopy and X-ray Microanalysis (Amsterdam: Elsevier), 169. [244]
HÁMOS, L. VON (1934). Nature, Lond. 134, 181. [6, 128]
HÁMOS, L. VON (1936). Metallwirtsch. Metallwiss. 15, 433. [128]

HÁMOS, L. VON (1938a). *Amer. Min.* **23**, 215. [128, 131]

HÁMOS, L. VON (1938b). *J. sci. Instrum.* **15**, 87. [127, 128, 129, 130, 131]

HÁMOS, L. VON (1939). *Z. Kristallogr.* **101**, 17. [128, 129]

HÁMOS, L. VON (1953). Thesis, Royal. Inst. of Technol. Stockh. [6, 128, 131]

HÁMOS, L. VON & ENGSTRÖM, A. (1944). *Acta Radiol., Stockh.* **25**, 325. [128]

HANAWALT, J. D. (1932). *J. Franklin Inst.* **214**, 569. [26]

HANSON, A. W., LIPSON, H. & TAYLOR, C. A. (1953). *Proc. Roy. Soc. A,* **218**, 371. [17]

HARDING, J. P. (1957). Cambridge Instrument Co., Cambridge, England. [249]

HELMCKE, J. G. (1954). *Optik,* **11**, 201. [264]

HENDEE, C. F. & FINE, S. (1954). *Phys. Rev.* **95**, 281. [193]

HENDRICK, R. (1957). *J. Opt. Soc. Amer.* **47**, 165. [91]

HENKE, B. L. (1957). *X-ray Microscopy and Microradiography* (New York: Academic Press), 72. [154, 239, 240, 263, 341]

HENKE, B. L. (1959). *Proc. 7th Conf. on X-ray Analysis,* 1958 (Denver: The University), 117. [80, 154, 264]

HENKE, B. L. (1960). *X-ray Microscopy and X-ray Microanalysis* (Amsterdam: Elsevier), 10. [91]

HENKE, B. L. & DuMOND, J. W. M. (1955). *J. Appl. Phys.* **26**, 912. [23, 107, 340]

HENKE, B. L., LUNDBERG, B. & ENGSTRÖM, A. (1957). *X-ray Microscopy and Microradiography* (New York: Academic Press), 248. [28, 162, 169, 171]

HENKE, B. L., WHITE, R. & LUNDBERG, B. (1957). *J. Appl. Phys.* **28**, 98. [149, 158, 375]

HERRNRING, G. & WEIDNER, W. (1952). Unpublished. (Quoted by Hildenbrand, 1956). [98, 111, 112]

HEVESY, G. VON (1932). *Chemical Analysis by X-rays and its Applications* (New York: McGraw-Hill). [209, 258]

HEWES, C. G., NIXON, W. C., BAEZ, A. V. & KAMPMEIER, J. F. (1956). *Science,* **124**, 129. [314]

HEYCOCK, C. T. & NEVILLE, F. H. (1898). *J. Chem. Soc.* **73**, 714. [1, 19, 322]

HEYCOCK, C. T. & NEVILLE, F. H. (1900). *Phil. Trans.* A, **194**, 201. [322]

HIBI, T. (1956). *J. Elect. Microscopy, Japan,* **4**, 11. [216]

HILDENBRAND, G. (1956). *Fortschritte der Physik,* **4**, 1. [90, 91, 92, 107, 108, 113, 114]

HILDENBRAND, G. (1958). See 'General Texts'.

HINK, W. (1957). *X-ray Microscopy and Microradiography* (New York: Academic Press), 151. [108, 116, 118, 119]

HINK, W. & PETZOLD, W. (1958). *Naturwissenschaften,* **45**, 107. [94]

HIRSCH, P. B. (1953). *Rev. Métall.* **50**, 333. [345]

HIRSCH, P. B. (1955). *X-ray Diffraction by Polycrystalline Materials.* Eds. H. S. Peiser, H. P. Rooksby, A. J. C. Wilson. (Inst. of Phys., Lond.), chap. 11, 278. [336, 344]

HOERLIN, H. (1949). *J. Opt. Soc. Amer.* **39**, 891. [162]
HOERLIN, H. (1951). *Z. Naturforsch.* **6a**, 344. [162]
HOH, F. C. & LINDSTRÖM, B. (1959). *J. Ultrastructure Res.* **2**, 512. [42, 160]
HOLLIDAY, J. E. & STERNGLASS, E. J. (1959). *J. Appl. Phys.* **30**, 1428. [369]
HOLMSTRAND, K. (1957). *Acta Orthopaed. Scand.*, Suppl. 26. [161, 316]
HOMES, G. A. & GOUZOU, J. (1951). *Rev. Métall.* **48**, 251. [322]
HONEYCOMBE, R. W. K. (1951). *J. Inst. Met.* **80**, 39 and 45. [16, 346]
HOOPER, K. (1958). M.Sc. Thesis, University of London. [333]
HOOPER, K. (1960). *X-ray Microscopy and X-ray Microanalysis* (Amsterdam: Elsevier), 216. [333]
HOPPE, W. & TRURNIT, H. J. (1947). *Z. Naturforsch.* **2a**, 608. [136, 137]
HUANG, L. Y. (1957). *Z. Physik*, **149**, 225. [357]
HULUBEI, H. (1946). *C.R. Acad. Sci.*, *Paris*, **198**, 79. [134]
HUXLEY, H. E. (1953). *Acta Cryst.*, *Camb.* **6**, 457. [337, 339]
HUYSEN, G. VAN, HODGE, H. C., WARREN, S. L. & BISHOP, F. W. (1933). *Dent. Cosmos*, **75**, 729. [145]
HUYSEN, G. VAN, BALE, W. F. & HODGE, H. C. (1934). *J. Dent. Res.* **14**, 168. [145]
HYDÉN, H. & LARSSON, S. (1960). *X-ray Microscopy and X-ray Microanalysis* (Amsterdam: Elsevier), 51. [182]
ISINGS, J., ONG, S. P., LE POOLE, J. B. & NEDERVEEN, G. VAN. (1957). *Proc. Stockh. Conf. Elect. Microscopy*, *1956* (Stockholm: Almqvist and Wiksell), 282. [331]
IVANOV, K. A. (1949). *J. Tech. Phys.*, *Moscow*, **19**, 1217. [331]
JACKSON, C. K. (1957a). *X-ray Microscopy and Microradiography* (New York: Academic Press), 487. [318]
JACKSON, C. K. (1957b). Unpublished. [319]
JACKSON, C. K. (1957c). *X-ray Microscopy and Microradiography* (New York: Academic Press), 623. [325]
JEFFERY, J. W. (1952). *J. Sci. Instrum.* **29**, 385. [349]
JEFFERY, J. W. & BULLEN, H. E. (1960). *X-ray Microscopy and X-ray Microanalysis* (Amsterdam: Elsevier), 440. [370]
JENTZSCH, F. (1929). *Phys. Z.* **30**, 268. [4, 94]
JONES, H. A. & LANGMUIR, I. (1927). *Gen. Elect. Rev.* **30**, 310. [214]
JOWSEY, J. (1955). *J. Sci. Instrum.* **32**, 159. [243]
JUSTER, M., FISCHGOLD, H. & METZGER, J. (1960). *X-ray Microscopy and X-ray Microanalysis* (Amsterdam: Elsevier), 191. [321]
KATO, Y., YAMANAKA, S. & SHINODA, G. (1953). *Technol. Rep.*, *Osaka Univ.* **3**, 5. [305]
KAZAN, B. & NICOLL, F. H. (1957). *J. Opt. Soc. Amer.* **47**, 887. [356]
KELLERMANN, K. (1943). *Ann. Phys.*, *Lpz.* **43**, 32. [94]
KELLSTRÖM, G. (1932). *Nova Acta Soc. Sci. Upsal.* (4), **8**, 61. [18, 58]
KELLY, A. (1953). *Proc. Phys. Soc.* A, **66**, 403. [349]
KENNARD, O. (1960). *X-ray Microscopy and X-ray Microanalysis.* (Amsterdam: Elsevier), 525. [370]
KIRKBY, H. W. & MORLEY, J. I. (1948). *J. Iron St. Inst.* **158**, 289. [323]
KIRKPATRICK, P. (1939). *Rev. Sci. Instrum.* **10**, 186. [236]

KIRKPATRICK, P. (1944). *Rev. Sci. Instrum.* **15**, 223. [236]

KIRKPATRICK, P. (1949). *J. Opt. Soc. Amer.* **39**, 796. [94]

KIRKPATRICK, P. (1957). *X-ray Microscopy and Microradiography* (New York: Academic Press), 17. [117, 119, 120]

KIRKPATRICK, P. & BAEZ, A. V. (1948). *J. Opt. Soc. Amer.* **38**, 766. [4, 95, 108, 116]

KIRKPATRICK, P. & PATTEE, H. H. (1953). *Advances in Biological and Medical Physics*, vol. 3 (New York: Academic Press), 247–83. [5, 117, 341, 342]

KIRKPATRICK, P. & PATTEE, H. H. (1957). See 'General Texts'.

KIRKPATRICK, P. & WIEDMANN, L. (1945). *Phys. Rev.* **67**, 321. [227, 232]

KLEMPERER, O. (1953). *Electron Optics* (Cambridge University Press). [272]

KLUG, H. P. & ALEXANDER, L. E. (1954). *X-ray Diffraction Procedures* (New York: Wiley). [336]

KOEHLER, W. F. (1953). *J. Opt. Soc. Amer.* **43**, 743. [119]

KRAMER, I. R. H. (1951). *Anat. Rec.* **111**, 91. [245]

KRATKY, O. (1931). *Z. Kristallogr.* **76**, 261. [336]

KRYNITSKY, A. I. & STERN, H. (1952). *Foundry*, **71**, 106. [324]

KUHLENKAMPFF, H. (1922). *Ann. Physik, Lpz.* **69**, 548. [226]

KURA, J. G., EASTWOOD, L. W. & DOIG, J. R. (1951). *Foundry*, **70**, 90 and 254. [327]

KURYLENKO, C. (1955). *Cah. Phys.* **54**, 1. [26]

KÜSTNER, H. (1931). *Z. Physik*, **70**, 324 and 468. [236]

KÜSTNER, H. (1932). *Z. Physik*, **77**, 52. [236]

LADD, W. A., HESS, W. M. & LADD, M. W. (1956). *Science*, **123**, 370. [365]

LADD, W. A. & LADD, M. W. (1957). *X-ray Microscopy and Microradiography* (New York: Academic Press), 383. [43, 365]

LAGERGREN, C. (1956). *Acta Radiol., Stockh.*, Suppl. 133. [313]

LAMARQUE, P. (1936). *C.R. Acad. Sci., Paris*, **202**, 684. [19]

LAMARQUE, P. (1938). *Brit. J. Radiol.* **11**, 425. [20, 244, 246]

LAMARQUE, P. & TURCHINI, J. (1936). *C.R. Soc. Biol., Paris*, **122**, 294. [20, 261]

LAMARQUE, P., TURCHINI, J. & CASTEL, P. (1937). *Arch. Soc. des sciences méd. et biol. de Montpellier et du Languedoc Méditerr.* **18**, 27. [20, 244]

LAMBOT, H. J. (1947). *Nature, Lond.* **159**, 676. [331]

LANE, R. O. & ZAFFARANO, D. J. (1954). *Phys. Rev.* **94**, 960. [222]

LANG, A. R. (1954). *Rev. Sci. Instrum.* **25**, 1032. [355]

LANG, A. R. (1956). *J. Sci. Instrum.* **33**, 96. [176, 198]

LANG, A. R. (1957). *Acta Metallurgica*, **5**, 358. [347]

LANG, A. R. (1958). *J. Appl. Phys.* **29**, 597. [347]

LANG, A. R. (1959). *J. Appl. Phys.* **30**, 1748. [347, 371]

LANGE, P. W. (1954). Unpublished; quoted by Engström, A. (1956). [332]

LANGE, P. W. & ENGSTRÖM, A. (1954). *Lab. Investigation*, **3**, 116. [142, 151]

LANGMUIR, D. B. (1937). *Proc. Inst. Radio Engrs, N.Y.* **25**, 977. [62]

LANGNER, G. (1957). *X-ray Microscopy and Microradiography* (New York: Academic Press), 293. [69, 70, 219]

C & N

LANGNER, G. (1960*a*). *X-ray Microscopy and X-ray Microanalysis* (Amsterdam: Elsevier), 31. [86]

LANGNER, G. (1960*b*). *X-ray Microscopy and X-ray Microanalysis* (Amsterdam: Elsevier), 90. [86]

LARSSON, S. (1957). *Exp. Cell Res.* 12, 666. [148]

LARSSON, A., SIEGBAHN, M. & WALLER, I. (1924). *Naturwissenschaften,* 52, 1212. [94]

LAUBERT, S. (1941). *Ann. Phys., Lpz.* 40, 553. [374]

LEGRAND, C. & SALMON, J. (1954). *Bull. de Microscopie Appliquée,* 2, 9. [317]

LE POOLE, J. B. & ONG, S. P. (1956). *Appl. Sci. Res.* B, 5, 454. [50, 81, 84, 85, 252]

LE POOLE, J. B. & ONG, S. P. (1957). *X-ray Microscopy and Microradiography* (New York: Academic Press), 91. [252, 281, 308]

LEROUX, G., FRANCOIS, J., COLLETTE, J. M. & NEETENS, A. (1960*a*). *X-ray Microscopy and X-ray Microanalysis* (Amsterdam: Elsevier), 233. [321]

LEROUX, G., FRANCOIS, J., COLLETTE, J. M. & NEETENS, A. (1960*b*), *X-ray Microscopy and X-ray Microanalysis* (Amsterdam: Elsevier), 239. [321]

LEWIS, D. (1955). *J. Sci. Instrum.* 32, 467. [349]

LIEBHAFSKY, H. A., PFEIFFER, H. G. & ZEMANY, P. D. (1960). *X-ray Microscopy and X-ray Microanalysis* (Amsterdam: Elsevier), 321. [368]

LIEBHAFSKY, H. A. & WINSLOW, E. A. (1956, 1958). See 'General Texts'.

LIEBMANN, G. (1949). *Proc. Phys. Soc.* B, 62, 213. [273, 285]

LIEBMANN, G. (1952). *Proc. Phys. Soc.* B, 65, 188. [277]

LIEBMANN, G. (1953). *Proc. Phys. Soc.* B, 66, 448. [281]

LIEBMANN, G. (1955). *Proc. Phys. Soc.* B, 68, 737. [273]

LIEBMANN, G. & GRAD, E. M. (1951). *Proc. Phys. Soc.* B, 64, 956. [281]

LINDBLOM, K. (1954). *Acta Radiol., Stockh.* 42, 465. [299]

LINDSLEY, C. H., FISHER, E. K. & BRANT, J. H. (1949). *Text. Res. (J.),* 19, 686. [333]

LINDSTRÖM, B. (1955). *Acta Radiol., Stockh.,* Suppl. 125. [27, 39, 146, 147, 151, 154, 156, 158, 162, 164, 247, 261, 263, 320]

LINDSTRÖM, B. (1957). *X-ray Microscopy and Microradiography* (New York: Academic Press), 443. [21]

LINDSTRÖM, B. (1960). *X-ray Microscopy and X-ray Microanalysis* (Amsterdam: Elsevier), 288. [182, 321]

LINDSTRÖM, B. & MOBERGER, G. (1954). *Exp. Cell. Res.* 6, 540. [316]

LION, K. S. & VANDERSCHMIDT, G. F. (1955). *J. Opt. Soc. Amer.* 45, 1024. [355]

LIPSON, H. & TAYLOR, C. A. (1951). *Acta Cryst., Camb.* 4, 458. [17]

LIQUIER-MILWARD, J. (1956). *Nature, Lond.* 177, 619. [40]

LOHODNY, A. & NONVEILLER, F. (1951). *Rev. Métall.* 48, 778. [326]

LONG, J. V. P. (1957). *X-ray Microscopy and Microradiography* (New York: Academic Press), 628. [170, 177, 199]

LONG, J. V. P. (1958). *J. Sci. Instrum.* 35, 323. [76, 170, 173, 174, 175, 176, 178, 180, 181, 321]

LONG, J. V. P. (1959). Ph.D. Thesis, Cambridge. [26, 72, 76, 125, 170, 173, 180, 186, 209, 221, 232, 334, 347, 368]

LONG, J. V. P. (1960*a*). *X-ray Microscopy and X-ray Microanalysis* (Amsterdam: Elsevier), 98. [182]

LONG, J. V. P. (1960*b*). *X-ray Microscopy and X-ray Microanalysis* (Amsterdam: Elsevier), 41. [26]

LONG, J. V. P. & COSSLETT, V. E. (1957). *X-ray Microscopy and Microradiography* (New York: Academic Press), 435. [170, 172, 173, 176, 186, 199, 207, 208, 211]

LONG, J. V. P. & McCONNELL, J. D. C. (1959). *Miner. Mag.* **32**, 117. [170, 333]

LONSDALE, K. (1947). *Phil. Trans.* **240**, 219. [347]

LUBLIN, P. (1960). *Proc. 8th Conf. on X-ray Analysis*, 1959 (Denver: The University), 1. [335]

LUCHT, C. M. & HARKER, D. (1951). *Rev. Sci. Instrum.* **22**, 392. [117]

LUCHT, C. M., MANN, M. & SMOLUCHOWSKI, R. (1946). *Phys. Rev.* **69**, 256 (abstract only). [330]

MACARTHUR, I. (1957). *X-ray Microscopy and Microradiography* (New York: Academic Press), 447. [321]

MACKAY, A. L. (1955). Private communication. [350]

MADDIGAN, S. E. & ZIMMERMAN, B. R. (1944). *Metals Tech.* **11**, 1. [322]

MALSCH, F. (1939). *Naturwissenschaften*, **27**, 854. [7, 49]

MANIGAULT, P. & SALMON, J. (1956). *J. Rech.* **37**, 319. [318]

MARINGER, R. E., RICHARD, N. A. & AUSTIN, A. E. (1959). *Trans. Met. Soc. A.I.M.E.* **215**, 56. [368]

MARTON, L. (1939). *Internal Report, RCA Laboratories, Princeton.* [7, 49]

MARTON, L. (1954). *Proc. Elect. Phys. Symp., Washington, 1951* (Nat. Bureau of Standards, Washington), 265. [49, 215]

MARTON, L., SCHRACK, P. A. & PLACIOUS, R. C. (1957). *X-ray Microscopy and Microradiography* (New York: Academic Press), 287. [284, 287]

McBRIAR, E. M., JOHNSON, W., ANDREWS, K. W. & DAVIES, W. (1954). *J. Iron St. Inst.* **177**, 316. [325]

McGEE, J. F. (1955). *Phys. Rev.* **98**, 282 (abstract only). [118]

McGEE, J. F. (1957). *X-ray Microscopy and Microradiography* (New York: Academic Press), 164. [4, 104, 108, 116, 118, 120]

McGEE, J. F. & MILTON, J. W. (1960). *X-ray Microscopy and X-ray Microanalysis* (Amsterdam: Elsevier), 118. [122]

McKAY, H. C. (1948). *Principles of Stereoscopy* (American Photo. Pub. Co., Boston). [266]

McMASTER, R. C. & GROVER, H. J. (1947). *Weld. J.*, **26**, 223. [329]

McMULLAN, D. (1953). *Proc. Instn Elect. Engrs*, **100** (11) 245. [202]

MELFORD, D. A. (1960). *X-ray Microscopy and X-ray Microanalysis* (Amsterdam: Elsevier), 407. [368]

MELFORD, D. A. & DUNCUMB, P. (1958). *Metallurgia*, **57**, 159. [205]

MILLER, J. (1954). *Brit. Dent. J.* **97**, 7. [316]

MILLER, D. C. & ZINGARO, W. (1960). *Proc. 8th Conf. on X-ray Analysis*, 1959 (Denver: The University), 49. [374]

MITCHELL, G. A. G. (1951). *Brit. J. Radiol. N.S.* **24**, 110. [311]

MITCHELL, G. A. G. (1954). *J. Photog. Sci.* **2**, 113. [20, 311, 317]

MOBERGER, G. & ENGSTRÖM, A. (1954). *J. Invest. Derm.* **22**, 477. [317]

MOLENAAR, I. (1960). *X-ray Microscopy and X-ray Microanalysis* (Amsterdam: Elsevier), 177. [244]

MÖLLENSTEDT, G. & HUANG, L. Y. (1957). *X-ray Microscopy and Microradiography* (New York: Academic Press), 392. [43, 356, 357, 358]

MONTEL, M. (1953). *Rev. Opt. théor. instrum.*, **32**, 585. [4, 98]

MONTEL, M. (1954). *Optica Acta*, **1**, 117. [4, 98, 108]

MONTEL, M. (1957). *X-ray Microscopy and Microradiography* (New York: Academic Press), 177. [109, 116, 118]

MONTEL, M. (1960). *X-ray Microscopy and X-ray Microanalysis* (Amsterdam: Elsevier), 129. [122]

MOON, R. J. (1950). *Science*, **112**, 389. [351]

MOSLEY, V. M., SCOTT, D. B. & WYCKOFF, R. W. G. (1956). *Science*, **124**, 683. [243, 302, 317]

MOSLEY, V. M., SCOTT, D. B. & WYCKOFF, R. W. G. (1957). *Biochim. Biophys. Acta*, **24**, 235. [247, 317]

MOSLEY, V. M. & WYCKOFF, R. W. G. (1958). *J. Ultrastructure Res.* **1**, 337. [334]

MÜLLER, D. & SANDRITTER, W. (1960). *X-ray Microscopy and X-ray Microanalysis* (Amsterdam: Elsevier), 263. [182, 321]

MULVEY, T. (1953). *Proc. Phys. Soc.* B, **66**, 441. [281]

MULVEY, T. (1957). *Brit. J. Appl. Phys.* **8**, 259 (summary only). [185]

MULVEY, T. (1959a), *J. Sci. Instrum.* **36**, 350. [185, 200, 368]

MULVEY, T. (1959b). *Rev. Mét.* **56**, 163. [368]

MULVEY, T. (1960a). *Proc. Internat. Conf. Elect. Microscopy, Berlin, 1958* (Berlin: Springer), 68 and 263. [185, 200, 201, 368]

MULVEY, T. (1960b). *X-ray Microscopy and X-ray Microanalysis* (Amsterdam: Elsevier), 372. [368]

MULVEY, T. & CAMPBELL, A. J. (1958). *Brit. J. Appl. Phys.* **9**, 406. [195]

NÄHRING, E. (1930). *Phys. Z.* **31**, 401 and 799. [89, 94]

NEWBERRY, S. P. & NIXON, W. C. (1953). *J. Appl. Phys.* **24**, 1415. [304] (abstract only).

NEWBERRY, S. P. & NORTON, J. F. (1958). *J. Appl. Phys.* **29**, 1618 (abstract only). [309]

NEWBERRY, S. P. & SUMMERS, S. E. (1956). *Proc. Internat. Conf. Electron Microscopy, Lond., 1954* (Roy. Microscopical Soc. Lond.), 305. [285, 304]

NEWBERRY, S. P. & SUMMERS, S. E. (1957). *X-ray Microscopy and Microradiography* (New York: Academic Press), 116. [304, 309]

NEWKIRK, J. B. (1958a). *Phys. Rev.* **110**, 1465. [346]

NEWKIRK, J. B. (1958b). *J. Appl. Phys.* **29**, 995. [346]

NEWKIRK, J. B. (1959) *Trans. Met. Soc. A.I.M.E.* **215**, 483. [371]

NIXON, W. C. (1952). Ph.D. Thesis, Cambridge Univ. [245, 250, 274, 301]

NIXON, W. C. (1955a). *Nature, Lond.* **175**, 1078. [8, 18, 359, 362]

NIXON, W. C. (1955b). *Proc. Roy. Soc.* A, **232**, 475. [8, 18, 49, 58, 74, 278, 301, 359, 362]

NIXON, W. C. (1956a). *Proc. Internat. Conf. Electron Microscopy, Lond., 1954* (Roy. Microscopical Soc., Lond.), 307. [326, 327]

NIXON, W. C. (1956b). *Brit. J. Radiol. N.S.* **29**, 657. [266]

NIXON, W. C. (1957a). Unpublished. [309]

NIXON, W. C. (1957b). *X-ray Microscopy and Microradiography* (New York: Academic Press), 336. [327, 339, 344, 345, 348, 349]

NIXON, W. C. (1960a). *Proc. Internat. Conf. Electron Microscopy, Berlin, 1958* (Berlin: Springer), 249. [362, 363]

NIXON, W. C. (1960b). *X-ray Microscopy and X-ray Microanalysis* (Amsterdam: Elsevier), 105. [362]

NIXON, W. C. & COSSLETT, V. E. (1955). *Brit. J. Radiol. N.S.* **28**, 532. [324, 327]

NIXON, W. C. & PATTEE, H. H. (1957). *X-ray Microscopy and Microradiography* (New York: Academic Press), 397. [354]

NÜRNBERGER, J. & ENGSTRÖM, A. (1956). *J. Cell. Comp. Physiol.* **39**, 215. [241]

O'CONNOR, D. T. & EAGLESON, E. W. (1950). *Non-Destr. Test.* **9**, 15. [335]

ODÉN, B., BELLMAN, S. & FRIES, B. (1958). *Brit. J. Radiol. N.S.* **31**, 70. [307]

ODERR, C. P. (1957). *Amer. J. Roentgenol.* **77**, 1071. [313]

ODERR, C. P. (1960). *X-ray Microscopy and X-ray Microanalysis* (Amsterdam: Elsevier), 290. [321]

OGBURN, F. & HILKERT, M. (1957). *Iron Age*, **179**, 123. [331]

OGILVIE, R. E. (1957). Private communication. [198]

OGILVIE, R. E. (1958). *Proc. 6th Conf. on X-ray Analysis, 1957* (Denver: The University), 439. [328]

OGILVIE, R. E. & LEWIS, R. K. (1960). *Proc. 8th Conf. on X-ray Analysis, 1959*, 376. (Denver: The University). [368]

OKAWA, C. & TROMBKA, J. I. (1956). *Amer. J. Clin. Path.* **26**, 758. [313]

ONG, S. P. (1959). D. Tech. Sc. Thesis, Delft.

ONG, S. P. & LE POOLE, J. B. (1957). *J. Appl. Phys.* **28**, 1368. [86, 359]

ONG, S. P. & LE POOLE, J. B. (1958a). *Appl. Sci. Res.* B, **7**, 233. [86, 360]

ONG, S. P. & LE POOLE, J. B. (1958b). *Appl. Sci. Res.* B, **7**, 265. [85]

ONG, S. P. & LE POOLE, J. B. (1959). *Appl. Sci. Res.* B, **7**, 343. [86]

OOSTERKAMP, W. J. (1948). *Philips Tech. Rep.* **3**, 58 and 303. [67, 217]

OOSTERKAMP, W. J. & PROPER, J. (1952). *Acta Radiol., Stockh.* **37**, 33. [232]

OSIPOV, K. A. & FEDOTOV, S. G. (1951). *Dokl. Akad. Nauk, SSSR*, **78**, 51. [331]

OSSWALD, E. (1949). *Z. Metallk.* **40**, 12. [326]

PAGE, R. S. & OPENSHAW, I. K. (1960). *X-ray Microscopy and X-ray Microanalysis* (Amsterdam: Elsevier), 385. [368]

PAIC, M. (1944). *Rev. Métall.* **41**, 169. [326]

PAIC, M. (1951a). *Metal Treatm.* **18**, 395. [328]

PAIC, M. (1951b). *Rev. Métall.* **48**, 116. [328]

PARRATT, L. G. & HEMPSTEAD, C. F. (1954). *Phys. Rev.* **94**, 1593. [89]

PARRISH, W. & KOHLER, T. R. (1956). *J. Appl. Phys.* **27**, 1215. [230]

PATTEE, H. H. (1953a). Ph.D. Thesis, Stanford Univ., U.S.A. [4, 98, 100, 103, 111, 112]

PATTEE, H. H. (1953b). *J. Opt. Soc. Amer.* **43**, 61. [9, 215, 299, 351]

PATTEE, H. H. (1957a). *X-ray Microscopy and Microradiography* (New York: Academic Press), 135. [6, 111, 112, 117]

PATTEE, H. H. (1957b). *X-ray Microscopy and Microradiography* (New York: Academic Press), 367. [9, 339, 352]

PATTEE, H. H. (1958). *Science*, **128**, 977. [86, 267]

PATTEE, H. H. (1960a). *X-ray Microscopy and X-ray Microanalysis* (Amsterdam: Elsevier), 56. [366]

PATTEE, H. H. (1960b). *X-ray Microscopy and X-ray Microanalysis* (Amsterdam: Elsevier), 61. [86, 366]

PATTEE, H. H. (1960c). *X-ray Microscopy and X-ray Microanalysis* (Amsterdam: Elsevier), 79. [269]

PATTEE, H. H., GARRON, L. K., McEWEN, W. K. & FEENEY, M. L. (1957). *X-ray Microscopy and Microradiography* (New York: Academic Press), 534. [314]

PATTEE, H. H. & WARNES, R. (1957). *X-ray Microscopy and Microradiography* (New York: Academic Press), 387. [365]

PAUL, W. & STEINWEDEL, H. (1955). *Beta and Gamma-ray spectroscopy*, ed. Siegbahn, K. (N. Holland Publ. Co. Amsterdam), 1. [223]

PEISER, H. S., ROOKSBY, H. P. & WILSON, A. J. C. (1955). *X-ray Diffraction by Polycrystalline Materials* (Inst. of Phys. Lond.). [236, 240, 336]

PELC, S. R. (1957). *Exp. Cell Res.*, Suppl. 4, 231. [45]

PELGROMS, J. D. (1951). *Papeterie*, 73, 2. [332]

PHILIBERT, J. & DE BEAULIEU, C. (1959). *Rev. Mét.* 56, 171. [368]

PHILIBERT, J. & BIZOUARD, H. (1959). *Rev. Mét.* 56, 187. [367, 368]

PHILIBERT, J. & BIZOUARD, H. (1960). *X-ray Microscopy and X-ray Microanalysis* (Amsterdam: Elsevier), 416. [367, 368]

PHILIBERT, J. & CRUSSARD, C. (1956). *J. Iron St. Inst.* 183, 42. [322]

POITTEVIN, M. (1947). *J. Phys. Radium*, 8, 102. [338]

POLLACK, H. C. & BRIDGMAN, C. F. (1954). *Radiology*, 62, 259. [332]

POLLACK, H. C., BRIDGMAN, C. F. & SPLETTSTOSSER, H. R. (1955). *Med. Radiogr. Photogr.*, 31, 74. [332]

PRINCE, E. (1950). *J. Appl. Phys.* 21, 698. [101, 117]

RAMACHANDRAN, G. N. & THATHACHARI, Y. T. (1951). *Curr. Sci.* 20, 314. [6, 135, 136]

RANWEZ, F. (1896). *C.R. Acad. Sci., Paris*, 122, 841. [1, 19, 317]

RATHENAU, G. W. & BAAS, G. (1951). *Physica*, 17, 117. [11]

RATHENAU, G. W. & BAAS, G. (1956). *Proc. Internat. Conf. Elect. Microscopy, London, 1954* (Royal Microscopical Soc., London), 387. [11]

RECOURT, A. (1957a). *X-ray Microscopy and Microradiography* (New York: Academic Press), 234. [40, 260]

RECOURT, A. (1957b). *X-ray Microscopy and Microradiography* (New York: Academic Press), 475. [244]

REGENSTREIF, E. (1951). *Ann. Radioélect.* 6, 31. [284]

REHDER, J. E. (1952a). *Amer. Foundrym.* 21, 44. [324]

REHDER, J. E. (1952b). *Amer. Foundrym.* 21, 87. [324]

REISDORF, B. G. (1960). *Proc. 8th Conf. on X-ray Analysis*, 1959. (Denver: The University), 163. [368]

RENNINGER, R. (1954). *Acta Cryst., Camb.* 7, 677. [125]

RICH, A. & DAVIES, H. (1957). Private communication. [350]

RIESER, L. M. (1951). Ph.D. Thesis, Stanford Univ., U.S.A. [120]

RIESER, L. M. (1957a). *X-ray Microscopy and Microradiography* (New York: Academic Press), 195. [91, 115, 120, 121]

RIESER, L. M. (1957b). *J. Opt. Soc. Amer.* 47, 987. [91, 115, 120, 121]

RIGGS, J. (1958). *Electron Probe Microanalyser Conf., Washington, 1958.* [186]

RÖCKERT, H. (1958). *Acta Odont. Scand.*, Suppl. 25. [170, 316, 321]

ROGERS, T. H. (1952). *J. Appl. Phys.* **23**, 881. [209, 258]

RÖHRL, W. (1951). *Klin. Wschr.* **29**, 307. [312]

RÖHRL, W. (1952). *Strahlentherapie*, **88**, 276. [312]

RÖNTGEN, W. C. (1895). *Sitzungsber. Phys.-Med. Ges. Würzburg*, 137–41. 'Über eine neue Art von Strahlen. 1 Mitteilung.' [1]

RÖNTGEN, W. C. (1896). *Sitzungsber. Phys.-Med. Ges. Würzburg*, 11–19. 'Über eine neue Art von Strahlen. 2 Mitteilung.' [48]

ROOKSBY, H. P. (1955). *Brit. J. Appl. Phys.* **6**, 272. [345]

ROSENGREN, B. H. O. (1956). *Nature, Lond.* **177**, 1127. [151, 170, 175]

ROVINSKY, B. M. & GAMBASHIDZE, N. D. (1951). *Dokl. Akad. Nauk, SSSR*, **76**, 399. [331]

ROVINSKY, B. M. & LUTSAU, V. G. (1957). *X-ray Microscopy and Microradiography* (New York: Academic Press), 128. [7, 58]

ROVINSKY, B. M., LUTSAU, V. G. & AVDEYENKO, A. T. (1957). *X-ray Microscopy and Microradiography* (New York: Academic Press), 269. [59, 60, 220, 331]

ROVINSKY, B. M., LUTSAU, V. G. & AVDEYENKO, A. T. (1960). *X-ray Microscopy and X-ray Microanalysis* (Amsterdam: Elsevier), 110. [59]

RUFF, A. W. & KUSHNER, L. M. (1960). *X-ray Microscopy and X-ray Microanalysis* (Amsterdam: Elsevier), 153. [335, 370]

SAGEL, K. (1959). See 'General Texts'.

SALMON, J. (1954). *Bull. Soc. Bot. Fr.* **101**, 429. [318]

SALMON, J. (1957 a). *X-ray Microscopy and Microradiography* (New York: Academic Press), 465. [317]

SALMON, J. (1957 b). *X-ray Microscopy and Microradiography* (New York: Academic Press), 484. [318]

SALMON, J. (1960). *X-ray Microscopy and X-ray Microanalysis* (Amsterdam: Elsevier), 311. [321]

SANDSTRÖM, A. E. (1957). *Handb. Physik*, **30**, 78–245. [224, 240]

SAUNDERS, R. L. DE C. H. (1957 a). *X-ray Microscopy and Microradiography* (New York: Academic Press), 561. [245, 307]

SAUNDERS, R. L. DE C. H. (1957 b). Private communication. [307, 311, 312, 313]

SAUNDERS, R. L. DE C. H. (1958). Private communication. [308]

SAUNDERS, R. L. DE C. H. (1960). *X-ray Microscopy and X-ray Microanalysis* (Amsterdam: Elsevier), 244. [321, 370]

SAUNDERS, R. L. DE C. H., LAWRENCE, J. & MACIVER, D. A. (1957). *X-ray Microscopy and Microradiography* (New York: Academic Press), 539. [306]

SAUNDERS, R. L. DE C. H. & VAN DER ZWAN, L. (1960). *X-ray Microscopy and X-ray Microanalysis* (Amsterdam: Elsevier), 293. [321, 370]

SAWKILL, J. (1959). Private communication. [348]

SCHAAFS, W. (1957). *Hand. Physik*, **30**, 1–77. [65, 226]

SCHERZER, O. (1947). *Optik*, **2**, 114. [65]

SCHMIDT, R. A. M. (1948). *Amer. J. Sci.* **246**, 615. [333]

SCHMIDT, R. A. M. (1952). *Science*, **115**, 94. [333]

SCHULZ, K. (1936). *Ann. Phys., Lpz.* **27**, 1. [375]

SCHUMACHER, B. W. (1958). *Electron Probe Microanalyser Conf., Washing-1958*. [186]

SCHWARZENBERGER, D. R. (1959). *Phil. Mag.* **4**, 1242. [349]

SCHWARZSCHILD, K. (1905). *Abhand. Kön. Ges. Wiss. Gött. Math-Phys. Kl.* 4, no. 1, 3; no. 2, 3; no. 3, 3. [113]

SEED, J. (1960). *Proc. Roy. Soc.* B, 152, 387. [310]

SEEMANN, H. E. (1937). *J. Appl. Phys.* 8, 836. [47]

SEEMANN, H. E. (1949). *J. Appl. Phys.* 20, 231. [47]

SEEMANN, H. E. (1950). *Rev. Sci. Instrum.* 21, 314. [165]

SENNETT, R. S. & SCOTT, G. D. (1950). *J. Opt. Soc. Amer.* 40, 203. [119]

SHARPE, R. S. (1957a). *X-ray Microscopy and Microradiography* (New York: Academic Press), 590. [251, 256, 258, 322, 325, 328, 330]

SHARPE, R. S. (1957b). *Industrial Image,* 2, 15. [256, 322]

SHERWOOD, H. F. (1937a). *J. Biol. Photogr. Ass.* 6, 78. [265]

SHERWOOD, H. F. (1937b). *J. Soc. Mot. Pict. Engrs,* 28, 614. [308]

SHERWOOD, H. F. (1947). *Rev. Sci. Instrum.* 18, 80. [252]

SHINODA, G., AMONO, Y., TOMURA, T. & SHIMIZU, R. (1960). *X-ray Microscopy and X-ray Microanalysis* (Amsterdam: Elsevier), 162. [335]

SIEGBAHN, M. (1925). *The Spectroscopy of X-rays* (Oxford University Press). [377]

SIEGEL, B. M. & KNOWLTON, K. C. (1955). *J. Appl. Phys.* 26, 1395 (abstract only). [305]

SIEGEL, B. M. & KNOWLTON, K. C. (1957). *X-ray Microscopy and Microradiography* (New York: Academic Press), 106. [305]

SIEVERT, R. M. (1936). *Acta Radiol., Stockh.* 17, 299. [7, 48]

SISSONS, H. A. (1950). *Brit. J. Radiol. N.S.* 23, 2. [313]

SISSONS, H. A., JOWSEY, J. & STEWART, L. (1960a). *X-ray Microscopy and Microanalysis* (Amsterdam: Elsevier), 199. [182, 321, 370]

SISSONS, H. A., JOWSEY, J. & STEWART, L. (1960b). *X-ray Microscopy and X-ray Microanalysis* (Amsterdam: Elsevier), 206. [321]

SKINNER, H. W. B. (1938). *Rep. Progr. Phys.* 5, 257. [240]

SMITH, A. U. (1958). *Nature, Lond.* 181, 1694. [242]

SMITH, D. S. (1957). *X-ray Microscopy and Microradiography* (New York: Academic Press), 492. [308]

SMITH, K. C. A. (1956). Ph.D. Thesis, Cambridge Univ. [276]

SMITH, K. C. A. & OATLEY, C. W. (1955). *Brit. J. Appl. Phys.* 6, 391. [202]

SOMMERFELD, A. (1931). *Ann. Phys., Lpz.* 11, 257. [227, 232]

SMOLUCHOWSKI, R., LUCHT, C. M. & MANN, M. (1946). *Phys. Rev.* 70, 318. [330, 347]

SPEAR, W. E. (1955). *Proc. Phys. Soc.* A, 68, 991. [222]

SPLETTSTOSSER, H. R. & SEEMANN, H. E. (1952). *J. Appl. Phys.* 23, 1217. [257, 258]

STANSFIELD, J. R. (1960). *X-ray Microscopy and X-ray Microanalysis* (Amsterdam: Elsevier), 483. [370]

STAUSS, H. E. (1930a). *Phys. Rev.* 36, 1101. [94]

STAUSS, H. E. (1930b). *J. Opt. Soc. Amer.* 20, 616. [94]

STEPHENSON, S. T. (1957). *Handb. Physik,* 30, 337–70. [226, 227, 232]

STERNGLASS, E. J. (1955). *Rev. Sci. Instrum.* 26, 1202. [357]

STODDARD, K. B. (1934). *Phys. Rev.* 46, 837. [229]

STURROCK, P. (1951). *Phil. Trans.* A, 243, 387. [277]

SWIFT, H. & RASCH, E. (1956). *Physical Techniques in Biolog. Res.,* ed. G. Oster and A. W. Pollister (New York: Academic Press), vol. 3, chap. 8. [147]

TASKER, H. S. & TOWERS, S. W. (1945). *Nature, Lond.* **156**, 50. [47]
TAYLOR, J. H. (1956). *Physical Techniques in Biolog. Res.* ed., G. Oster
  and A. W. Pollister (New York: Academic Press), vol. 3, chap. 11. [45]
TERRILL, H. M. (1924). *Phys. Rev.* **24**, 616. [222]
TERRILL, H. M. (1930). *X-ray Technology* (New York: van Nostrand).
  [222]
TEVES, M. C. & TOL, T. (1952). *Philips Tech. Rev.* **14**, 33. [354]
THATHACHARI, Y. T. (1953). *Proc. Ind. Acad. Sci.* A, **37**, 41. [126, 135]
THATHACHARI, Y. T. & RAMACHANDRAN, G. N. (1952). *J. Indian Inst. Sci.*
  34, 67. [135]
THEVENARD, P. (1955). *Med. Biol. Illustration*, **5**, 66. [309]
THEVENARD, P. (1956). *J. Phot. Sci.* **4**, 29. [309]
THEWLIS, J. (1940). *Med. Res. Council. Special Rep.*, no. 238. [316]
THORP, J. S. (1951). *J. Sci. Instrum.* **28**, 239. [344]
TIRMAN, W. S., CAYLOR, C. E., BANKER, H. W. & CAYLOR, T. E. (1951).
  Radiology, **57**, 70. [311]
TOL, T., OOSTERKAMP, W. J. & PROPER, J. (1955). *Philips Res. Rep.* **10**,
  141. [356]
TOLANSKY, S. (1948). *Multiple-beam Interferometry* (Oxford: Clarendon
  Press). [119]
TOMBOULIAN, D. H. (1957). *Handb. Physik*, **30**, 246–304. [240]
TRILLAT, J. J. (1940). *Rev. Sci.*, Paris, **78**, 212. [19, 46]
TRILLAT, J. J. (1945a). *Rev. Gén. Caoutch.* **22**, 3. [335]
TRILLAT, J. J. (1945b). *Peint-Pigm.-Vern.* **21**, 178. [335]
TRILLAT, J. J. (1948). *J. Appl. Phys.* **19**, 844. [46, 325]
TRILLAT, J. J. (1949). *Rev. Métall.* **46**, 79. [325]
TRILLAT, J. J. (1956). *Metallurgical Rev.* **1**, 3–30. [322]
TRILLAT, J. J. & LEGRAND, C. (1947). *C.R. Acad. Sci.*, Paris, **224**, 1000.
  [325]
TRILLAT, J. J. & LEGRAND, C. (1948). *Peint-Pigm.-Vern.* **24**, 153. [335]
TRILLAT, J. J. & URBAIN, P. (1943). *C.R. Acad. Sci.*, Paris, **216**, 534. [325]
TRURNIT, H. J. & HOPPE, W. (1946). *Nachr. Akad. Wiss. Göttingen*, 29.
  [136]
USPENSKI, N. (1914). *Phys. Z.* **15**, 717. [7, 48]
VICTOREEN, J. A. (1949). *J. Appl. Phys.* **20**, 1141. [149]
VICTOREEN, J. A. (1950). *Med. Phys.*, vol. II, 887 (Chicago: The Year
  Book Publishers). [149]
VINCENT, J. (1956). *Arch. Biol.* **65**, 531. [316]
VIRTAMA, P. (1958). Private communication. [160]
VOTAVA, E., BERGHEZAN, A. & GILLETTE, R. H. (1957). *X-ray Microscopy
  and Microradiography* (New York: Academic Press), 603. [330, 347]
WALLGREN, G. (1957). *Acta Paediat.*, Stockh., Suppl. 113. [159,162,237,
  316]
WALLGREN, G. & HOLMSTRAND, K. (1957). *Exp. Cell Res.* **12**, 188. [152,
  159]
WEISSMANN, S. (1956). *J. Appl. Phys.* **27**, 389 and 1335. [16, 346]
WEISSMANN, S. (1959). *Proc. 7th Conf. on X-ray Analysis*, 1958 (Denver:
  The University), 21 and 47. [371]
WEISSMANN, S. (1960). *X-ray Microscopy and X-ray Microanalysis*
  (Amsterdam: Elsevier), 488 and 497. [371]

WEISSMANN, S. & TURNER, K. A. (1959). *Proc. 7th Conf. on X-ray Analysis*, 1958 (Denver: The University), 37. [371]

WHIDDINGTON, R. (1912). *Proc. Roy. Soc.* A, **86**, 365. [222]

WHIDDINGTON, R. (1914). *Proc. Roy. Soc.* A, **89**, 554. [222]

WHITE, M. (1958). Ph.D. Thesis, London Univ. [39]

WIGGLESWORTH, V. B. (1950). *Quart. J. Micr. Sci.* **91**, 217. [245]

WILLIAMS, F. E. & CUSANO, D. A. (1955). *J. Appl. Phys.* **26**, 358. [356]

WILLIAMS, W. M. & SMITH, C. S. (1952). *J. Metals*, **4**, 755. [266, 326, 327]

WITTRY, D. B. (1957). Ph.D. Thesis, Cal. Inst. Tech., U.S.A. [199]

WITTRY, D. B. (1958). *J. Appl. Phys.* **29**, 1543. [367]

WITTRY, D. B. (1960a). *Proc. 8th Conf. on X-ray Analysis*, 1959 (Denver: The University), 185. [367]

WITTRY, D. B. (1960b). *Proc. 8th Conf. on X-ray Analysis*, 1959 (Denver: The University),197. [368]

WOLFE, K. J. B. & ROBINSON, I. R. (1950). *Metal Treatm.* **16**, 209. [323]

WOLFE, K. J. B. & ROBINSON, I. R. (1951). *Metallurgia*, **43**, 3. [323]

WOLTER, H. (1952a). *Ann. Phys., Lpz.* **10**, 94. [98, 107, 111, 112]

WOLTER, H. (1952b). *Ann. Phys., Lpz.* **10**, 286. [98, 107, 111, 112]

WOO, Y. H. & SUN, C. P. (1947). *Sci. Rep. Nat. Tsing. Hua Univ.* **4**, 398. [372, 374]

WORTHINGTON, C. R. & TOMLIN, S. G. (1956). *Proc. Phys. Soc.* A, **69**, 401. [231, 369]

WYLIE, T. S. (1960). *X-ray Microscopy and X-ray Microanalysis* (Amsterdam: Elsevier), 510. [321]

YARWOOD, J. (1956). *High Vacuum Technique* (London: Chapman and Hall). 3rd ed. [291]

YOUNG, J. R. (1956). *J. Appl. Phys.* **27**, 1. [222]

YOUNG, J. R. (1957). *J. Appl. Phys.* **28**, 524. [222]

ZEITLER, E. & BAHR, G. F. (1957). *Exp. Cell Res.* **12**, 44. [45]

ZEITZ, L. & BAEZ, A. V. (1957). *X-ray Microscopy and Microradiography* (New York: Academic Press), 417. [170, 186, 210, 211]

ZINGARO, P. W. (1954a). *Norelco Reporter*, **1**, 67 and 78. [224, 375]

ZINGARO, P. W. (1954b). *Norelco Reporter*, **2**, 92. [224, 377]

ZIRKLE, R. E. (1957). *Advances in Biological and Medical Physics*, ed. J. H. Lawrence and C. A. Tobias (New York: Academic Press), 103. [310]

ZWORYKIN, V. K., MORTON, G. A., RAMBERG, E. G., HILLIER, J. & VANCE, A. W. (1945). *Electron Optics and the Electron Microscope* (New York: Wiley). [283]

INDEX

Printed in the United States
By Bookmasters